Nanoparticle Toxicity and Compatibility

Edited by

Jorddy N. Cruz

Institute of Biological Sciences, Federal University of Pará, Belém 66075-110, PA, Brazil

Published by **Materials Research Forum LLC**
Millersville, PA 17551, USA

Published as part of the book series
Materials Research Foundations
Volume 161 (2024)
ISSN 2471-8890 (Print)
ISSN 2471-8904 (Online)

Print ISBN 978-1-64490-298-1
eBook ISBN 978-1-64490-299-8

Distributed worldwide by

Materials Research Forum LLC
105 Springdale Lane
Millersville, PA 17551
USA
https://www.mrforum.com

Manufactured in the United States of America
10 9 8 7 6 5 4 3 2 1

Table of Contents

Preface

In the realm of nanotechnology, the exploration of nanoparticles has emerged as a transformative frontier, ushering in unprecedented possibilities in fields ranging from medicine to environmental science. As we delve into the intricate world of these minuscule structures, the paramount concern lies in understanding their dual nature—potentially groundbreaking yet delicately poised on the precipice of unforeseen consequences. This compendium, "Nanoparticle Toxicity and Compatibility," seeks to unravel the complexities surrounding the interplay between nanoparticles and biological systems.

The chapters curated within this volume span a diverse spectrum of research avenues, each contributing to a comprehensive understanding of nanoparticle behavior and impact. "Future Directions in Nanomaterials Research for Biological Applications" sets the stage by delineating the evolving landscape of nanomaterials, illuminating the pathways for future exploration.

The synthesis, characterization, and application of nanomaterials in tissue engineering take center stage in "Nanomaterials for Tissue Engineering," offering insights into the promising strides made in this interdisciplinary domain. "Interaction of Nanoparticles with Macromolecules" extends the narrative into biomedical and food science applications, exploring the intricate dance between nanoparticles and the macromolecular milieu.

Turning the focus to specific health concerns, "Cardiovascular Toxicity of Nanoparticles" scrutinizes the potential implications of nanoparticle exposure on the cardiovascular system. "Colon Targeted Nano Drug Delivery Systems" then navigates the terrain of innovative drug delivery strategies, with an emphasis on precision and efficacy. "Bioinspired Nano-engineering for Plasmon-enhanced Biosensing Applications" explores the amalgamation of nature-inspired design principles with nanotechnology, paving the way for advanced biosensing capabilities.

"Strategies for Enhancing Biocompatibility of Nanoparticles" encapsulates an array of methodologies aimed at refining the compatibility of nanoparticles, ensuring their seamless integration into biological systems. Finally, "Ecotoxicology of Nanoparticles" scrutinizes the environmental impact of nanoparticles, completing the circle of exploration from the laboratory bench to the broader ecosystem.

This compilation stands as a testament to the collective efforts of researchers and scholars dedicated to demystifying the intricate dynamics between nanoparticles and living organisms. As we embark on this intellectual journey, may the insights gleaned from these chapters propel us towards responsible and sustainable applications of nanotechnology, fostering a future where innovation harmonizes with the intricate dance of life.

Nanoparticle Toxicity and Compatibility
Materials Research Foundations 161 (2024) 1-26

Materials Research Forum LLC
https://doi.org/10.21741/9781644902998-1

Chapter 1

Future Directions in Nanomaterials Research for Biological Applications

Rahul Das[1]*, Manab Deb Adhikari[2], Pratap Singh Chauhan[3]

[1]Department of Physics, The University of Burdwan, Bardhaman 713104, India

[2]Department of Biotecnology, University of North Bengal, Darjeeling 734013, India

[3]Department of Physics, Government Dau Kalyan Arts and Commerce College, Chhattisgarh 493332, India

rdas@phys.buruniv.ac.in

Abstract

Nowadays, nanomaterial-based technologies have reached a great height from the application point of view. However, this chapter is mainly focused on nanomaterials research for biological applications and its future directions. Depending on the shape, size and elemental composition, nanomaterials are capable of exhibiting some unique and remarkable functional properties. Due to such functional properties, nanomaterials have attracted much attention for biomedical applications and are being tested for easier treatment and diagnosis without any side effects. This chapter not only provides the expected biological applications of nanomaterials but also points out some information starting from key properties of biologically applicable nanomaterials to characterization through fabrication processes.

Keywords

Nanomaterials, Nanofabrications, Biological Applications, Diagnostic Tools, Drug Delivery, Tissue Engineering, Nanorobots

Contents

1.   Introduction

Due to their unique physical and chemical properties, nanomaterials are becoming very promising in various applications. In a word, the application of nanomaterial-based technologies is a very fast growing research field that has great potential to fascinate humanity with revolutionary breakthroughs. This technology will soon have a significant impact on every aspect of life. Recently, various nanomaterials/nanotechnology-based advances and enhancements are utilizing in the field of biology. Nano sized materials can be elegantly incorporated into the tiny devices that could be used for biological purposes because most biological systems belong to the nano dimension [1]. Various biological applications of nanomaterials are drug carriers, cancer treatment, tissue engineering, biological markers/tracking agents, vectors for gene therapy, hyperthermia treatments, diagnosis, detection of antibodies and proteins, etc. [2-5]. Additionally, individuals experiencing major traumatic injuries and impaired organ functions could profit from the utilization of nano sized medicines [6]. Continuous advancements in small-scale medicine using nanomaterials have opened up the appropriate time for implementation in an assortment of therapeutic disciplines. These biologically applicable nano scale materials (some of which are listed in Table 1) either from nature or fabricated in the laboratory through top-down and bottom-up approaches can be composed of a variety of materials, including noble metals, semiconductors, insulators, magnetic compounds, and combining form of two or more of them [7-19]. A brief discussion about the scope of nanomaterials in various biological applications is presented in the following sections.

Table 1. Different nanomaterials with their biological activities/applications.

Sl. No.	Nanomaterials	Biological activities/applications	Ref.
1	Silver-based	Antimicrobial, drug delivery, gene therapies, tissue developments, imaging, plastered on surgical nets for pelvic reconstruction, contrasts on contact lenses, coating on the tracking catheter, treatments for dermatitis, etc.	[20, 21]
2	Iron-based	Magnetic imaging, hyperthermia-based therapy, controlled drug delivery, environmental remediation, etc.	[22, 23]
3	Gold-based	Antimicrobial, drug delivery, cellular imaging, surface plasmon resonance (SPR) sensing, Potential usage in Surface-enhanced Raman scattering (SERS) detection of biomolecules and biomarkers, cancer treatment, surgical devices, etc.	[23, 24]
4	Cerium oxide	Redox activity, biofilm inhibition, antibacterial activity, anti-inflammatory activity, bioimaging and surgical devices, etc.	[23, 26]
5	Silica oxide	Production of thermal and electric insulators gene delivery, catalyst applications, drug carriers, efficient adsorbents, serve as filler materials, etc.	[23]
6	Zinc oxide	Antimicrobial activity, bio-imaging, wound healing, implant coating, tissue engineering, skin protectant, creation of anticancer medicines, etc.	[27, 28]
7	Titanium dioxide	Antibacterial, antioxidant, anticancer, coating material, food colorants, dyes, pharmaceuticals, cosmetics, use in toothpaste, etc.	[29, 30]
8	Nickel-based	Biomedical applications such as anticancer, antibacterial, antifungal, antileishmanial, etc.	[31, 32]

9	Copper oxide	CuO can work as antibiotic, antimicrobial, antifouling, antibacterial, antioxidant, anticancer, antiparasitic, antidiabetic, antiviral, antifungal, etc.	[23, 33]
10	Aluminum oxide	Antifungal, antibacterial, antiviral, cytotoxicity against the breast cancer cell line (MCF-7), etc.	[34, 35]
11	Calcium oxide	Strong antimicrobial activity connected to active oxygen species and alkalinity, other biomedical applications like effective drug delivery, endodontic treatment, etc.	[36, 37]
12	Magnesium oxide	Effective antibacterial and anti-larvicidal agents	[36, 38]
13	Bismuth-based	Cancer therapy, bioimaging, drugs delivery systems, etc.	[39, 40]
14	Manganese dioxide	Biocatalysis, fluorescence sensing, controlled drug delivery, gene therapy, stimuli-activated imaging, It can also be combined with reactive oxygen species in the tumor microenvironment to relief tumor hypoxia.	[7, 41]

2. Key characteristics of nanomaterials for biological applications

Due to the outstanding characteristic properties, nano sized materials have expanded the boundaries of science and technology and opened up new avenues for a variety of biological applications. Several kye characteristics of nanomaterials, such as size, shape, chemical/elemental composition, surface morphology, surface charge, aggregation, agglomeration, solubility, etc., can have a significant impact on how they interact with biomolecules and cells. Some key characteristics of these biologically active materials are discussed below:

2.1 Size

The ability of nanomaterials to perform biological functions is mostly attributed to their small size. Because of this small size, nanomaterials can easily adapt themselves to the biological components like cells, tissues, subcellular structures, etc. If necessary, these tiny particles can even pass through the cell membrane to deliver medications inside the cells. After its medicinal activity, nanomaterials must be biocompatible with our body or removed from the body to reduce the toxicity. Due to the smaller size, the nanomaterials can easily be removed from the biological system via the kidney. On the other hand, as we know smaller size particles provide larger surface-to-volume ratio. Considering a spherical shaped nanomaterial of radius r, we can see that the surface-to-volume ratio is $3/r$ i. e. surface-to-volume ratio is inversely proportional to its radius. When the diameter of a spherical shaped nanomaterials reduced from 100 nm to 1 nm, its surface-to-volume ratio

Materials Research Forum LLC
https://doi.org/10.21741/9781644902998-1

raises from 6×10^{-2} nm^{-1} to 6 nm^{-1}, which is 100 times bigger. The similar kind phenomenon (increase of surface-to-volume ratio with decrease of material size) is also occurred for other shape of nanomaterials. Because of this larger surface area with lesser volume, nano sized materials become more interactive in the biological system, more capacitive for drug-loading (drug delivery) on the surface (from the surface) and more sensitive for applications. Highly drug-loaded nanoparticles get optimum size for longer circulation times in the bloodstream due to reduced renal clearance. This longer time of circulation can prolong therapeutic process and provides the larger time for maximum drug delivery.

2.2 Biocompatibility

The term "the biocompatibility of nanomaterials" may be associated with serious safety issues. Nanoparticles designed for biological applications must be examined for cytotoxicity and biocompatibility. Cytotoxicity is the term used to describe how they affect cell viability and functionality. According to a consensus conference of the European Society for Biomaterials in 1986, the ability of a material to elicit an adequate host response in a given application is referred to as biocompatibility [42]. It is important to keep in mind that a variety of factors, including the host factors, the substance's qualities, location and length of the exposure, might affect how a material interacts with a host. Nanomaterials should be investigated for possible toxicity in a variety of in vitro and in vivo situations to better understand the nature and scope of these interactions. However, there are currently no standardized methods for assessing the toxicity and biocompatibility of nanomaterials in biological systems [43].

2.3 Surface chemistry

It is another important characteristic of biologically active nanomaterials. This characteristic is not only related with the chemical composition, structure, surface area and size/shape of the nanomaterials but also related with the attached biomolecules, such as proteins, antibodies, DNA, enzymes, targeting ligands, etc. Depending on size, composition and surface morphology, nanomaterials can have different amounts of surface charges which are responsible for different surface chemistry. As a result, they are responding differently to the biological systems. The attachment of biomolecules to the surface of nanomaterial can enhance their surface chemistry for specific purposes, such as the development of customized drug delivery systems, biosensors, platforms for biological imaging and diagnostics, etc. Depending on this surface chemistry characteristic, the surface reactivity, as well as other functionalities of the nanomaterials in various environments can be determined and then improved by modifications. Surface chemistry plays a significant role for determination of toxicity and biocompatibility of nano range materials. Modifications of surface chemistry can reduce the toxicity of these nano sized materials and can help to improve their biocompatibility for the use in biological applications. Thus, we can develop an advanced nanosystem for imaging, biomedicine, diagnostics, and other biological applications by modifying the surface chemistry of nanomaterials.

2.4 Magnetic properties

Nanomaterials should have some suitable magnetic properties for biological applications such as drug administration, imaging, magnetic separation, hyperthermia, manipulation of cells/tissues, etc. One appropriate property is superparamagnetism. Due to this property, nanomaterials can exhibit strong magnetic response in presence of external applied magnetic field. Superparamagnetic nano sized materials are desirable for targeted drug delivery, magnetic resonance imaging, hyperthermia treatment of tumors, etc. High magnetization is another required magnetic property of nanomaterials. In this case, nano sized materials should hold a huge number of magnetic atoms per unit volume to guarantee a strong magnetic response. This property is preferred for efficient magnetic separation, cell sorting, imaging, etc.

2.5   Stability

In case of biological applications, biologically responding nanomaterials should be stable under the physiological conditions to maintain their desired properties. When nanomaterials interact with various biological fluids or physiological environments, this stability feature should shield them from oxidation, rapid breakdown, and aggregation.

2.6   Optical properties

Biomedically active nanomaterials should have unique optical properties. These optical characteristics can be exploited for various diagnostic and therapeutic applications. For instance, metallic nanoparticles with strong surface plasmon resonance, like gold or silver, can absorb and scatter light, which make them valuable for imaging, diagnostics, and photothermal therapy.

2.7   pH and thermal sensitivity

Nanomaterials also have to have highly sensitive to pH and/or temperature. This type of highly sensitive extremely delicate nanomaterial can be designed to carry and release medicinal chemicals under precise control. This regulated release can be accomplished using stimuli-responsive coatings or by utilizing the pH and/or temperature sensitivity of nanoparticles.

Bio-medically active nanomaterials can be fabricated by different methods such that they have the desired characteristic mentioned in the above section.

3. Fabrication of nanomaterials for biomedical application

Depending on Top-down and Bottom-up approaches, there are three main nano fabrication methods named chemical method, physical method and biological method (shown in the Fig. 1). Sol-gel process, Coprecipitation method, Electrochemical method, Microemulsion method, Hydrothermal process, Polyol process, Chemical vapour process, Plasma enhanced chemical vapour deposition method, Pyrolysis, Sonochemical method, etc. which involves the use of various chemical reactions to prepare nano size materials belongs to chemical method. Physical methods involve the techniques such as Arc discharge method, Electron beam lithography, Ion implantation, Inert gar condensation, Mechanical grinding, Milling, Spray pyrolysis, Vacuum sputtering, etc. which use specific energy sources (mechanical energy applying pressure, radiations energy, thermal energy as well as electrical energy) to break, abrasion, melting, evaporation as well as condensation the materials, leading to the formation of nanoparticles. On the other hand, biological methods include plant-based synthesis, microorganism-based synthesis, enzyme-based synthesis, synthesis using algae, synthesis from industrial and agricultural wastes [44]. Due to the use of natural resources, this method provides non-toxic, biocompatible nanomaterials in a more environmentally friendly way. Here are some common methods have been discussed:

3.1 Sol-gel process

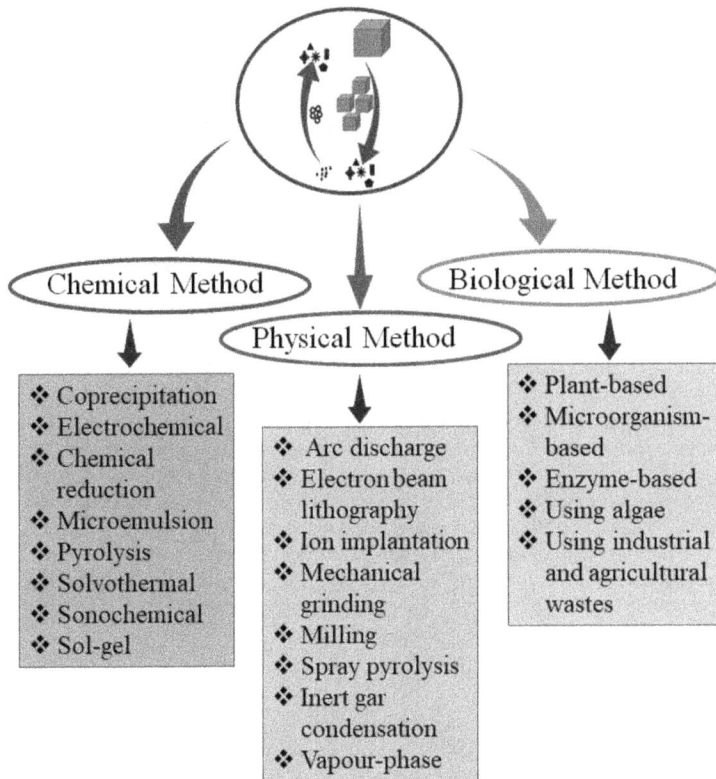

Figure 1. Different fabrication methods for biomedically applicable nanomaterials.

Sol-gel process is a bottom-up wet chemical process for the synthesis of different nano size materials, particularly metal-based nanomaterials [45]. Here two important things are "sol" and "gel". The colloidal suspension of solid particles in a liquid is known as "sol," and polymers containing liquid is known as "gel" [46]. So, as part of this processing, "sols" are formed in the liquid by dissolving or dispersing the appropriate molecular precursors (usually metal alkoxide) in a suitable solvent. This precursor solution aids in the creation of "gel," which is a network of distinct particles. The main stages of the sol-gel process (shown in Fig. 2) include hydrolysis, condensation, aging, drying, and calcination. The precursor goes through hydrolysis, which is the initial step in converting the precursor into the gel phase. In the hydrolysis process, water that was previously used to make the precursor solution helps to break the molecular bonds. Condensation, which occurs when the hydrolyzed precursor species start to condense, causes the development of a three-dimensional network or gel. During this condensation process, the hydroxyl groups on the surface of the hydrolyzed precursors react with one another, resulting in the creation of bonds between the metal atoms. Here, a variety of factors including acid and basic catalysts, can modify the gel structure by affecting the rates of hydrolysis and condensation [47]. After that, the gel is allowed to undergo aging, during which more polymerization and network rearrangement take place. This process aids in enhancing the gel's mechanical characteristics and structural integrity. Drying methods, such as thermal drying, freeze-drying, or supercritical drying, can be used to dry this wet gel at room temperature or under controlled conditions. The removal of solvent molecules leads to the shrinkage and solidification of the gel into a porous or non-porous nanomaterial. In some cases, the as-prepared nanomaterial may require additional thermal treatment at high temperatures. Calcination or annealing contributes to the enhancement of nanomaterial's desirable properties, the elimination of residual organic/inorganic species, and the crystallization of amorphous structures.

3.2   Ball milling

In this top-down physical process, the material powders kept in the ball mill are subjected to high-energy collisions from the balls. The milled material is exposed to the tremendous kinetic energy of the moving balls during these collisions. As a result, the powdered components are broken down into tiny pieces, disrupting their chemical bonds with freshly formed surfaces. In this sense, a ball mill is a specific type of grinder used to crush the micro material powder into extremely fine powders. The amount of energy transferred between the balls and the material powders during the milling process can be controlled by various factors such as milling speed, milling media, milling time, type of milling (dry or wet), ball-to-powder weight ratio, type of high energy ball mill (vibrator mill, planetary mill, attritor mill, tumbler ball mill, etc.), etc. This in turn influences the physical and morphological properties of the resulting nanomaterials [46].

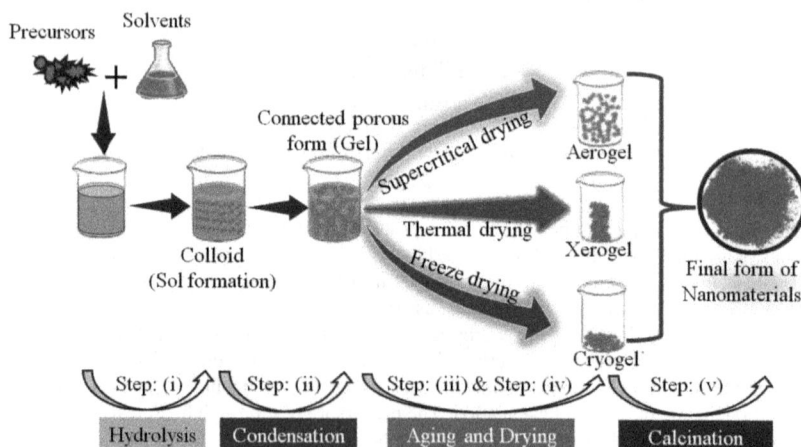

Figure 2. Schematic representation of different stages of sol-gel process [45].

3.3　Plant extract-based synthesis

In contrast to common traditional chemical and physical preparation methods, there is a plant-based cost-effective energy-saving easy eco-friendly synthesis method for the fabrication of different nano-size materials. To reduce the input plug energy, number of complicated steps and prevent the use of undesirable costly hazard chemicals which are typically utilized during the conventional physical and chemical synthesis processes, this type of bottom-up, energy-saving, straightforward green route technique has been developed. Here plant extract-based biocompatible phytochemicals act as reducing, capping as well as stabilizing agents during nanomaterial synthesis (mechanism shown in Fig. 3) [48-52]. Another benefit of this green route method is the fact that fabricated nanomaterials do not require any further functionalization agent [53]. Fig. 4 illustrates the mechanism of the synthesis of plant extract-mediated Au and Ag nanomaterials for biomedical applications in wound healing [54]. Some examples of plant extract-based green synthesized nanomaterials and their biomedical activities are presented in Table 2.

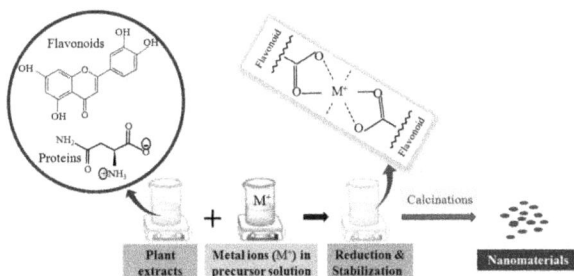

Figure 3. Proposed mechanism of plant-based synthesis of nanomaterials [48-52].

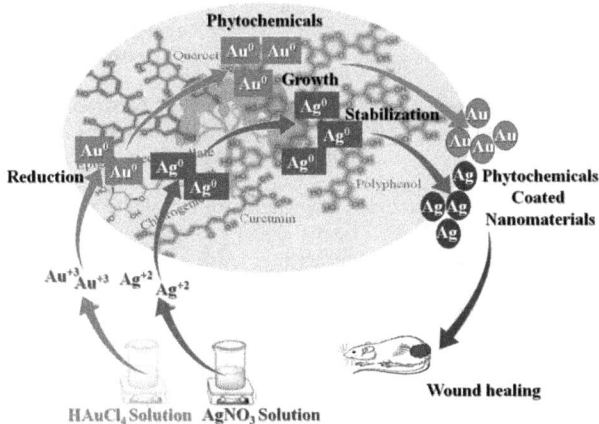

Figure 4. Plant extract-based Synthesis mechanism of Au and Ag nanomaterials for wound healing biomedical application [54].

Table 2. Some green synthesised nanomaterials using different plant and their biomedical activities.

Sl. No.	Nanomaterial	Precursors	Plant	Biological applications	Ref.
1	Au	Tetrachlo roauric acid	Leaf Extract of Ziziphus zizyphus	Applicable for antimicrobial and antifungal purposes	[55]
2	NiO	Nickel chloride hexahydr ate	Punica granatum peel extract	Purify the water by methyl orange degradation	[56]
3	AgCl	Silver nitrate hexahydr ate	Calotropis gigantea	Antimicrobial activities against the Gram-positive Listeria monocytogenes MTCC 657 and the Gram-negative Salmonella typhimurium MTCC 3224 bacteria	[57]
3	Iron oxide	Iron chloride tetrahydr ate	Hibiscus rosa sinensis Flowers	Applicable for antibacterial purposes	[58]
4	Cu	Copper (II) Sulfate pentahyd rate	Celastrus paniculatus leaves extract	Degradation efficiency on organic dye Methylene blue under sunlight and antifungal activity against F. oxysporum	[59
5	Ag	Silver nitrate	Cucumis prophetarum Leaf Extract	Antibacterial and Antiproliferative Activity Against Cancer Cell Lines	[60]
6	ZnO	Zinc acetate dihydrate	Leaf extracts of Cassia fistula/Melia azadarach	Applicable for antibacterial purpose	[61]

4. Characterization of biologically active nano-particles

Different characterization techniques can be used to determine the size distribution, shape, chemical compositions, structural properties, magnetic properties, optical properties, electrical properties, etc., of the biologically active nanomaterials and infer their mode of action during biological applications. Some of these techniques are X-ray diffraction (XRD), transmission electron microscopy (TEM), scanning electron microscopy equipped with the energy dispersive X-ray detector (SEM-EDX), dynamic light scattering (DLS), atomic force microscopy (AFM), thermal gravimetric analysis (TGA), vibrating-sample magnetometer (VSM), physical property measurement system (PPMS), Fourier transform infrared spectroscopy (FTIR), UV/VIS spectroscopy, gas-phase electrophoretic mobility molecular analyzer (GEMMA), inductively coupled plasma optical emission spectrometry (ICP-OES), liquid chromatography (HPLC), matrix-assisted laser desorption/ionization–time-of-flight mass spectrometry (MALDI-TOF MS), etc. [62-64]. A detailed description of the importance of some techniques is discussed below.

4.1 X-ray diffraction

X-ray diffraction (XRD) technique is commonly used to study crystallographic structure and atomic positions of any sample. By analyzing the diffraction patterns, it is possible to know the internal arrangement of building blocks within the materials. Additionally, it can also reveal the crystallographic flaws like vacancies, defects, dislocations, etc. These crystallographic information data are crucial for understanding how nanomaterials behave and interact in biological systems. Under specific circumstances, many nano size biological systems, including proteins and viruses, can display crystalline ordering [65]. XRD can also be used to collect the crystallographic information of these nano size biological systems, which can give insights into their functions and behaviours. In addition, XRD can be used to track structural modifications in biological nanosystems in response to various external stimuli, such as adjustments in pH, pressure, temperature, etc. It is possible to monitor the structural transformations, phase transitions, and conformational changes of any materials by accumulating X-ray diffraction patterns under various conditions. This information is necessary to know the reactivity and adaptability of these nanomaterials in biological environments.

4.2 Scanning electron microscopy equipped with the energy dispersive X-ray detector

Scanning electron microscopy equipped with the energy dispersive X-ray detector (SEM-EDX) is an analytical technique that provides the valuable insights into the morphological and compositional information of the materials. SEM-EDX's significance for biologically active nanomaterials is based on its ability to: (a) *Characterize nanomaterial's morphology:* SEM provides imaging, allowing for the observation and analysis of the size, shape, surface properties, and distribution of nanoparticles. Knowing this information is necessary to comprehend the physical properties and potential interactions of nanoparticles with biological systems; (b) *Identify Elemental Composition:* EDX makes it possible to recognize and measure the elements that are present in nanomaterials. The atomic composition and elemental distribution of the sample can be determined using EDX by analysing the sample emitted X-rays. This information can be used to validate the existence of desired elements and identify impurities or contaminants. This data is essential for ensuring the quality and purity of nanomaterials useable in biological applications; (c) *Investigation of Nanomaterial-Biological Interactions:* The interaction of nanomaterials with biological systems can be better understood with the help of SEM-EDX. By studying the location and distribution of nanomaterials within cells or tissues, researchers can gain a better understanding of their uptake, internalization mechanisms, and potential toxicity. Additionally, elemental analysis performed by EDX can assist to know the connection of certain nanoparticles to cellular or tissue response; (d) *Optimize Nanomaterial Synthesis and Modification:* SEM-EDX allows researchers to characterize and optimize the synthesis and modification processes of nanomaterials. By observing the morphology and elemental composition of nanomaterials at different stages of the synthesis process, researchers can identify the potential issues, such as agglomeration or incomplete synthesis, and can then take the necessary corrective action to improve the efficacy, stability, and bioactivity of the nanomaterials; *(e) Quality Control:* SEM-EDX is a quality control tool that may be used to evaluate the reliability, reproducibility or consistency of nanomaterial synthesis and functionalization. Researchers may make sure that nanomaterials intended for biological applications are trustworthy, consistent, and reproducible from batch to batch by comparing the size, shape and elemental content of different nanomaterials which produced by individual batches.

4.3   Vibrating sample magnetometer

Vibrating Sample Magnetometer (VSM) is a characterization technique which frequently used in the field of magnetism to measure the magnetic properties of a material based on Faraday's Law of Induction. VSM has a variety of uses in the context of biologically active nanomaterials, including (a) *Magnetic characterization*: It is possible to measure the magnetic moments, coercivity, remanence, and hysteresis of biologically active nanomaterials. This makes it simpler to understand how nano size materials respond to an external magnetic field; (b) *Quality control:* The magnetic properties of biologically active nanomaterials can be evaluated using VSM to determine their reliability and quality. This is crucial for uses like targeted medication administration or magnetic hyperthermia where precise control of the magnetic behavior is essential; (c) *Optimization:* By shedding light on the correlation between synthesis variables and magnetic characteristics, VSM data can aid in the optimization of the synthesis of biologically active nanomaterials. This could result in the development of nano size materials with enhanced magnetic properties; (d) *Guidance applications*: VSM measurements can provide information about the saturation magnetization of biologically active magnetic nanomaterials and required magnetic field to guiding these nanomaterials effectively. This aids in designing and optimizing their utilization in several biomedical applications like magnetic targeting or magnetic resonance imaging (MRI).

4.4   Fourier transform infrared spectroscopy

Fourier transform infrared spectroscopy (FTIR) is a useful characterization technique which is generally used to identify and characterize material based on their molecular bonds (functional groups). FTIR has various significant roles for analysis of physiologically relevant nanomaterials, including the following: (a) *Identification of the material*: FTIR can help to determine the nature and composition of nanomaterials, allowing researchers to confirm the presence of desired biomolecules or functional groups. This is crucial for quality control purposes and to ensure that the nanomaterials meet the required specifications; (b) *Structural analysis*: FTIR can be used to determine the molecular structure of nanomaterials, such as the presence of bonds, functional groups, and chemical interactions. This helps in comprehending how nanoparticles behave and function in biological systems; (c) *Stability assessment*: FTIR can track and monitor changes in the molecular composition or structure of nanomaterials over time. This makes it possible for scientists to evaluate how stable the nanomaterials are under a variety of environmental factors, such as temperature, humidity, or contact with biological fluids. This information is essential for the long-term preservation and application of nanomaterials; (d) *Interaction with biomolecules*: FTIR can shed light on how biological molecules like proteins or DNA interact with nanomaterials.

Researchers can understand the binding mechanisms, affinities, and conformational changes that take place during such interactions by examining the changes in the FTIR spectra. This is essential for making and designing nanomaterials with the necessary bioactivity and understanding their potential uses in tissue engineering, diagnostics, drug delivery, etc.

5.　Futuristic applications of nanomaterials in the field of biology

The development of technology based on nanomaterials is undeniable. Due to the vast advantages of nanomaterials, these are also very much positive as well as promising for future technologies. Only this nanomaterial-based technology can lead to significant advances in various fields of biology. Some of the potential developments based on nanotechnology and their future applications in biology are discussed below:

5.1　Diagnostic tools

The sensitivity and specificity of diagnostic techniques can be enhanced by nanomaterials, allowing early stage and more accurate disease detection. Nanomaterials can be functionalized with specific ligands or targeting moieties that have high affinity towards the target biomarkers or molecules. By attaching these targeting ligands to the surface of nano size materials, their binding specificity can be increased, leading to enhanced efficiency of the diagnostic tools. Such a way, nanomaterials can be utilized for efficient cellular imaging. Encapsulating the enhanced green fluorescent proteins inside the nano scale silica by multiple covalent bonding, fluorescence intensity and stabilities against protease, denaturant and heat have been increased. Due to this silica nanomaterials attached with enhanced green fluorescent proteins can be used as a potential fluorescence probe for bio-imaging [66]. Nanostructure with definite shape and large number porosities show high surface areas compared to their volume. This increased surface area allows for more binding sites and strong interactions with the target molecules, resulting in enhanced sensitivity for detection. Sensitivity of the diagnostic tools also can be improved by combining the nanomaterials with the various signal transduction, such as electrochemical sensors or surface-enhanced Raman scattering (SERS) techniques. Based on the produced signals by the interaction between the nanoparticles and the target, these techniques enable the sensitive detection and quantification of target molecules. To provide quick and precise diagnostic testing, nanosensors can be engineered to detect specific and smaller amounts of biomarkers associated with specific diseases. Gold nanomaterial-based nanosensors can be functionalized with multiple biomarkers or antibodies, resulting in amplified signals when the target molecule binds to these nanomaterials. This amplification procedure increases the sensitivity of the diagnostic tool to detect biomarkers at low concentrations. Diagnostic tools can use nano size magnetic materials to boost the sensitivity through magnetic manipulation. Magnetic nanomaterials can capture target biomarkers by functionalizing them with particular ligands. The captured nanomaterials can easily be concentrated and isolated from the sample using a magnetic field. Some nano range materials have special optical characteristics like plasmon resonance, fluorescence, or phosphorescence that can be exploited to increase the sensitivity of diagnostic devices. For instance, localized surface plasmon resonance in gold nanoparticles might result in amplified scattering or absorption signals that can be used for detection. The visibility of biological structures (tissues and organs) can be improved with imaging agents carried by suitably engineered nanostructured materials, assisting in the precise diagnosis and monitoring of disorders. Sometimes, nanomaterials synthesized by different compositions itself can be used as contrast agents in biomedical imaging techniques like magnetic resonance imaging (MRI), computed tomography (CT) scan, near infrared fluorescence molecular tomography (NIR FMT), X-ray, and ultrasonogram (USG). A few examples of nanosized materials used as contrast agents in bio-imaging are listed in Table 3. Generally, paramagnetic chelated gadolinium or superparamagnetic iron oxide nanoparticles are used as nonspecific or targeted probes in MRI [67, 68]. But in the near future, the improvement of MRI process would be done by combining both properties (paramagnetic as well as superparamagnetic) which will help to create the dual contrast agents [69, 70]. These factors make nanomaterial-based technologies more likely to be used to create cutting-edge biomedical diagnostic instruments in the future.

Table 3. Few examples of nano sized materials which can be used as contrast agents for advanced diagnostic tools.

Nano sized materials	Applications	Refs.
Gadolinium-based nanopolymers	MRI imaging	[71-77]
Gold nanoparticles	X-ray/CT scan	[78-80]
Superparamagnetic iron-oxide nanostructures coated with dextran	MRI imaging	[81-88]
Quantum Dots	Florescence or optical imaging	[89-92]
Silicon nanoparticles coated with conductive layers	MRI imaging	[93]
Nano sized Bismuth sulphide	CT scan	[94]
Iodinated liposomal carriers, inorganic nanomaterials	CT scan	[95-101]

5.2   Drug delivery

An ideal drug delivery system should have two major functions: the ability to target the specific location and control over drug release. Side effects can be reduced significantly, and drug efficiency can be enhanced by specifically targeting, controlled drug releasing and destroying the sick cells [102]. The nanomaterial-based development of drug formulations has opened up new avenues for delivering the drugs and treating the complicated diseases. Nanotechnology can be used to design more efficient drug delivery systems by encapsulating or incorporating the drugs within nanomaterials. For example, without affecting the heart or kidneys, highly toxic doxorubicin drug can be delivered directly to tumour cells using liposomes (shown in Fig. 5) [103]. On the other hand, for the purpose of chemotherapeutic treatment of metastatic breast cancers, paclitaxel incorporated Methoxypoly(ethylene glycol)-poly(dl-lactic acid) or mPEG-PLA micelles can be utilized [81,104-105]. Beside of these, some other nanomaterials can be coated with some unique ligands that specifically response and bind to infected cells, allowing for targeted delivery of drugs directly to the infected site. These specially engineered nanomaterials can enhance drug

Materials Research Forum LLC

https://doi.org/10.21741/9781644902998-1

stability, target specific tissues or cells, and increase drug release kinetics, leading to improved therapeutic outcomes. Nanomaterials could play a significant role in personalized medicine by allowing for individualized drug delivery based on a patient's specific needs or genetic makeup. By tailoring nanomaterials to respond to specific stimuli or genetic markers, drugs can be delivered in a customized manner. Thus, nanotechnology can enable targeted drug delivery to specific sites within the body, reducing the side effects.

5.3 Cancer treatment

With the advancement of nanotechnology, we might soon be able to provide novel strategies for treating cancer. Imaging methods based on nanotechnology can help with early cancer detection and treatment response monitoring. Anticancer treatments are often observed as more useful if the therapeutic agent can reach and influence the exact target site while ensuring zero side effects. Nano sized materials can be used to deliver chemotherapy drugs directly to cancer cell, reducing systemic toxicity and improving treatment effectiveness. Chemically modified surface of nano sized carriers can improve this required targeted delivery. The incorporation of Polyethylene glycol or polyethylene oxide at the surface of nanomaterials is the one of the best examples of this kind of modified nanocarriers. Since the body's immune system cannot detect these modified nanocarriers as the foreign objects, they can easily circulate in the bloodstream until their fixed destination i.e., cancer cells. This type of surface incorporation not only can enhance the specificity of drug delivery, but also the cancer cell targeting ability [106]. Specially engineered nano scale devices can also be utilized to select the target cancer cells and deliver therapeutic payloads. Additionally, with the help of external stimuli, certain nanomaterials can convert their magnetic or other forms of energy into heat which can help to destroy the cancer cells. By these process nanomaterials can selectively kill infected cells while sparing healthy tissue, making it a promising alternative to traditional cancer treatments like radiation therapy. Nanomaterials can be utilized to enhance the efficacy of immunotherapy, which uses the body's immune system to combat with cancer. Thus, in the near future, nanoscale materials could dramatically improve cancer treatment.

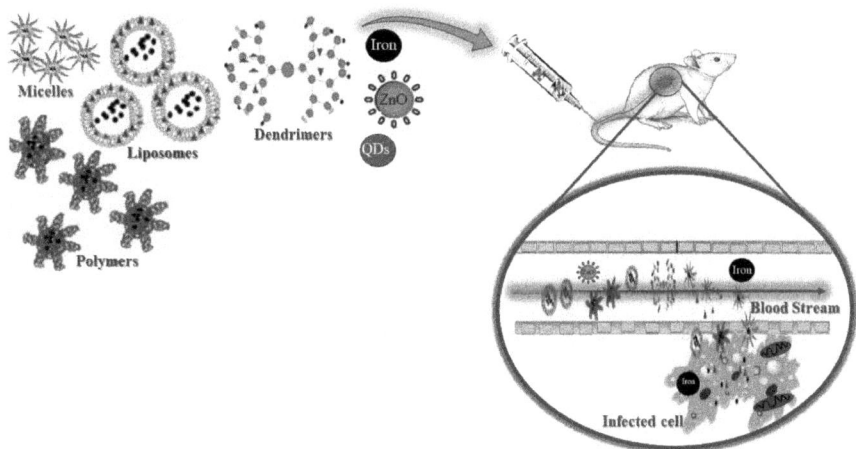

Figure 5. Schemetic diagram of drug delivery using nanomaterials [103].

5.4 Tissue engineering and regenerative medicine

There is a growing demand for tissue engineering and regenerative medicine (TERM) because there are many drawbacks to tissue and organ transplantation, including the scarcity of donors and transplant rejection. The influence of nanomaterial-based technology has changed the conventional and straightforward methods in TERM towards more sophisticated and effective systems. Various materials, including ceramics, metals, and organic or synthetic polymers, can be used to make nano range materials. Their compositions and unique properties, such as high surface area and tunable surface properties, make them one of the most popular candidates in the TERM field for imaging, mechanical strength enhancement, supplements to bio-ink, and carriers for microbes and bioactive agents [107]. Nanomaterials such as nanoparticles, nanofibers, and nanopatterned surfaces have been applied in the TERM field to control cell behavior. Nano scale structures can impact cellular activities like adhesion, proliferation, and segregation. Most significantly, nanomaterials have unique optical and magnetic properties, making them suitable agents for observing in vivo cellular function after a transplant. Applications of nanomaterials in TERM include using therapeutic and imaging systems, controlling the release of several bioactive agents, reducing toxicity and increasing biocompatibility through tissue-specific delivery, particularly growth factors to direct stem cell fate and morphogenesis, embedding novel biomaterials with superior control within scaffolds, and adjusting the mechanical strength of scaffolds for hard tissue applications [108]. Nanomaterial scaffolds can be used to resemble the extracellular matrix in nature. These scaffolds could be created with particular mechanical, structural, and biochemical characteristics for improved tissue regeneration. By incorporating nanomaterials with self-healing properties into tissue engineering constructs, it may be possible to create materials that can repair themselves which will allow prolonged functionality and improved lifespan of implants.

5.5   Environmental applications

The use of nanotechnology can help to solve environmental problems. Highly sensitive and selective sensors for environmental monitoring can be created using nanomaterials. For example, fluorescent probes made of quantum dots (nanoscale semiconductors) can be used to detect and measure contaminants in the air, water, and soil. Additionally, nano scale materials can be functionalized to precisely bind with certain pollutants, enabling quick and effective real-time detection. After detecting, nanomaterials can be used as a smart and efficient filtration medium for removing contaminants, bacteria, and heavy metals from water (mechanisms are shown in Fig. 6). Carbon nanotubes and graphene oxide have shown promise in water treatment procedures due to their large surface area and ion adsorption properties. Nano sized materials can be added to air filtration systems to capture and neutralize pollutants such as volatile organic compounds, nitrogen oxides, and other particulate contaminants. When exposed to natural or artificial light, nanostructured materials with photocatalytic characteristics, such as titanium dioxide, nickel oxide and zinc oxide, can degrade dangerous air and water pollutants [56]. In the case of soil, persistent organic pollutants like polychlorinated biphenyls and pesticides can be targeted and degraded by nanoparticles coated with specific chemicals. Nanomaterials can therefore aid in the removal of contaminants from a variety of sources, resulting in cleaner and safer environments.

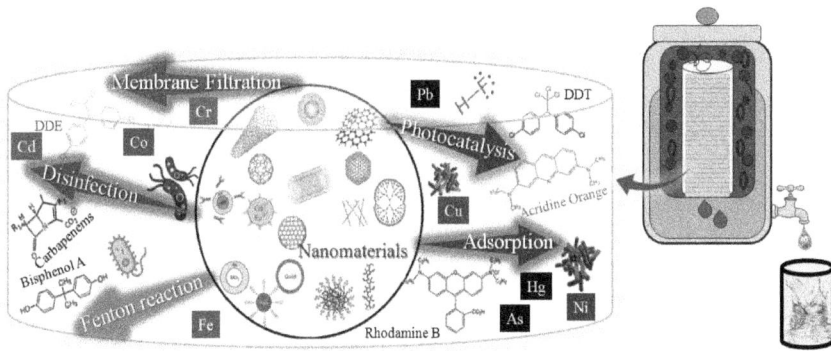

Figure 6. A proposed mechanisms of water filtration by nanomaterials [109].

5.6   Nanorobots

Nano-biotechnology has opened up new possibilities for biomedical using robotics or nanobots. Nanorobots can carry out intracellular surgery, and the elimination of lesions [110]. Although they are still in the research and development stages, they are capable of carrying out their specific task at the molecular and cellular levels [111]. With their advancements in drug delivery and cancer diagnosis, nano-robots have recently interrupted the biomedicine industry. Nanorobots, also known as nano scale machines or gadgets, may be used to precisely manipulate damaged tissues and repair them at the cellular level (shown in Fig. 7). These nanorobots may be able to perform tasks including precise therapeutic drug delivery, cellular debris cleaning, or even tissue regeneration.

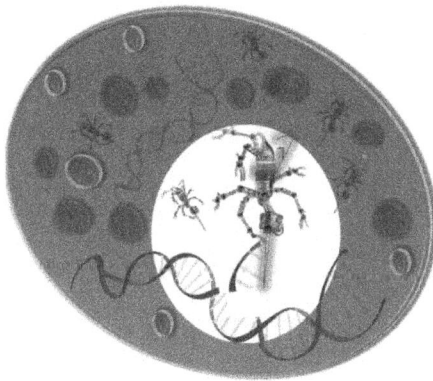

Figure 7. Propose schematic representation of tissue repairing at the cellular level by nanorobots [110].

Conclusion

The field of nanomaterial-based biomedical applications is growing so drastically that some of these have already succeeded in the implementation of nanomedicine with diagnostic tools. Most expected biomedical applications are the development of advanced drug carrier and release systems, enhanced sensing and imaging techniques, and the designing of novel biomaterials for tissue engineering and regenerative medicine. Future researches are going on to investigate and optimize the properties of nanomaterials such as size, shape, surface chemistry, and biocompatibility to ensure their safe and effective application for biological purposes. Additionally, focusing on improving the scalability and cost-effectiveness of nanomaterial synthesis and manufacturing processes will ensure their widespread adoption in biological applications.

References

[1]   A.P. Ramos, M.A.E. Cruz, C.B. Tovani, P. Ciancaglini, Biomedical applications of nanotechnology, Biophys. Rev. 9 (2017) 79–89. https://doi.org/10.1007/s12551-016-0246-2.

[2]   E. Boisselier, D. Astruc, Gold nanoparticles in nanomedicine: preparations, imaging, diagnostics, therapies and toxicity, Chem. Soc. Rev. 38 (2009) 1759–1782. https://doi.org/10.1039/b806051g

[3]   Q.A. Pankhurst, N.K.T. Thanh, S.K. Jones, J. Dobson, Progress in applications of magnetic nanoparticles in biomedicine, J. Phys. D. Appl. Phys. 42 (2009) 224001. https://doi.org/10.1088/0022-3727/42/22/224001

[4]   T. Jamieson, R. Bakhshi, D. Petrova, R. Pocock, M. Imani, A.M. Seifalian, Biological applications of quantum dots, Biomaterials. 28 (2007) 4717–4732. https://doi.org/10.1016/j.biomaterials.2007.07.014

[5]   Q.A. Pankhurst, J. Connolly, S.K. Jones, J. Dobson, Applications of magnetic nanoparticles in biomedicine, J. Phys. D. Appl. Phys. 36 (2003) R167–R181. https://doi.org/10.1088/0022-3727/36/13/201

[6]   R. Kramer, D. Cohen, Functional genomics to new drug targets, Nat. Rev. Drug Discov. 3 (2004) 965–972. https://doi.org/10.1038/nrd1552.

[7]   Y. Chen, H. Cong, Y. Shen, B. Yu, Biomedical application of manganese dioxide nanomaterials, Nanotechnology. 31 (2020) 202001. https://doi.org/10.1088/1361-6528/ab6fe1

[8]   F.S. Alves, J.N. Cruz, I.N. de Farias Ramos, D.L. do Nascimento Brandão, R.N. Queiroz, G.V. da Silva, G.V. da Silva, M.F. Dolabela, M.L. da Costa, A.S. Khayat, J. de Arimatéia Rodrigues do Rego, D. do Socorro Barros Brasil, Evaluation of Antimicrobial Activity and Cytotoxicity Effects of Extracts of Piper nigrum L. and Piperine, Separations. 10 (2023) 21. https://doi.org/10.3390/separations10010021

[9]   P. Podsiadlo, S. Paternel, J.M. Rouillard, Z. Zhang, J. Lee, J.W. Lee, E. Gulari, N.A. Kotov, Layer-by-layer assembly of nacre-like nanostructured composites with antimicrobial properties, AIChE Annu. Meet. Conf. Proc. 21 (2005) 5198–5210. https://doi.org/10.1021/la051284+

[10] S. Murthy, P. Effiong, C.C. Fei, Metal oxide nanoparticles in biomedical applications, Met. Oxide Powder Technol. Fundam. Process. Methods Appl. (2020) 233–251. https://doi.org/10.1016/B978-0-12-817505-7.00011-7.

[11] D. Ung, L.D. Tung, G. Caruntu, D. Delaportas, I. Alexandrou, I.A. Prior, N.T.K. Thanh, Variant shape growth of nanoparticles of metallic Fe-Pt, Fe-Pd and Fe-Pt-Pd alloys, CrystEngComm. 11 (2009) 1309–1316. https://doi.org/10.1039/b823290n

[12] T. Jamieson, R. Bakhshi, D. Petrova, R. Pocock, M. Imani, A.M. Seifalian, Biological applications of quantum dots, Biomaterials. 28 (2007) 4717–4732. https://doi.org/10.1016/j.biomaterials.2007.07.014

[13] J. Drbohlavova, V. Adam, R. Kizek, J. Hubalek, Quantum dots - characterization, preparation and usage in biological systems, Int. J. Mol. Sci. 10 (2009) 656–673. https://doi.org/10.3390/ijms10020656

[14] A.L. Rogach, A. Eychmüller, S.G. Hickey, S. V. Kershaw, Infrared-emitting colloidal nanocrystals: Synthesis, assembly, spectroscopy, and applications, Small. 3 (2007) 536–557. https://doi.org/10.1002/smll.200600625

[15] K. Shahane, M. Kshirsagar, S. Tambe, D. Jain, S. Rout, M.K.M. Ferreira, S. Mali, P. Amin, P.P. Srivastav, J. Cruz, R.R. Lima, An Updated Review on the Multifaceted Therapeutic Potential of Calendula officinalis L., Pharmaceuticals. 16 (2023) 611. https://doi.org/10.3390/ph16040611

[16] A.G. Roca, R. Costo, A.F. Rebolledo, S. Veintemillas-Verdaguer, P. Tartaj, T. González-Carreño, M.P. Morales, C.J. Serna, Progress in the preparation of magnetic nanoparticles for applications in biomedicine, J. Phys. D. Appl. Phys. 42 (2009) 224002. https://doi.org/10.1088/0022-3727/42/22/224002

[17] N.T.K. Thanh, L.A.W. Green, Functionalisation of nanoparticles for biomedical applications, Nano Today. 5 (2010) 213–230. https://doi.org/10.1016/j.nantod.2010.05.003

[18] Scientific and Clinical Applications of Magnetic Carriers, Sci. Clin. Appl. Magn. Carriers. (1997). https://doi.org/10.1007/978-1-4757-6482-6

[19] M. Chen, D.E. Nikles, Synthesis of spherical FePd and CoPt nanoparticles, J. Appl. Phys. 91 (2002) 8477–8479. https://doi.org/10.1063/1.1456406

[20] R. Shanmuganathan, I. Karuppusamy, M. Saravanan, H. Muthukumar, K. Ponnuchamy, V.S. Ramkumar, A. Pugazhendhi, Synthesis of Silver Nanoparticles and their Biomedical Applications - A Comprehensive Review, Curr. Pharm. Des. 25 (2019) 2650–2660. https://doi.org/10.2174/1381612825666190708185506

[21] I. Rezić, Nanoparticles for Biomedical Application and Their Synthesis, Polymers (Basel). 14 (2022) 4961. https://doi.org/10.3390/polym14224961

[22] C.S.S.R. Kumar, F. Mohammad, Magnetic nanomaterials for hyperthermia-based therapy and controlled drug delivery, Adv. Drug Deliv. Rev. 63 (2011) 789–808. https://doi.org/10.1016/j.addr.2011.03.008

[23] A.M. Schrand, M.F. Rahman, S.M. Hussain, J.J. Schlager, D.A. Smith, A.F. Syed, Metal-based nanoparticles and their toxicity assessment, Wiley Interdisc. Rev. Nanomedicine Nanobiotechnology. 2 (2010) 544–568. https://doi.org/10.1002/wnan.103

Materials Research Forum LLC
https://doi.org/10.21741/9781644902998-1

[24] M.I. Anik, N. Mahmud, A. Al Masud, M. Hasan, Gold nanoparticles (GNPs) in biomedical and clinical applications: A review, Nano Sel. 3 (2022) 792–828. https://doi.org/10.1002/nano.202100255

[25] L. S Jairam, A. Chandrashekar, T.N. Prabhu, S.B. Kotha, M.S. Girish, I.M. Devraj, M. Dhanya Shri, K. Prashantha, A review on biomedical and dental applications of cerium oxide nanoparticles — Unearthing the potential of this rare earth metal, J. Rare Earths. 41 (2023) 1645–1661. https://doi.org/10.1016/j.jre.2023.04.009

[26] S. Pantic, S.R. Skodric, Z. Loncar, I. Pantic, Zinc oxide nanoparticles: Potential novel applications in cellular physiology, pathology, neurosciences and cancer research, Rev. Adv. Mater. Sci. 58 (2019) 17–21. https://doi.org/10.1515/rams-2019-0002

[27] S. Jha, R. Rani, S. Singh, Biogenic Zinc Oxide Nanoparticles and Their Biomedical Applications: A Review, J. Inorg. Organomet. Polym. Mater. 33 (2023) 1437–1452. https://doi.org/10.1007/s10904-023-02550-x

[28] A. Esmaeilnejad, P. Mahmoudi, A. Zamanian, M. Mozafari, Synthesis of titanium oxide nanotubes and their decoration by MnO nanoparticles for biomedical applications, Ceram. Int. 45 (2019) 19275–19282. https://doi.org/10.1016/j.ceramint.2019.06.177

[29] S. Jafari, B. Mahyad, H. Hashemzadeh, S. Janfaza, T. Gholikhani, L. Tayebi, Biomedical applications of TiO2 nanostructures: Recent advances, Int. J. Nanomedicine. 15 (2020) 3447–3470. https://doi.org/10.2147/IJN.S249441

[30] S. Khan, A.A. Ansari, A. Malik, A.A. Chaudhary, J.B. Syed, A.A. Khan, Preparation, characterizations and in vitro cytotoxic activity of nickel oxide nanoparticles on HT-29 and SW620 colon cancer cell lines, J. Trace Elem. Med. Biol. 52 (2019) 12–17. https://doi.org/10.1016/j.jtemb.2018.11.003

[31] N. Behera, M. Arakha, M. Priyadarshinee, B.S. Pattanayak, S. Soren, S. Jha, B.C. Mallick, Oxidative stress generated at nickel oxide nanoparticle interface results in bacterial membrane damage leading to cell death, RSC Adv. 9 (2019) 24888–24894. https://doi.org/10.1039/c9ra02082a

[32] S. Naz, A. Gul, M. Zia, R. Javed, Synthesis, biomedical applications, and toxicity of CuO nanoparticles, Appl. Microbiol. Biotechnol. 107 (2023) 1039–1061. https://doi.org/10.1007/s00253-023-12364-z

[33] P. Paramita, V.D. Subramaniam, R. Murugesan, M. Gopinath, I. Ramachandran, S. Ramalingam, X.F. Sun, A. Banerjee, F. Marotta, S. Pathak, Evaluation of potential anti-cancer activity of cationic liposomal nanoformulated Lycopodium clavatum in colon cancer cells, IET Nanobiotechnology. 12 (2018) 727–732. https://doi.org/10.1049/iet-nbt.2017.0106

[34] T. Sabah, K.H. Jawad, N. Al-attar, Synthesis and Biomedical Activity of Aluminium Oxide Nanoparticles by Laser Ablation Technique, Res. J. Pharm. Technol. 16 (2023) 1267–1273. https://doi.org/10.52711/0974-360X.2023.00209

[35] S.M. Dizaj, F. Lotfipour, M. Barzegar-Jalali, M.H. Zarrintan, K. Adibkia, Antimicrobial activity of the metals and metal oxide nanoparticles, Mater. Sci. Eng. C. 44 (2014) 278–284. https://doi.org/10.1016/j.msec.2014.08.031

[36] M. Kumari, B. Sarkar, K. Mukherjee, Nanoscale calcium oxide and its biomedical applications: A comprehensive review, Biocatal. Agric. Biotechnol. 47 (2023) 102506. https://doi.org/10.1016/j.bcab.2022.102506

[37] S. Abinaya, H.P. Kavitha, Magnesium Oxide Nanoparticles: Effective Antilarvicidal and Antibacterial Agents, ACS Omega. 8 (2023) 5225–5233. https://doi.org/10.1021/acsomega.2c01450

[38] K. Szostak, P. Ostaszewski, J. Pulit-Prociak, M. Banach, Bismuth Oxide Nanoparticles in Drug Delivery Systems, Pharm. Chem. J. 53 (2019) 48–51. https://doi.org/10.1007/s11094-019-01954-9

[39] M.A. Shahbazi, L. Faghfouri, M.P.A. Ferreira, P. Figueiredo, H. Maleki, F. Sefat, J. Hirvonen, H.A. Santos, The versatile biomedical applications of bismuth-based nanoparticles and composites: Therapeutic, diagnostic, biosensing, and regenerative properties, Chem. Soc. Rev. 49 (2020) 1253–1321. https://doi.org/10.1039/c9cs00283a

[40] M. Wu, P. Hou, L. Dong, L. Cai, Z. Chen, M. Zhao, J. Li, Manganese dioxide nanosheets: From preparation to biomedical applications, Int. J. Nanomedicine. 14 (2019) 4781–4800. https://doi.org/10.2147/IJN.S207666

[41] D.F. Williams, A model for biocompatibility and its evaluation, J. Biomed. Eng. 11 (1989) 185–191. https://doi.org/10.1016/0141-5425(89)90138-6

[42] M.A. Dobrovolskaia, S.E. McNeil, Immunological properties of engineered nanomaterials, Nat. Nanotechnol. 2 (2007) 469–478. https://doi.org/10.1038/nnano.2007.223

[43] M. Huston, M. Debella, M. Dibella, A. Gupta, Green synthesis of nanomaterials, Nanomaterials. 11 (2021) 2130. https://doi.org/10.3390/nano11082130

[44] D. Bokov, A. Turki Jalil, S. Chupradit, W. Suksatan, M. Javed Ansari, I.H. Shewael, G.H. Valiev, E. Kianfar, Nanomaterial by Sol-Gel Method: Synthesis and Application, Adv. Mater. Sci. Eng. 2021 (2021) 1–21. https://doi.org/10.1155/2021/5102014

[45] C. Dhand, N. Dwivedi, X.J. Loh, A.N. Jie Ying, N.K. Verma, R.W. Beuerman, R. Lakshminarayanan, S. Ramakrishna, Methods and strategies for the synthesis of diverse nanoparticles and their applications: A comprehensive overview, RSC Adv. 5 (2015) 105003–105037. https://doi.org/10.1039/c5ra19388e

[46] A.E. Danks, S.R. Hall, Z. Schnepp, The evolution of "sol-gel" chemistry as a technique for materials synthesis, Mater. Horizons. 3 (2016) 91–112. https://doi.org/10.1039/c5mh00260e

[47] S. Barman, S. Sikdar, A. Biswas, B.K. Mandal, R. Das, Structural microanalysis of green synthesized AlxZn(1-x)O nanoparticles, Nano Express. 1 (2020) 20003. https://doi.org/10.1088/2632-959X/ab9f54

[48] Y. Huang, C.Y. Haw, Z. Zheng, J. Kang, J.C. Zheng, H.Q. Wang, Biosynthesis of Zinc Oxide Nanomaterials from Plant Extracts and Future Green Prospects: A Topical Review, Adv. Sustain. Syst. 5 (2021). https://doi.org/10.1002/adsu.202000266

[49] S. Barman, S. Sikdar, A. Biswas, A. Islam, R. Das, Green synthesis of MnxZn(1-x)O nanostructure using Azadirachta indica leaf extract and its microstructural and optical study, Phys. Scr. 97 (2022) 45002. https://doi.org/10.1088/1402-4896/ac520c

[50] M.H. Sarfraz, M. Zubair, B. Aslam, A. Ashraf, M.H. Siddique, S. Hayat, J.N. Cruz, S. Muzammil, M. Khurshid, M.F. Sarfraz, A. Hashem, T.M. Dawoud, G.D. Avila-Quezada, E.F. Abd_Allah, Comparative analysis of phyto-fabricated chitosan, copper oxide, and chitosan-based CuO nanoparticles: antibacterial potential against Acinetobacter baumannii isolates and anticancer activity against HepG2 cell lines, Front. Microbiol. 14 (2023) 1188743. https://doi.org/10.3389/fmicb.2023.1188743

[51] S. Barman, S. Sikdar, R. Das, A comprehensive study on ZrO2-ZnO nanocomposites synthesized by the plant-mediated green method, Phys. Scr. 98 (2023) 85947. https://doi.org/10.1088/1402-4896/ace857

[52] B. Shkodra-Pula, A. Vollrath, U.S. Schubert, S. Schubert, Polymer-based nanoparticles for biomedical applications, Front. Nanosci. 16 (2020) 233–252. https://doi.org/10.1016/B978-0-08-102828-5.00009-7

[53] M. Ovais, I. Ahmad, A.T. Khalil, S. Mukherjee, R. Javed, M. Ayaz, A. Raza, Z.K. Shinwari, Wound healing applications of biogenic colloidal silver and gold nanoparticles: recent trends and future prospects, Appl. Microbiol. Biotechnol. 102 (2018) 4305–4318. https://doi.org/10.1007/s00253-018-8939-z

[54] A.A.A. Aljabali, Y. Akkam, M.S. Al Zoubi, K.M. Al-Batayneh, B. Al-Trad, O.A. Alrob, A.M. Alkilany, M. Benamara, D.J. Evans, Synthesis of gold nanoparticles using leaf extract of ziziphus zizyphus and their antimicrobial activity, Nanomaterials. 8 (2018) 174. https://doi.org/10.3390/nano8030174

[55] B.K. Mandal, R. Mandal, S. Sikdar, S. Sarma, A. Srinivasan, S.R. Chowdhury, B. Das, R. Das, Green synthesis of NiO nanoparticle using Punica granatum peel extract and its characterization for methyl orange degradation, Mater. Today Commun. 34 (2023) 105302. https://doi.org/10.1016/j.mtcomm.2022.105302

[56] S.A. Razack, A. Suresh, S. Sriram, G. Ramakrishnan, S. Sadanandham, M. Veerasamy, R.B. Nagalamadaka, R. Sahadevan, Green synthesis of iron oxide nanoparticles using Hibiscus rosa-sinensis for fortifying wheat biscuits, SN Appl. Sci. 2 (2020). https://doi.org/10.1007/s42452-020-2477-x

[57] S.C. Mali, A. Dhaka, C.K. Githala, R. Trivedi, Green synthesis of copper nanoparticles using Celastrus paniculatus Willd. leaf extract and their photocatalytic and antifungal properties, Biotechnol. Reports. 27 (2020) e00518. https://doi.org/10.1016/j.btre.2020.e00518

[58] Hemlata, P.R. Meena, A.P. Singh, K.K. Tejavath, Biosynthesis of Silver Nanoparticles Using Cucumis prophetarum Aqueous Leaf Extract and Their Antibacterial and Antiproliferative Activity against Cancer Cell Lines, ACS Omega. 5 (2020) 5520–5528. https://doi.org/10.1021/acsomega.0c00155

[59] M. Naseer, U. Aslam, B. Khalid, B. Chen, Green route to synthesize Zinc Oxide Nanoparticles using leaf extracts of Cassia fistula and Melia azadarach and their antibacterial potential, Sci. Rep. 10 (2020). https://doi.org/10.1038/s41598-020-65949-3

[60] S.E. Sandler, B. Fellows, O. Thompson Mefford, Best Practices for Characterization of Magnetic Nanoparticles for Biomedical Applications, Anal. Chem. 91 (2019) 14159–14169. https://doi.org/10.1021/acs.analchem.9b03518

[61] I. Rezić, Nanoparticles for Biomedical Application and Their Synthesis, Polymers (Basel). 14 (2022) 4961. https://doi.org/10.3390/polym14224961

[62] M. Mabrouk, D.B. Das, Z.A. Salem, H.H. Beherei, Nanomaterials for biomedical applications: Production, characterisations, recent trends and difficulties, Molecules. 26 (2021) 1077. https://doi.org/10.3390/molecules26041077

[63] S. Muzammil, J. Neves Cruz, R. Mumtaz, I. Rasul, S. Hayat, M.A. Khan, A.M. Khan, M.U. Ijaz, R.R. Lima, M. Zubair, Effects of Drying Temperature and Solvents on In Vitro Diabetic Wound Healing Potential of Moringa oleifera Leaf Extracts, Molecules. 28 (2023) 710. https://doi.org/10.3390/molecules28020710

[64] Z. Cai, Z. Ye, X. Yang, Y. Chang, H. Wang, Y. Liu, A. Cao, Encapsulated enhanced green fluorescence protein in silica nanoparticle for cellular imaging, Nanoscale. 3 (2011) 1974–1976. https://doi.org/10.1039/c0nr00956c

[65] B.K. Mandal, R. Mandal, D. Limbu, M.D. Adhikari, P.S. Chauhan, R. Das, Green synthesis of AgCl nanoparticles using Calotropis gigantea: Characterization and their enhanced antibacterial activities, Chem. Phys. Lett. 801 (2022) 139699. https://doi.org/10.1016/j.cplett.2022.139699

[66] R. Weissleder, Scaling down imaging: Molecular mapping of cancer in mice, Nat. Rev. Cancer. 2 (2002) 11–18. https://doi.org/10.1038/nrc701

[67] S. Flacke, S. Fischer, M.J. Scott, R.J. Fuhrhop, J.S. Allen, M. McLean, P. Winter, G.A. Sicard, P.J. Gaffney, S.A. Wickline, G.M. Lanza, Novel MRI contrast agent for molecular imaging of fibrin implications for detecting vulnerable plaques, Circulation. 104 (2001) 1280–1285. https://doi.org/10.1161/hc3601.094303

[68] R.B.M. de Almeida, D.B. Barbosa, M.R. do Bomfim, J.A.O. Amparo, B.S. Andrade, S.L. Costa, J.M. Campos, J.N. Cruz, C.B.R. Santos, F.H.A. Leite, M.B. Botura, Identification of a Novel Dual Inhibitor of Acetylcholinesterase and Butyrylcholinesterase: In Vitro and In Silico Studies, Pharmaceuticals. 16 (2023) 95. https://doi.org/10.3390/ph16010095

[69] N. Arndt, H.D.N. Tran, R. Zhang, Z.P. Xu, H.T. Ta, Different Approaches to Develop Nanosensors for Diagnosis of Diseases, Adv. Sci. 7 (2020). https://doi.org/10.1002/advs.202001476

[70] L. Helm, Optimization of gadolinium-based MRI contrast agents for high magnetic-field applications, Future Med. Chem. 2 (2010) 385–396. https://doi.org/10.4155/fmc.09.174

[71] J.A. Park, J.J. Lee, J.C. Jung, D.Y. Yu, C. Oh, S. Ha, T.J. Kim, Y. Chang, Gd-DOTA conjugate of RGD as a potential tumor-targeting MRI contrast agent., Chembiochem. 9 (2008) 2811–2813. https://doi.org/10.1002/cbic.200800529

[72] W.L. Zhang, D.W. Yong, J. Huang, J.H. Yu, S.Y. Liu, M.X. Fan, Fabrication of polymer-gadolinium (III) complex nanomicelle from poly(ethylene glycol)-polysuccinimide conjugate and diethylenetriaminetetraacetic acid-gadolinium as magnetic resonance imaging contrast agents, J. Appl. Polym. Sci. 120 (2011) 2596–2605. https://doi.org/10.1002/app.33464

[73] L.S. Karfeld-Sulzer, E.A. Waters, N.E. Davis, T.J. Meade, A.E. Barron, Multivalent protein polymer mri contrast agents: Controlling relaxivity via modulation of amino acid sequence, Biomacromolecules. 11 (2010) 1429–1436 https://doi.org/10.1021/bm901378a

[74] N. Kamaly, A.D. Miller, Paramagnetic liposome nanoparticles for cellular and tumour imaging, Int. J. Mol. Sci. 11 (2010) 1759–1776. https://doi.org/10.3390/ijms11041759

[75] H. Bin Na, I.C. Song, T. Hyeon, Inorganic nanoparticles for MRI contrast agents, Adv. Mater. 21 (2009) 2133–2148. https://doi.org/10.1002/adma.200802366

[76] Y.Z. Shao, L.Z. Liu, S.Q. Song, R.H. Cao, H. Liu, C.Y. Cui, X. Li, M.J. Bie, L. Li, A novel one-step synthesis of Gd3+-incorporated mesoporous SiO2 nanoparticles for use as an efficient MRI contrast agent, Contrast Media Mol. Imaging. 6 (2011) 110–118. https://doi.org/10.1002/cmmi.412

[77] L. Nie, F. Liu, P. Ma, X. Xiao, Applications of gold nanoparticles in optical biosensors, J. Biomed. Nanotechnol. 10 (2014) 2700–2721. https://doi.org/10.1166/jbn.2014.1987

[78] H. Daraee, A. Eatemadi, E. Abbasi, S.F. Aval, M. Kouhi, A. Akbarzadeh, Application of gold nanoparticles in biomedical and drug delivery, Artif. Cells, Nanomedicine Biotechnol. 44 (2016) 410–422. https://doi.org/10.3109/21691401.2014.955107

[79] L.E. Cole, R.D. Ross, J.M. Tilley, T. Vargo-Gogola, R.K. Roeder, Gold nanoparticles as contrast agents in X-ray imaging and computed tomography, Nanomedicine. 10 (2015) 321–341. https://doi.org/10.2217/nnm.14.171

[80] D. Lombardo, M.A. Kiselev, M.T. Caccamo, Smart Nanoparticles for Drug Delivery Application: Development of Versatile Nanocarrier Platforms in Biotechnology and Nanomedicine, J. Nanomater. 2019 (2019) 1–26. https://doi.org/10.1155/2019/3702518

[81] D. Bobo, K.J. Robinson, J. Islam, K.J. Thurecht, S.R. Corrie, Nanoparticle-Based Medicines: A Review of FDA-Approved Materials and Clinical Trials to Date, Pharm. Res. 33 (2016) 2373–2387. https://doi.org/10.1007/s11095-016-1958-5

[82] R. Qiao, C. Yang, M. Gao, Superparamagnetic iron oxide nanoparticles: From preparations to in vivo MRI applications, J. Mater. Chem. 19 (2009) 6274–6293. https://doi.org/10.1039/b902394a

[83] A.S. Teja, P.Y. Koh, Synthesis, properties, and applications of magnetic iron oxide nanoparticles, Prog. Cryst. Growth Charact. Mater. 55 (2009) 22–45. https://doi.org/10.1016/j.pcrysgrow.2008.08.003

[84] J. Mürbe, A. Rechtenbach, J. Töpfer, Synthesis and physical characterization of magnetite nanoparticles for biomedical applications, Mater. Chem. Phys. 110 (2008) 426–433. https://doi.org/10.1016/j.matchemphys.2008.02.037

[85] C. Sun, J.S.H. Lee, M. Zhang, Magnetic nanoparticles in MR imaging and drug delivery, Adv. Drug Deliv. Rev. 60 (2008) 1252–1265. https://doi.org/10.1016/j.addr.2008.03.018

[86] J. Hong, P. Gong, D. Xu, H. Sun, S. Yao, Synthesis and characterization of carboxyl-functionalized magnetic nanogel via "Green" photochemical method, J. Appl. Polym. Sci. 105 (2007) 1882–1887. https://doi.org/10.1002/app.25655

[87] A.K. Gupta, M. Gupta, Synthesis and surface engineering of iron oxide nanoparticles for biomedical applications, Biomaterials. 26 (2005) 3995–4021. https://doi.org/10.1016/j.biomaterials.2004.10.012

[88] J. Pan, Y. Wang, S.S. Feng, Formulation, characterization, and in vitro evaluation of quantum dots loaded in poly(Lactide)-Vitamin E TPGS nanoparticles for cellular and

molecular imaging, Biotechnol. Bioeng. 101 (2008) 622–633.
https://doi.org/10.1002/bit.21924

[89] L. Yang, H. Mao, Y. Andrew Wang, Z. Cao, X. Peng, X. Wang, H. Duan, C. Ni, Q. Yuan, G. Adams, M.Q. Smith, W.C. Wood, X. Gao, S. Nie, Single chain epidermal growth factor receptor antibody conjugated nanoparticles for in vivo tumor targeting and imaging, Small. 5 (2009) 235–243. https://doi.org/10.1002/smll.200800714

[90] B. Ballou, L. Ernst, A. Waggoner, Fluorescence Imaging of Tumors In Vivo, Curr. Med. Chem. 12 (2005) 795–805. https://doi.org/10.2174/0929867053507324

[91] B. Ballou, L.A. Ernst, S. Andreko, T. Harper, J.A.J. Fitzpatrick, A.S. Waggoner, M.P. Bruchez, Sentinel lymph node imaging using quantum dots in mouse tumor models, Bioconjug. Chem. 18 (2007) 389–396. https://doi.org/10.1021/bc060261j

[92] P. Wunderbaldinger, L. Josephson, C. Bremer, A. Moore, R. Weissleder, Detection of lymph node metastases by contrast-enhanced MRI in an experimental model, Magn. Reson. Med. 47 (2002) 292–297. https://doi.org/10.1002/mrm.10068

[93] O. Rabin, J.M. Perez, J. Grimm, G. Wojtkiewicz, R. Weissleder, An X-ray computed tomography imaging agent based on long-circulating bismuth sulphide nanoparticles, Nat. Mater. 5 (2006) 118–122. https://doi.org/10.1038/nmat1571

[94] M.A. Hahn, A.K. Singh, P. Sharma, S.C. Brown, B.M. Moudgil, Nanoparticles as contrast agents for in-vivo bioimaging: Current status and future perspectives, Anal. Bioanal. Chem. 399 (2011) 3–27. https://doi.org/10.1007/s00216-010-4207-5

[95] D.B. Elrod, R. Partha, D. Danila, S.W. Casscells, J.L. Conyers, An iodinated liposomal computed tomographic contrast agent prepared from a diiodophosphatidylcholine lipid, Nanomedicine Nanotechnology, Biol. Med. 5 (2009) 42–45. https://doi.org/10.1016/j.nano.2008.06.007

[96] S. Kweon, H.J. Lee, W.J. Hyung, J. Suh, J.S. Lim, S.J. Lim, Liposomes coloaded with iopamidol/lipiodol as a res-targeted contrast agent for computed tomography imaging, Pharm. Res. 27 (2010) 1408–1415. https://doi.org/10.1007/s11095-010-0135-5

[97] J. Zheng, C. Allen, S. Serra, D. Vines, M. Charrong, D.A. Jaffray, Liposome contrast agent for CT-based detection and localization of neoplastic and inflammatory lesions in rabbits: Validation with FDG-PET and histology, Contrast Media Mol. Imaging. 5 (2010) 147–154. https://doi.org/10.1002/cmmi.378

[98] A. Chrastina, J.E. Schnitzer, Iodine-125 radiolabeling of silver nanoparticles for in vivo SPECT imaging, Int. J. Nanomedicine. 5 (2010) 653–659. https://doi.org/10.2147/IJN.S11677

[99] J.L. Van Herck, G.R.Y. De Meyer, W. Martinet, R.A. Salgado, B. Shivalkar, R. De Mondt, H. Van De Ven, A. Ludwlg, P. Van Der Veken, L. Van Vaeck, H. Bult, A G. Herman, C.J. Vrints, Multi-slice computed tomography with N1177 identifies ruptured atherosclerotic plaques in rabbits, Basic Res. Cardiol. 105 (2010) 51–59. https://doi.org/10.1007/s00395-009-0052-0

[100] K.L. Aillon, N. El-Gendy, C. Dennis, J.P. Norenberg, J. McDonald, C. Berkland, Iodinated NanoClusters as an inhaled computed tomography contrast agent for lung visualization, Mol. Pharm. 7 (2010) 1274–1282. https://doi.org/10.1021/mp1000718

[101] S. Zalipsky, Chemistry of polyethylene glycol conjugates with biologically active molecules, Adv. Drug Deliv. Rev. 16 (1995) 157–182. https://doi.org/10.1016/0169-409X(95)00023-Z

[102] Z. Li, S. Tan, S. Li, Q. Shen, K. Wang, Cancer drug delivery in the nano era: An overview and perspectives (Review), Oncol. Rep. 38 (2017) 611–624. https://doi.org/10.3892/or.2017.5718

[103] S. Katsuki, T. Matoba, J.I. Koga, K. Nakano, K. Egashira, Anti-inflammatory Nanomedicine for Cardiovascular Disease, Front. Cardiovasc. Med. 4 (2017). https://doi.org/10.3389/fcvm.2017.00087

[104] N.A. Ochekpe, P.O. Olorunfemi, N.C. Ngwuluka, Nanotechnology and drug delivery part 1: Background and applications, Trop. J. Pharm. Res. 8 (2009) 265–274. https://doi.org/10.4314/tjpr.v8i3.44546

[105] S. Sim, N.K. Wong, Nanotechnology and its use in imaging and drug delivery (Review), Biomed. Reports. 14 (2021). https://doi.org/10.3892/br.2021.1418

[106] M. Fathi-Achachelouei, H. Knopf-Marques, C.E. Ribeiro da Silva, J. Barthès, E. Bat, A. Tezcaner, N.E. Vrana, Use of Nanoparticles in Tissue Engineering and Regenerative Medicine, Front. Bioeng. Biotechnol. 7 (2019). https://doi.org/10.3389/fbioe.2019.00113

[107] Y.L. Colson, M.W. Grinstaff, Biologically responsive polymeric nanoparticles for drug delivery, Adv. Mater. 24 (2012) 3878–3886. https://doi.org/10.1002/adma.201200420

[108] M.P. Ajith, M. Aswathi, E. Priyadarshini, P. Rajamani, Recent innovations of nanotechnology in water treatment: A comprehensive review, Bioresour. Technol. 342 (2021) 126000. https://doi.org/10.1016/j.biortech.2021.126000

[109] M. Rai, J.C. Dos Santos, M.F. Soler, P.R. Franco Marcelino, L.P. Brumano, A.P. Ingle, S. Gaikwad, A. Gade, S.S. Da Silva, Strategic role of nanotechnology for production of bioethanol and biodiesel, Nanotechnol. Rev. 5 (2016) 231–250. https://doi.org/10.1515/ntrev-2015-0069

[110] A.A.G. Requicha, Nanorobots, NEMS, and nanoassembly, Proc. IEEE. 91 (2003) 1922–1933. https://doi.org/10.1109/JPROC.2003.818333

[111] S.M. Asil, J. Ahlawat, G.G. Barroso, M. Narayan, Application of nanotechnology in stem-cell-based therapy of neurodegenerative diseases, Appl. Sci. 10 (2020) 4852. https://doi.org/10.3390/app10144852

Nanoparticle Toxicity and Compatibility
Materials Research Foundations 161 (2024) 27-63

Materials Research Forum LLC
https://doi.org/10.21741/9781644902998-2

Chapter 2

Nanomaterials for Tissue Engineering: Synthesis, Characterisation and Application

Drishya Prakashan[1,2], Riya Sharma[1,2], Sayanti Halder[1], Sonu Gandhi[1,2,*]

[1] DBT-National Institute of Animal Biotechnology (NIAB), Hyderabad-500032, Telangana, India

[2] DBT-Regional Centre for Biotechnology (RCB), Faridabad-121001, Haryana, India

gandhi@niab.org.in

Abstract

Tissue engineering has recently become an effective method for restoring and rebuilding injured tissues and organs. Scaffolds for tissue engineering are essential because they not only give targeted cells structural support, but also act as templates to direct regeneration of tissue and regulate structure of tissue. Nanomaterials of various types have gradually grown and attracted a wide spectrum of research interests over the previous few years due to their distinctive physicochemical properties and exceptional biocompatibility, allowing remarkable advancements in the repair of wounds, wound healing, regeneration of neural tissue, and cardiac tissue engineering. This chapter focuses on the most recent different types of nanomaterials, its synthesis method, functionalisation and characterisation method for the different application in tissue regeneration and engineering. The chapter also focusses on the developments in the usage of scaffolds, nanosheets, or hydrogels based on different nanomaterials that are designed to repair cartilage, bone, and skin tissues. We have also summarised the difficulties and potential of nanomaterial applications in tissue engineering.

Keywords

Tissue Engineering, Nanoparticles, Surface Modifications, Functionalisation, Drug Delivery

Contents

1. Introduction

A relatively recent area of "tissue engineering" (TE) combines engineering methods with those of cell biology and material science to produce artificial tissues which can mimic natural tissues both physically and physiologically [1,2]. To repair, replace, retain, or increase tissue- and organ-level functions, TE seeks to create biological alternatives. The limitations of current TE techniques include the absence of suitable biomaterials for treatment, unsuccessful cell growth on these materials, unavailability of suitable methods for making suitable physiological architectures, and unstable and inadequate formation of growth factor which are needed to promote cell communication and the right response [3]. The existing limits in this sector are also exacerbated by the incapability to regulate cellular operations and their varied qualities (biological, mechanical, electrochemical, and others), as well as problems with biomolecular detection and biosensors [4]. When trying to replicate the extracellular matrix (ECM) composition of a tissue by constructing a three-dimension (3D) scaffold for cells that has sufficient mechanical strength, ease for tracking cellular activities, and delivery of bioactive agents, a nanoscale approach instead of a macroscopic one is needed to achieve satisfactory results. Nanotechnology involves the synthesis of nanoparticles of sizes ranging from 10 to 1,000 [5]. Nanoparticles are used widely in the biomedical industry, including as vaccine adjuvants, medicine delivery systems, imaging [6,7], therapeutics [8]and biosensors [9–11]. Due to their unique physicochemical characteristics and excellent biocompatibility, various nanomaterials have been investigated and explored as candidates for the fabrication of tissue engineering scaffolds over the past few years. This has led to significant improvements in bone repair, healing of wounds, regeneration of neural tissues, and cardiac tissue engineering [12,13]. They have been employed in TE to achieve better mechanical and biological characteristics [14]. In a wide range of TE applications, gold nanoparticles (GNPs), silver and other metallic nanoparticles and metal oxides, quantum dots, and carbon nanotubes (CNTs) have all proven to be very beneficial due to their surface conjugation and conducting properties, antimicrobial properties, fluorescence properties, and unique electromechanical properties. Additionally, the research of cell mechanotransduction, gene transfer, regulating cell patterning, and the creation of complex 3D tissues have all been used using magnetic nanoparticles (MNPs) [15,16]. Nanoparticles have an advantage over peptides and small proteins in TE because of their size in nanometer range and the high surface to volume ratio that goes along with it. They are simple to diffuse across membranes, which aids cell uptake. Nanoparticles can also be produced in customised sizes and surface properties to suit any need; thus, one is not constrained by a determined size. They also mimic the extracellular matrix (ECM) constituents of tissues inherent nanometer size scale. In this chapter, we describe an outline of the different applications of nanomaterials for tissue engineering. This chapter discusses about the different types of nanomaterials, their synthesis, surface modifications and characterisation for various biomedical applications. the insights in this chapter will be advantageous for the future researchers studying with nanomaterials for TE, permitting them to navigate through requests that fill fit their requirements.

Materials Research Forum LLC
https://doi.org/10.21741/9781644902998-2

2. Types of nanomaterials

To address the rising need for replacement tissues and organs, tissue engineering has attracted a lot of attention to maintain, improve, and restore tissue function. To increase the effectiveness and adaptability of tissue regeneration, researchers have recently been actively investigating new nanomaterials for tissue engineering. Nanomaterials are essential in tissue engineering because they have special qualities that can improve tissue regeneration and repair. The types of nanomaterials utilised most frequently in tissue engineering are listed below (Figure 1) [17].

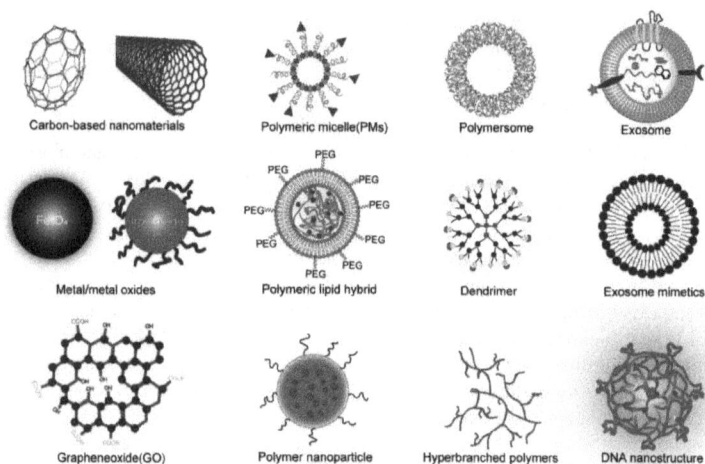

Figure 1: Different nanomaterials used in tissue engineering. Recently reported nanoparticles include metal and metal oxide nanoparticles, inorganic nanoparticles, polymeric nanomaterials exosomes, carbon-based nanomaterials etc. [18].

2.1 Metallic based nanoparticle

Metal nanoparticles are particularly well suited for topical usage in healing due to their wide range of beneficial features for skin tissue engineering, including improved mechanical strength, precise release, and antibacterial action against almost all microbes. The materials that have been most extensively explored for this purpose include Au, Cu, Ag, Fe, Al, Ni, Zr, etc. [19].

2.1.1 Gold nanoparticles

The clinical applications of nanomedicine have created an excessive deal of attention in gold nanoparticles (AuNPs). Due to their special qualities and wide range of uses, gold nanoparticles are important in tissue engineering. Using a combination of cells, biomaterials, and signaling molecules, tissue engineering strives to repair, replace, or regenerate damaged tissues and organs.

A biocompatible and tenable surface chemistry makes gold nanoparticles one of the biomaterials employed in this field. The cardiac excitability and cellular adhesion can be improved by adding gold nanoparticles to scaffold structures. The fabrication and testing of microporous thiol-hydroxyethyl methacrylate scaffolds with immobilised gold nanoparticles for cardiac tissue engineering [20].

For tissue engineering applications, an efficient technique to create nanostructures that transport growth factors and therapeutics directly to their intended sites is by incorporating gold nanoparticles (Au NP) into biopolymer scaffolds. A biodegradable, photocurable polymer resin called polypropylene fumarate can be functionalized with gold nanoparticles to create a hybrid material that can be used directly in stereolithography (SL) advancements. PVP may be coated on the Au NP nanofillers to increase the colloidal particle stability. Using a mask projection excimer laser SL and the manufactured resin, a novel type of complicated construction was created [21].

Additionally, the development of osteoclasts may be impacted by gold nanoparticles produced by citrate reduction of HAuCl4 [22]. A novel method for regeneration of bone tissue has recently been developed in this direction, using a biodegradable hydrogel filled with Au nanoparticles [23]. In order to demonstrate the feasibility of using hydrogels loaded with gold nanoparticles in regeneration techniques for bone tissue engineering, the scientists used photo-curable gelatin hydrogels. Further, Electrospinning was also successful in incorporating peptide-functionalized gold nanoparticles into polymethylglutarimide nanofibers. Grafting functionalized gold nanoparticles increased the adherence of HeLa cells to the nanofiber and potentiated the differentiation of human pluripotent stem cells into cardiomyocytes. This was accomplished by localising cell-adhesive peptides on the nanofiber in high density. Additionally, Li et al. found that mouse embryonic fibroblast cells proliferated more quickly when exposed to gold nanoparticles immobilised on a silica substrate to form a 500 nm SiO2/gold nanoparticles core-shell structure. This phenomenon was explained by the silica substrate's ability to keep gold nanoparticles outside of cells and the concavo-convex gold shell's ability to facilitate cell attachment and proliferation at the nanoscale. Additionally, it was demonstrated that the anti-inflammatory and antioxidative capabilities of the gold nanoparticles in the final core-shell nanomaterials can aid in the promotion of wound healing [24]. Finally, for any successful clinical use, it is crucial to address the problem and detailed study concerning the toxicity and adverse effects of gold nanoparticles used in TE.8T

2.1.2 Silver nanoparticles

The use of silver nanoparticles (AgNPs) in regenerative medicine and TE has attracted a lot of interest. They are strong prospects for a variety of uses because of their special qualities, particularly in TE. To increase the antibacterial activity of implants, silver (Ag), either as a metal or as Ag salts, is being employed as an addition in implanted devices. The penetration of the bacterial cell wall by Ag ions, incorporation into the cell, and subsequent DNA condensation result in the microorganisms' death and prevention of growth, according to one mechanism proposed for the antibacterial activity of silver [25]. AgNPs and hydroxyapatite have been included into the TiO2 nanofibres. Their potent anti-bacterial properties made them an effective implant material against *S. aureus* and *E. coli* [26]. The scaffolds made of chitin and AgNP are frequently employed in tissue engineering applications. These scaffolds are employed in applications for wound healing and have good blood clotting properties. In tissue engineering, the antibacterial activity and in vitro cell compatibility of silver nanoparticles loaded on scaffolds made of poly 3-hydroxybutyrate co-

3-hydroxyvalerate (PHBV) nanofibers are significant. Potential applications for these bioengineered scaffolds include joint arthroplasty [27].

2.1.3 Other metallic nanoparticles

Several other metallic nanoparticles have been researched for their possible use in tissue engineering in addition to gold and silver nanoparticles. Magnetic scaffolds can help in regeneration of bone by absorbing the *in vivo* growth factors, stem cells, and other grafted biomolecular agents to the surface of magnetic nanoparticles. Kim and coworkers created a magnetic scaffold for bone restoration in 2014 using polycaprolactone (PCL) and functionalized magnetic nanoparticles (MNPs) [28]. De Santis et al. created a 3D-scaffold using a poly(-caprolactone) (PCL) matrix reinforced with iron-doped HAp (FeHAp) nanoparticles, and they assessed the mechanical and magnetic properties of this nanocomposite magnetic scaffold. The mechanical strength of the PCL/FeHAp and PCL scaffolds was assessed using the Brazilian test, an indirect tensile test. When bone marrow stem cells (BMSCs) were loaded into magnetic scaffolds in the presence of a static magnetic field, the ensuing cell proliferation was 36 percent and 2.2 times more than it would have been in the absence of the magnetic field [29]. Comparing nickel alloy to conventional scaffolding materials reveals certain advantages. It has multiple functions, great resilience to stress, decent strength, a low Young's modulus, shape memory effects, and super elasticity, for instance. In 2015, nano-hydroxyapatite (nHAp) was synthesised as a control, and different Ni^{2+} doped nano-hydroxyapatite MG63 cell viability responded to Ni^{2+} doped nHAp similarly to the control group, without any appreciable toxicity, according to Anu Priya et al.'s claim that Ni^{2+} doped nHAp might aid as a proangiogenic-osteogenic material. N10-HAp enhanced Runx2 expression, which is a controlling factor involved in osteoblastic differentiation and skeletal morphogenesis, according to RT-PCR data. In the same way that Ni^{2+} acts as an agonist to Ca^{2+} ions, increased Runx2 expression activated the Ca^{2+}-dependent Wnt5 signaling pathway. According to an ELISA investigation, Ni^{2+}-doped nHAp (at a low concentration) may be able to stimulate the expression of VEGF [30]. Antimicrobial and angiogenic properties of copper nanoparticles make them useful in applications for tissue regeneration and wound repair. The capacity of copper to stimulate collagen fibre deposition and angiogenesis, as well as its industrial catalytic properties, antibacterial and antifungal activities, has drawn a significant lot of interest in bone TE in recent years [31]. Copper is necessary for the catalytic activity of lysyl oxidase. The third and final steps in the production of collagen, elastin, and keratin can be mediated by this Cu-dependent enzyme. It can also control how calcium and phosphorus are normally deposited in bones [32].

2.2 Polymeric based nanoparticle

Polymers, which are big molecules formed of repeated monomeric building blocks, are used to create polymeric nanoparticles, which are minuscule particles. These nanoparticles are incredibly small and suited for a variety of biological and industrial applications. Their usual size ranges from 1 to 1000 nanometers (nm). Polymeric nanoparticles are used in tissue engineering as carriers for medicinal medicines, scaffolding materials, or imaging agents. They come with a number of benefits like regulated drug release, biocompatibility, and the capacity to shield delicate components from oxidation. Researchers have discovered and created a variety of synthetic polymer materials, such as polylactic acid (PLA), polyglycolic acid (PGA), polyamide, PLA-glycolic acid copolymer (PLGA), polycaprolactone polyester, polyanhydride, polyacrylate, polysaccharides, polyurethane, and proteins, among others, all of which have excellent biocompatibility, in order to attain the required efficiency of PNP [33].

2.2.1 Poly(lactic-co-glycolic acid) (PLGA)

Due to its biodegradability and biocompatibility, this copolymer of lactic acid and glycolic acid is frequently employed in biomedical applications. Poly (lactic-co-glycolic acid) nanoparticles, also known as PLGA nanoparticles, are an illustration of a polymeric nanoparticle utilised in tissue engineering. Lactic acid and glycolic acid units are the building blocks of PLGA, a biodegradable and biocompatible copolymer. PLGA modification has thus been used extensively in the area of TE. For instance, bioactive glass can enhance the biological activity of PLGA-based scaffolds. PLGA and poly-L-lactic acid often considerably enhance the material's physical compression capabilities and biodegradability of PGA-based fibre tubes (PLLA). Additionally, the silica-based scaffold functionalized with PLGA and gelatin significantly improved its hydrophilicity and tensile mechanical characteristics. Additionally, *in vitro* findings show that the scaffold greatly enhances cell adhesion and proliferation, demonstrating a broad range of application potential for stem cell culture and TE [34]. One of the initial steps in tissue regeneration in TE is vascularization. For bladder reconstruction therapy, VEGF-loaded PLGA nanoparticles have recently been developed, optimised, and integrated into a thermosensitive hydrogel in porcine bladder acellular matrix. By integrating PLGA nanoparticles, the distribution of angiogenic elements from implants has been extensively studied for building a vascular network within the developing tissue [35]. Wei et al. used HAMA hydrogel as scaffolds to create a matrix for cell homing and cartilage regeneration that contained kartogenin (KGN)-loaded nanoparticles. A small-molecule substance called katogenin (KGN) can convert mesenchymal stem cells (BMSCs) from the bone marrow into chondrocytes. Bone marrow-derived mesenchymal stem cells (BMSCs) and synovium-derived mesenchymal stem cells (SDMSCs) were used for cell homing using an ultraviolet-reactive, quickly cross-linkable scaffold combined with kartogenin-loaded nanoparticles (SMSCs) [36].

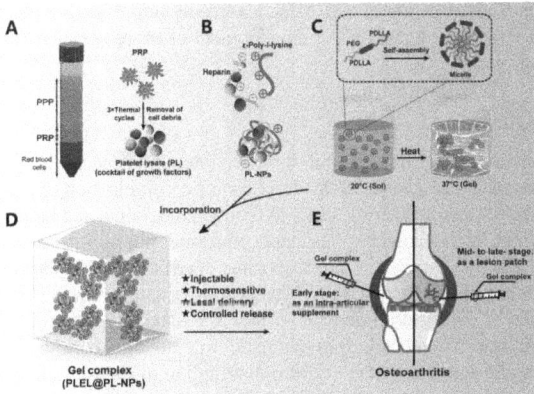

Figure 2: Representation of the steps involved in PL extraction, PL-loaded Hep/EPL NP preparation, PLEL hydrogel thermogelation, PLEL@PL-NP gel complex creation, and PL-NP gel complex treatment for osteoarthritis [37].

2.2.2 Polyethylene glycol (PEG)

Polyethylene glycol (PEG) plays a remarkable role in the structure of PEG-based polymeric nanoparticles. A synthetic, water-soluble polymer made of ethylene glycol repeating units is known as polyethylene glycol. Due to its great biocompatibility and capacity to minimise unintended interactions with proteins and cells, it is widely employed in the biomedical industry and is a extensively used material for tissue engineering. PEG esters have been used to enhance the nano formulation's pharmacokinetics. The PLEL@PL-NPs (poly(d,l-lactide)-poly(ethylene glycol)-poly(d,l-lactide) platelet lysate nanoparticle) technology improved early cartilage degeneration in *in vivo* applications and encouraged cartilage repair in the advanced stages of osteoarthritis (Figure 2) [37]. Additionally, studies on the transport of oligonucleotides, DNA, proteins, and medicines have been conducted using natural PNP, such as albumin, chitosan, polysaccharides, and heparin. While this is going on, small molecule medications' stability and biodistribution can be greatly enhanced when they are combined with polysaccharides like chitosan or human serum albumin. PNP forms include nanospheres, nanogels, polymersomes, polymeric micelles, dendrimers, and nanocapsules are frequently utilised in scientific study. The choice of a particular PNP type depends on its composition, dimensions, shape, purpose, physical/chemical characteristics, and application [38,39]. Zhiting Sun et. al developed L-arginine (LArg), a NO donor, and indocyanine green (ICG), a photosensitizer and co-encapsulated in poly (lactic-glycolic acid) (PLGA) nanoparticles before being added to the hydrogel composed of poly(-caprolactone)-poly (ethylene glycol)-poly(-caprolactone) (PCLPEGPCL). In addition to causing cancer cells to undergo apoptosis, reactive oxygen species (ROS) generated by PLGA@ICG@L-Arg/Gel under near-infrared (NIR) light irradiation CAN also oxidise L-Arg to produce NO, which inhibit the growth of cancer cells. Further oxidation of NO by ROS could result in the production of peroxynitrite anions (ONOO). The tumour microenvironment suppresses by ONOO by activating matrix metalloproteinases (MMPs), which are known for degrading collagen in extracellular matrix [40].

However, polymeric-based nanoparticles offer a versatile and effective platform for enhancing tissue engineering strategies, enabling the development of more advanced and successful tissue regenerative therapies.

2. Nanocomposites

Nanocomposites are extremely important in tissue engineering because they have special benefits and properties that make them useful in this field. Using a combination of cells, biomaterials, and bioactive chemicals, tissue engineering seeks to produce functional and biocompatible substitutes for tissue engineering. As the name implies, nanocomposites are substances made of nanoparticles scattered within a matrix material. Nanocomposites' matrix materials can be polymers, metals, ceramics, or any other kind of material. When creating a nanocomposite, the homogeneity and overall performance of the material depend on how the nanoparticles are distributed throughout the matrix. Better reinforcement and synergistic interactions between the matrix and the nanoparticles are made possible by the uniform distribution of nanoparticles, which improves the material's characteristics. The biocompatibility of nanocomposites can be enhanced, resulting in better interactions with cells and a decrease in unfavourable responses, inflammation, or rejection. It is also possible to achieve targeted interactions with particular cell types by functionalizing nanoparticles. There are several different forms of nanocomposite hydrogels that are frequently

Materials Research Forum LLC
https://doi.org/10.21741/9781644902998-2

utilised to heal cartilage tissue, including carbon-based materials, polymer nanoparticles, metal and metal oxide nanocomposites, inorganic non-metallic nanoparticles, and newly modified materials like cold-air plasma [41].

By combining the advantages of organic and inorganic synthesis, nanocomposite materials have relatively improved mechanical, biodegradability, and high dimensional stability compared to the original materials. Inorganic nanomaterials with considerable biological activity include calcium phosphate nanoparticles, metal oxide nanoparticles, bioglass, and HAp (Figure 3). These materials are derived from substances prevalent in biological tissues. While polymers give nanocomposite materials flexibility and durability, NPs impart and improve signal conductivity (mechanics, bioelectricity, etc.). These nanocomposites are perfect for tissue engineering scaffolds because they have surface chemistry that supports protein adsorption and increased matrix stiffness. Gold (Au) may be used to stimulate the development of apatite, and hydroxyapatite (HAP) is appropriate for bone tissue engineering. The hydrothermal method was used by Prakash et al. to create a GO/HAP/Au nanocomposite. The ternary nanocomposites' superior mechanical capabilities, outstanding antibacterial qualities, great chemical stability, and ability to increase osteoblast cell survival were demonstrated in vitro, pointing to a possible use in bone tissue regeneration [42]. High mechanical strength and highly porous architectures of GO-based composites make them excellent candidates for use as scaffolds for bone regeneration. According to Liang et al., composite scaffolds made of HAp/collagen (C)/poly(lactic-co-glycolic acid)/GO (nHAp/C/PLGA/GO) could promote the proliferation of MC3T3-E1 cells. For the purpose of creating scaffolds, they created nHAp/C/PLGA/GO nanomaterials with varying GO weight percentages and assessed the mechanical characteristics of the scaffold. The findings demonstrated that adding 1.5 weight percent of GO might strengthen the scaffold's structure and give the cells a good surface on which to adhere and multiply. In addition to these benefits, the inclusion of GO to (nHAp/C/PLGA/GO) enhanced the scaffolds' hydrophilic characteristics, which can aid in cell attachment [19].

When compared to scaffolds without nanoparticle reinforcement, nanoparticle-impregnated nanocomposite polymers, which are in the form of electrospun fibers and hydrogels, show improved mechanical characteristics for TE applications [43,44]. It is crucial to thoroughly research the potential hazards connected to the use of nanocomposites in tissue engineering, especially in regard to nanoparticle toxicity and long-term impacts, as with any newly developed technology. For tissue engineering structures based on nanocomposite materials to be safe and functional, proper characterization and testing are required.

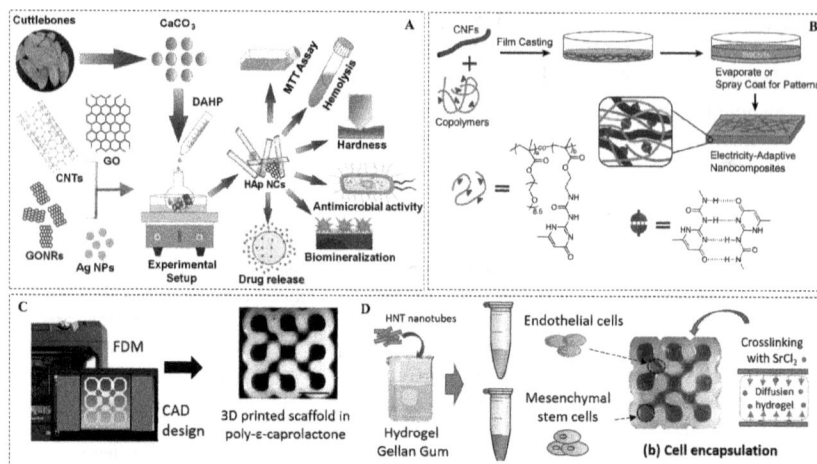

Figure 3: A. Diagram showing the creation of Hydroxyapatite (Hap) nanocomposites (and evaluations of their biological and mechanical properties [45]; B. Aqueous casting of CNFs and EG-UPy29 polymers bearing fourfold hydrogen bonding dimers and subsequent deposition of single walled carbon nanotubes (SWNTs) on the cellulose nanofibril/poly(oligo ethylene glycol methacrylate-ureidopyrimidinone (CNF/EG-UPy29) nanocomposites is shown in the schematic fabrication of bioinspired nanocomposites [46]; The preparation of the 3D build is shown schematically. C. Starting with the CAD design using Autodesk Inventor Professional software and ending with the MakerBot Replicator 2X 3D printed PCL scaffold; D. Cell encapsulation in hydrogels based on gellan gum [47,48]

3. Synthesis methods, functionalisation and characterisation of nanomaterials

3.1 Synthesis methods

Two major approaches followed for the synthesis of nanomaterials are: (1) Top-down and (2) Bottom-up approaches (Figure 4). As evident from their names, top-down is the destructive approach while bottom-up is the constructive approach. This means that larger materials are degraded in a sequential arrangement of steps to finally generate nanostructured materials in the top-down approach. Contrarily, synthesis begins at atomic or molecular level in bottom-up approach and makes its way up to nanoparticles (NPs) synthesis. Based on which process is being followed during the synthesis in these two approaches, the methods are further classified as- physical, chemical or biological methods of nanomaterial synthesis.

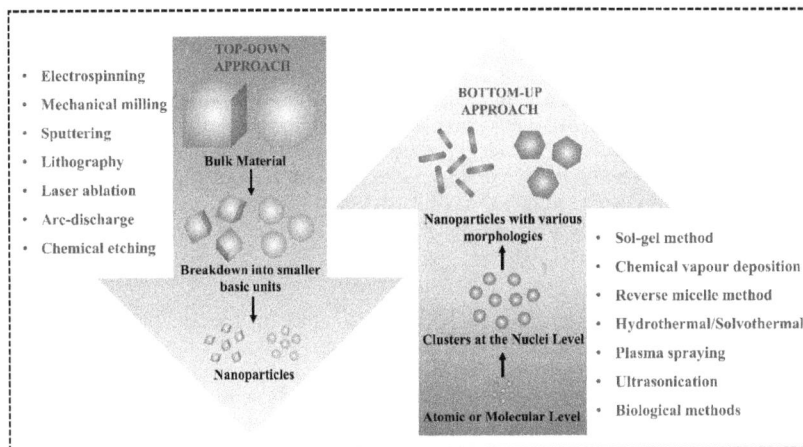

Figure 4: Diagram showing the different approaches for the synthesis of nanoparticles for various biomedical applications.

3.1.1 Synthesis of nanomaterials via top-down approach

Most of the techniques under the top-down approach follow the use of physical processes in nanomaterials generation. Electrospinning, sputtering, pulsed laser ablation, mechanical or powder ball milling, electron beam lithography, mechanochemical processing, electroexplosion and aerosol-based methods are some of the techniques categorized as top-down approaches.

3.1.1.1 Electrospinning

Electrospinning is one of the most popular methods for producing specific fibres at versatile scales. It uses predominantly organic polymers as a solution or a melt to produce nanofibers. Under the influence of a high electric field, the typically spherical droplet of the melt or polymer at the tip of a blunt needle becomes conical. This occurs as a result of the body of the polymer fluid being charged as the voltage rises, and electrostatic repulsion directly competing with surface tension. The electrostatic attraction between the polymer solution and the collector causes a jet of polymer solution to come out from the cone tip in the direction of the grounded collector surface. As the solvent evaporates the stream of nanofibers gets deposited onto the collector surface. Chris et al have highlighted the generation of electrospun iron oxide based nanofibre composites for tissue engineering applications [49,50]. One disadvantage with electrospinning is that basic systems are capable of flow rate of only 0.5ml/h or even less. To overcome this needless electrospinning technology where Taylor cone randomly generate several polymer jets have been employed [51]. For instance, Michal et al created electrospun gelatin and poly-ε-caprolactone (PCL) nanofibers for use as scaffolds in wound healing [52].

3.1.1.2 Mechanical or powder ball milling

Bulk material can be mechanically milled down to a nanoscale in size. This approach can be used to combine many phases and is better suited for large-scale production. The essential idea underlying this procedure is the transfer of energy from the balls during the operation to the sample. Plastic deformation, particle fracture, and cold welding are some significant aspects that have an impact on the manufacturing process and assist in maintaining the quality of nanomaterials. NPs are kept in their desired shapes by plastic deformation, their sizes are decreased by fracture, and their sizes are further reduced by cold welding [53]. Khandan et al demonstrated use of nanocomposite of tri-calcium phosphates (TCPs) prepared by mechanical activation for tissue engineering applications [54].

3.1.1.3 Sputtering

Pislova et al used sputtering for successful creation of round gold nanoparticles (AuNPs) and rod platinum nanoparticles (PtNPs) below the size of 10nm [54]. Their approach didn't involve the use of any harmful reducing agents or additional stabilizers. The sample surface is bombarded with energetic gaseous ions causing release of surface atoms as clusters. One advantage is that there is no vaporisation or melting of the sample unlike other methods and thin films of even thickness are generated. However, it requires skilled operator and is a costly technique compared to the time invested and yield obtained.

3.1.1.4 Lithography

It utilizes a focused beam of light or electrons for development of nanoarchitectures. It can be masked or maskless depending on whether molds are used in fabrication process or not. High yield can be achieved by incorporating deposition and etching along with lithography. Soft lithography, photolithography nanoimprint lithography are a few types of masked lithography methods. Electron ion beam, focused ion beam, scanning probe are some of the maskless approaches. Recently, Ryan et al have manufactured β-TCP scaffolds designed via lithography [55]. They have shown that although scaffolds alone might not be able to bear all the load, still they have enough mechanical strength for potential use in bone regeneration and tissue engineering.

3.1.1.5 Laser ablation

Powerful laser irradiation penetrates the sample forming free electrons which then collide with the atoms of the bulk sample. This transfer of energy leads to heating of the surface and ultimately sample vaporisation forming NPs. It occurs either in vacuum or gas conditions. Pulsed laser ablation in liquids involves a tube furnace and has proven quite beneficial for developing monodispersed solutions of NPs. A study reviewed the synthesis of bare AuNPs and silicon nanoparticles (SiNPs) by laser ablation and their usage as functional additives onto electrospun chitosan-polyethylene oxide wherein they concluded it was convenient to control their size, dissolution behaviour and that they could be exploited for advanced therapeutic applications [56].

3.1.2 Synthesis of nanomaterials via bottom-up approach

Techniques such as sol-gel method, chemical vapour deposition, reverse micelle method, plasma arcing, wet chemical methods and solvothermal or hydrothermal synthesis come under the bottom-up approach and most of them involve the use of chemical processes for nanomaterial synthesis.

Materials Research Forum LLC
https://doi.org/10.21741/9781644902998-2

3.1.2.1 Sol-gel method

A typical sol-gel procedure includes hydrolysis of the metal precursor to produce metal hydroxide which then condensates to form a gel. Once dried the gel gets converted to an aerogel or a xerogel. This wet chemical technique is popular because of its advantages like low processing temperature and the development of extensively diverse high-quality metal-oxide based homogenous nanomaterials. Fan et al produced spherical bioactive glass (BG) NPs (<60nm) and checked their response to human osteoblasts for use in bone tissue engineering. They reported generated NPs were not cytotoxic, induced cell proliferation and differentiation [57]. Studies suggest the use of CNTs as filler material in bone replacement. For instance, Shokri et al utilized BG prepared by sol-gel method in combination with carbon nanotubes (CNTs) and chitosan (Ch) in a variety of ratios to develop nanocomposite scaffold for bone tissue engineering [58,59].

3.1.2.2 Chemical vapour deposition

The manufacturing of two-dimensional (2D) NPs via chemical vapour deposition (CVD) is a well-known and good approach for creating high-quality carbon-based nanomaterials. Through the chemical reaction of vapour phase precursors, a thin coating is created on the substrate surface. The morphology and kind of nanomaterial produced are significantly influenced by the catalyst selection. The benefits of CVD include selective deposition on anticipated pattern, thin films with consistent thickness and low porosity % and great throwing power [60]. Kwon et al deposited a layer of polymer glycidyl methacrylate (pGMA) at the rate of 100nm/15min onto the surface of bare titanium (Ti) discs using initiated CVD (iCVD). They compared the amount of recombinant human bone morphogenic protein-2 (rhBMP2) attached to bare Ti discs versus pGMA fabricated Ti discs and reported 40.2 ± 32.7 ng and 243.9 ± 25.7 ng deposition respectively. Hence their study depicted improved osteogenic expression in human adipose-derived stem cells (hASCs) by utilizing iCVD technique.

3.1.2.3 Reverse micelle method

Contrary to the idea of normal micelle formation in an oil-in-water emulsion, reverse micelle (RM) method involves the presence of water droplet within the range of 5-100nm towards the centre and all hydrophilic heads surrounding it in a water-in-oil emulsion. This core system acts as a "nanoreactor" creating different size NPs based on the ratio of water to surfactant. The basic principle here is that the lesser the water concentration, the smaller will be the central droplet and hence the size of the NPs synthesized will also be small. One advantage is generation of fine monodispersed NPs. A few disadvantages are the synthesized NPs require excessive washing followed by some chemical treatment for better stabilization. Recently, Aniruddha et al. developed sugar-glass NPs (SGNPs) via inverse micelle method. They tested horseradish peroxidase (HRP) encapsulation within SGNPs and reported 93% protection efficiency. Furthermore, characterization of prepared SGNPs by Fourier transform infrared (FTIR), scanning electron microscope (SEM), differential scanning calorimeter (DSC) and Native polyacrylamide gel electrophoresis (Native-PAGE) proved them as capable protein carriers for use in tissue engineering [61]. Orellano et al obtained chitosan NPs (Ch-NPs) and studied their cross-linking in two RM systems: 1,4-bis-2-ethylhexylsulfosuccinate (AOT) and benzyl-n-hexadecyldimethylammonium chloride (BHDC). Various characterization studies showed better results with AOT as RM system. They further reported successful curcumin encapsulation within prepared Ch-NPs with improved solubility in water for theragnostic purposes [62].

Materials Research Forum LLC
https://doi.org/10.21741/9781644902998-2

3.1.2.4 Hydrothermal/Solvothermal method

The hydrothermal/solvothermal process refers to the reaction that takes place when solvents that are sealed in vessels reach their critical temperature under their own internal pressure and substances usually less soluble get dissolved. Solvothermal method differs from hydrothermal method in utilizing organic or non-aqueous solvent for nanomaterial extraction. By adjusting reaction parameters such as-pH, temperature and reactant concentrations, the shape of NPs obtained can be managed. A large variety of nanomaterials like-nanorods, nanosheets, nanowires and nanospheres have thus been constructed using this technique. The process essentially involves the nucleation of crystals followed by their development. For example, polycaprolactone (PCL) nanocomposites containing hydroxyapatite (HA) nanorods have been created using in-situ solvothermal methods. The study also highlighted an increase in size ratio (from 2 to 7) of obtained nanorod structures with the increase in reaction temperature conditions [63]. Another recent approach combines the advantages of hydrothermal and microwave processes, is faster and hence has garnered a lot of attention for engineering nanomaterials. Very recently, Zong et al have reviewed the role of solvothermal/hydrothermal developed carbon dots (CDs) in wound healing and various organ tissue engineering applications [64].

3.1.2.5 Biological methods

A diverse range of nanomaterials including carbon-based materials [65], metal oxides, sulfoxides and metal NPs [66] etc. have been synthesized using biological systems like yeast, bacteria, viruses, fungi, plant extracts [67], and actinomycetes. Compared to other top-down and bottom-up approaches, biological methods are more environment friendly, cheaper and less hazardous. NPs obtained through this process are influenced by various factors such as temperature, pressure, pH, reactant concentration, nature of the solvent, reaction time and exhibit high biocompatibility making them particularly suitable for medical applications. The synthesis of NPs through microbes follows an approach of bottom-up, wherein NPs are designed as a part of the microorganism's defence mechanism-based detoxification. It is considered a fundamental survival procedure that involves the oxidation/reduction of metal ions, generating various forms of metal phosphates, carbonates, and sulfides, or even metal ions volatilization. Through these mechanisms, microorganisms can convert toxic metal ions into less harmful or inert nanomaterials, thereby protecting themselves from potential harm caused by these metal ions. NP synthesis using microorganisms is slow, expensive, and challenging to maintain due to the need for various species. Additionally, large-scale application is limited. However, an alternative approach involving plant-mediated synthesis of metal nanoparticles offers several advantages. This method is rapid, easily reproducible, eco-friendly, and cost-effective. Furthermore, it can be readily applied on an industrial scale. This biologically driven process has gained significant attention due to its potential applications in synthesis of NPs via green nanotechnology.

For instance, Sahoo et al developed graphene oxide (GO) fabricated magnesium oxide (MgO) nanomaterial using *Azadirachta indica* (neem) leaf extract for removing chromium ions from contaminated water [68]. Similarly, another group employed *Trachyspermum ammi* (carom seeds) leaf extract for generating MgO NPs [69]. One study reported AgNPs (50nm) can be accumulated in the roots, stems, and leaves of *Brassica juncea* and *Medicago sativa* when these plants are grown in Hoagland's solution [67]. Various types of chemical, physical and biological methods used for nanomaterials synthesis are summarized below in Table1.

Nanoparticle Toxicity and Compatibility   Materials Research Forum LLC
Materials Research Foundations 161 (2024) 27-63   https://doi.org/10.21741/9781644902998-2

Table 1. Various top-down and bottom-up synthesis methods for nanomaterials and their applications.

S.No.	Method Approach	Method Name	Feature	Reference
1.	Chemical	Pyrolysis	High temperature decomposition involving the breakdown of precursor molecules to form NPs.	[70]
2.	Chemical	Coprecipitation	Mixing two or more precursor solutions to cause a precipitation reaction yielding nanomaterials.	[71]
3.	Chemical	Sonochemical	Ultrasound waves induce cavitation and create reactive conditions within a solution or a mixture to promote the formation of nanomaterials.	[72]
4.	Chemical	Electrochemical	Electrochemical reduction or oxidation of metal ions leads to the formation of nanomaterials.	[73]
5.	Physical	Arc discharge	High-voltage electric arcs between electrodes vaporize and condense materials, yielding nanomaterials.	[74]
6.	Physical	Evaporation-condensation	Materials are evaporated and then condensed to produce nanomaterials in a controlled environment or substrate.	[75]
7.	Physical	Pulse wire discharge	Rapid discharge of electrical pulses to vaporize a wire, forming NPs.	[76]
8.	Physical	Solvent evaporation	Dissolving nanomaterial precursors in a solvent, subsequently evaporating the solvent to obtain NPs with controlled properties and sizes.	[77]
9.	Biological	Bacterial based	Anaerobic magnetotactic bacteria at the seabed use magnetosomes to form magnetic particles.	[78]
10.	Biological	Fungal based	Living fungi or their extracts mediate the reduction and stabilization of metal ions, leading to the formation of NPs with various properties.	[79]
11.	Biological	Plant extract based	Phytochemicals act as effective reducing agents, facilitating the reduction of metal NPs.	[80]

3.2　Functionalization methods for synthesized nanomaterials

The modification of NPs has a significant impact on their basic properties such as colloidal stability, dispersion, and controlled assembly. Functionalizing the surface of NPs with biomolecules can confer several benefits like increased cellular internalization, better intracellular delivery, and an overall improved biocompatibility. NPs functionalization is achieved through a variety of synthesis strategies and employs the use of diverse ligands which may include the use of small molecules, dendrimers, polymers, biomolecules, surfactants and inorganic materials. For example, Addition of hydrophobic substances like oleic acid creates hydrophobic NPs whereas using hydrophilic substances like polyethylene glycol (PEG) will generate hydrophilic NPs. Other major mechanisms involved in carrying out NPs functionalization involve the use of electrostatic interactions, phenomena like chemisorption and physisorption, intrinsic surface engineering, non-covalent or supramolecular interactions etc. For example, chemisorption modifies the surface properties by attaching thiol groups to the NP surface while physisorption employs milling or mixing to bind molecules. Non-covalent interactions offer the advantage of simplicity, do not alter the structure of the molecules involved or their interactions with target biological materials but can be sensitive to various factors, such as pH and ionic strength, which can easily influence their stability and behaviour [81,82].

Liu et al. demonstrated the one-step functionalization process wherein CNTs were combined with water soluble ssDNA to form a negatively charged complex. Simultaneously, 3D-printed polypropylene fumarate scaffolds were treated to introduce positively charged amine groups. The negatively charged complex was then coated onto the positively charged 3D-printed scaffolds using electrostatic force [83]. The developed even surface non-toxic CNTs coating improved pre-osteoblast cell's attachment, growth and differentiation for use in bone tissue engineering. One study reported that human corneal epithelial cells were able to grow and adhere successfully on nanofunctionalized gels made from polypropyleneimine octaamine dendrimers crosslinked via carbodiimide EDC with highly concentrated collagen. There was no observed cell toxicity, suggesting that these gels could be promising scaffolds for corneal tissue engineering [84]. Surface modification for functionalization can also be achieved by using either inorganic polymers or organic polymers. For instance, gold is commonly used for surface coating, particularly on magnetic iron oxide nanoparticles (IONPs). There are two methods for obtaining a gold-shell coating on the IONPs: direct and indirect (Figure 5). When gold is coated with strong sulphur conjugation, it provides significant advantages. Gold-coated MNPs (magnetic nanoparticles) are highly stable due to the chemical inertness of gold. Similarly, another study suggested spherical MNPs implanted with organic polymer chitosan enhanced MRI image contrast [85].

Pourmadadi et al. reviewed the application of functionalized molybdenum disulphide (MoS_2) NPs for drug delivery and tissue engineering. They reported that using surface modifying agents having groups like dithiolenes led to new covalent bonds formation with MoS_2 and further enhanced photochemical and electrochemical properties of synthesized NPs [87].

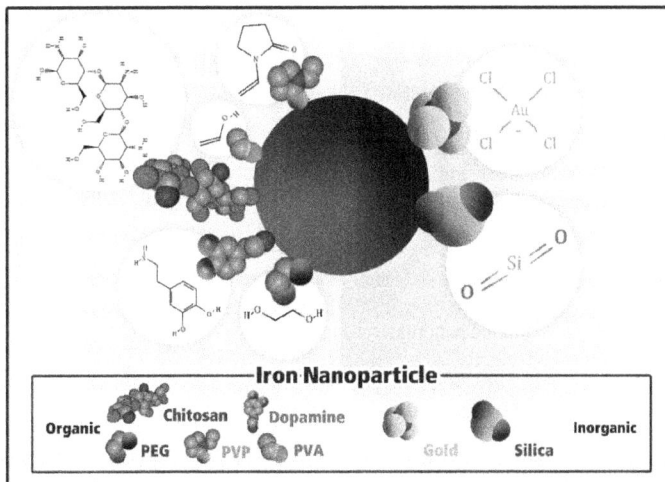

Figure 5: Variety of distinct hydrophilic ligand molecules on an iron oxide nanoparticle. Iron oxide nanoparticles' surface alteration is shown schematically. Polyethylene glycol (PEG), polydopamine (PDA), chitosan, polyvinyl alcohol (PVA), polyvinyl pyrrolidone (PVP), gold, and silica are listed from left to right, from organic to inorganic [86].

Nanomaterials based on 3D self-assembled peptide hydrogels are being used widely because of their convenient functionalization amongst other properties such as-porous composition, robust mechanical stability and excellent biocompatibility. Functionalization of these hydrogels is achieved by the incorporation of some other nanoscale components which may be certain NPs, proteins, DNA, polymers, etc. For instance, the use of PEG incorporated peptide hydrogel as a suitable substrate for human mesenchymal stem cells (hMSCs) has been shown to improve its scaffolding ability. Similarly, some research literature has depicted better hydrophilicity in surface functionalized peptide hydrogel coated PCL scaffolds compared to normal PCL scaffolds [88,89].

3.3   Characterization methods for synthesized nanomaterials

Upon successful functionalization of the synthesized nanomaterials, the next step is to perform their desired characterization. Before choosing their appropriate applications, several parameters are taken into consideration while characterizing the developed nanomaterials such as determining their size, shape, structure, morphology, elemental & mineral composition, existing bonds and checking for the presence of any special characters like magnetic properties etc. Most of the characterization is performed by the help of various advanced microscopic techniques and spectroscopic analysis including - UV-Visible (UV-Vis) spectroscopy, Dynamic light scattering (DLS), Fourier Transform Infrared (FT-IR) spectroscopy, Raman spectroscopy (RS), X-ray Photoelectron Spectroscopy (XPS), Energy Dispersive X-ray Spectroscopy (EDS), Mossbauer Spectroscopy (MS), X-ray powder diffraction (XRD), Electron Plasmon Resonance (EPR), Atomic Force Microscopy (AFM), Scanning electron microscopy (SEM), Transmission Electron

Microscopy (TEM), Magnetic Force Microscopy (MFM), [90] etc. These procedures are performed either individually or sometimes in combination to reveal the best possible surface and internal characteristics of synthesized NPs. For instance, the surface morphology study is conducted by using SEM, TEM and AFM [91]. While SEM provides analysis of the surface structure, TEM aids further in analyzing the crystal orientation of the NPs and their structure at electron level and AFM does the same but it is used particularly for dry samples. The obtained images act as a guide about the overall shape of the NPs, whether they are spherical, or rod shaped and give an idea about the potential number of pores seen in their structure which helps in deciding if there is scope for any further modifications to improve their biomedical applications. Techniques like DLS help in calculating average or mean size of the synthesized NPs [92] in comparison with the existing reference range whereas XRD analysis tells even the unit cell composition existing at the elemental and mineral composition levels [93]. The presence of metal oxygen bonds and other functional groups occurring in the nanomaterial structure is performed by methods like FT-IR, RS and XPS [94,95]. Lastly, methods such as MFM, EPS and MS are helpful in checking for the occurrence of any magnetic properties in synthesized nanomaterials [96]. These investigate the presence of any paramagnetic centres and free radicals in the structure, check for the existing oxidation state, spin state, structural bonding and orientation of constituent elements, their covalence and electronegativity among other properties. All these techniques ultimately act as a baseline to determine the quality of the synthesized nanomaterials for use in potential biomedical, theranostic and tissue engineering applications.

4.   Application of nanomaterials in tissue engineering

4.1   Skin tissue engineering

Skin tissue damage puts the patient's life in danger, adds to the social and personal burden of therapy, and disrupts the body's balance. Angiogenesis, a crucial stage in the healing of wounds, supplies enough nutrients and oxygen to the affected area [97,98]. Traditional clinical therapies, however, are insufficient to control the development of the vascular system and facilitate wound healing [99,100]. Nanomaterials have achieved significant advancements in the field of medicine because of their distinctive qualities and variety of uses [101,102]. Through the stimulation of essential elements during the vascular regeneration phase, nanomaterials offer a more efficient method of treating wounds and accelerating angiogenesis [103,104]. In recent years, researchers have conducted in-depth and extensive research in this sector and have documented a number of successes by using several innovative 2D nanomaterials in wound infection management. Due to their great qualities and similar cellular characteristics to those of human skin tissues, such as those relevant to adhesion and infiltration, a variety of natural materials have been used in skin tissue engineering. Collagen [105,106], gelatin [107,108], cellulose [109–111], and chitosan [112,113], are some of the natural materials, have been employed as electrospun scaffolds for tissue engineering. These materials incorporated further nanomaterials such as gold nanoparticles, silver nanoparticles, silica, copper etc. which have antibacterial, antioxidant properties etc. enabling the material to be more effective in tissue regeneration. Due to AuNPs great biocompatibility, they are able to minimise inflammation, encourage the creation of granulation tissue, and prevent skin tissues from rejecting them. By obstructing the mechanisms that keep keratocytes and fibroblasts alive, AgNPs can promote their proliferation. Additionally, they have the capacity to dampen the innate immune system, which has been linked to faster wound healing and slower scarring [114].

Materials Research Forum LLC
https://doi.org/10.21741/9781644902998-2

Further, endothelial cell migration is stimulated by another nanoparticle copper nanoparticles (20, 40, and 80 nm, all spherical in shape), and keratinocyte and fibroblast cell proliferation are induced at particular sizes and concentrations. More effectively than smaller CuNPs (40 nm), larger CuNPs (80 nm) stimulate collagen 1A1 expression in cultured fibroblast cells. Additionally, in rat models, copper nanoparticles can promote the development of new blood vessels and hasten the healing of full-thickness skin wounds without causing any negative side effects [115]. Since they dissolve as new tissue forms, biodegradable materials are generally preferable. However, because to poor mechanical qualities, their usage as such in skin tissue regeneration is somewhat constrained. Hydrogels can be made into multifunctional nanocomposites with a variety of other benefits by using nano- and biomaterials that can strengthen their structure. In reality, their formulation can be appropriately changed to perform many tasks at once, as is frequently the case with other nanomaterials for skin regeneration [116].

4.2   Bone tissue engineering

Bone is a nanocomposite with a hierarchical structure that spans the nano- to macroscale, made up of organic (mostly collagen) and inorganic (nanocrystalline hydroxyapatite) components. For bone tissue engineering, several NPs can be applied, with a focus on enhancing scaffolds and enhancing drug delivery [117,118]. Scaffolds made of nanofibers, nanotubes, nanoparticles, and hydrogel have lately come to light as viable possibilities in the past decade to better imitate the nanostructure in natural ECM and produce scaffolds that effectively replace damaged tissues [119]. Due to their capacity to encourage cell differentiation, AuNPs have been employed as scaffolds for promoting bone regeneration among NPs (both organic and inorganic) [120]. Heo et al. demonstrated improved bone regeneration by combining AuNPs and gelatin scaffolds [121]. Adipose-derived stem cells are differentiated into osteoblasts both in vitro and in vivo as a result of this combination [122]. Human mesenchymal stem cells (MSCs) have been shown to efficiently take up 2,2,6,6-tetramethylpiperidine-N-oxyl (TEMPO) coupled AuNPs and prevent the excessive generation of reactive oxygen species in them at modest dosages of TEMPO [123]. Through many signalling pathways, Au-NPs encourage MSC osteogenesis. Choi et al. investigated the Wnt/-catenin pathway for osteogenic differentiation of stem cells. By turning on the Wnt/-catenin signalling route and the p38 MAPK signalling pathway, respectively, osteogenesis of bone-derived MSCs and AD-MSCs was facilitated [124]. Au-NPs coated with artificial oligonucleotides in a study by Rosi et al. shown excellent biocompatibility, stability, and effective uptake by the cells. Au-NPs customised with oligonucleotides make it easier to detect a particular mRNA with precision and can be utilised to separate out distinct populations of skeletal cells from the mix of bone marrow stromal cells. Analysis using a transmission electron microscope showed that the Au-NPs activated osteogenic genes and downregulated adipogenic genes by interacting with the cytoplasmic proteins of particular signalling pathways [125]. Ag-NPs are utilised in bone engineering to create antibacterial nanocomposite-based scaffolds and implants to repair significant flaws. They successfully control bacterial infection and inflammation, which promotes bone regeneration and wound healing. Due to its excellent antimicrobial, antioxidant, and anti-inflammatory properties, this silver nanoparticle (AgNP) and melatonin biomolecules loaded hydrogel enhances re-epithelialization and collagen deposition in the wound tissue and holds promise as a more effective therapeutic material for the treatment of infected burns and diabetic wounds (Figure 6). Research in vitro demonstrates that Ag-NPs can effectively promote MSC proliferation, osteogenic differentiation, and ECM mineralization [48,126]. In a matrix of poly-L-lactic acid (PLA) nanofibers, a newly studied novel scaffold was made up of nanohydroxyapatite mixed with chitosan (CS) and was enriched with Pt-NPs, Au-NPs, and TiO2-NPs. In contrast to scaffolds enriched with TiO2-NPs, better the biomineralization, cell adhesion, proliferation, and differentiation of MG-63 cells were seen in the presence of Pt-NPs and Au-NPs [127]. Similarly, in the realm of bone tissue engineering, nano silica has a significant role as a bone regeneration and hierarchical bone structure promoter through an ossification process. Recently, not only in the area of bone regeneration, but also in other fields, nanoparticles based on a variety of various metals and their oxides have been examined.

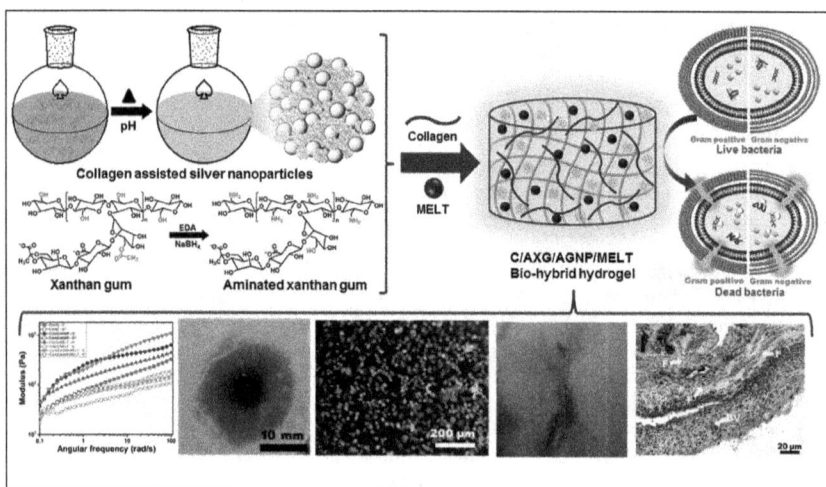

Figure 6: Synthesis of C/AXG/AGNP/MELT (cellulose/xanthum gum/silver nanoparticles/melatonin) hydrogel and its in vitro cell cytotoxicity and fluorescence photomicrographs of bio-hybrid hydrogel systems [128] The hydrogel promoted re-epithelialization and collagen deposition in the wound tissue in animal models.

4.3 Neural tissue engineering

The most structurally intricate cells in the human body, neurons are made up of electrically excitable components that control both the organism's voluntary and involuntary processes. As a result, the loss of neuronal connection is the fundamental cause of a wide range of nervous system illnesses, which adds to the significant load on the global health system [129]. In this regard, neuroregeneration processes continue to be the only solution that has the potential to produce new neurons and neuronal components as well as restore the connectivity that has been disrupted. Hence, numerous engineering techniques based on nanoparticles have been created and are being researched to prevent or treat nerve damage [130]. several nanomaterial-based studies produced encouraging findings, it was demonstrated that these nanoparticles were able to boost cell adhesion, proliferation, and differentiation of neuronal cells, and improved neuron regeneration

Materials Research Forum LLC
https://doi.org/10.21741/9781644902998-2

[131]. There are many commercially available nanotechnology-based products on the market, including the Avance® nerve graft and the Axoguard® nerve protector [132]. It is becoming more common to use nanotechnology and tools like confocal microscopy, MRI, and two-phonon techniques to better understand the structure and operations of biological systems at the nanoscale scale [133,134]. Nanomaterials are currently a prospective contender to be used in the management of nerve damage. Changing the physicochemical characteristics of nanoparticles may make it possible to stop or reverse neurodegeneration [135]. At the nanoscale, metals display excellent physicochemical characteristics. They are widely employed for many applications, such as bio imaging, sensing, and tagging, because of their special features [136]. The optical characteristics of metallic (Au and Ag) nanoparticles can be tailored by varying the size and shape of the particles. In order to facilitate the targeted delivery of genes (DNA and RNA) and growth factors (nerve growth factor, or NGF), biomolecules can be conjugated to the surface of metallic nanoparticles [137]. Another material that is widely used in tissue engineering, drug delivery, nanomedicine, and catalysis are silica nanostructures [138]. According to Shah et al., siRNA and miRNA were delivered into neural stem cells to promote neuronal development [139]. In relation to the brain, silica nanostructures are gaining interest due to their exceptional biocompatibility. Functional silica nanoparticles containing DNA-encoding EGFP were injected into mice' midbrains in a study by Alvarez-Buylla et al., and it was shown that the particles were not hazardous for up to 4 weeks [140]. In spite of these, several organic nanomaterials such as liposomes, nanofibers, carbon-based nanomaterials, micelles etc. are also used for neural tissue engineering. Gene treatment for age-related illnesses like Parkinson's disease uses liposomes. To deliver dopamine-depleted rat striatum, a PEG immunoliposome bound with the transferrin receptor is employed [141,142]. Nerve growth factors like NGF and GDNF are delivered using liposomes and micelles, which aid in neural regeneration [143,144]. Over the years, researchers have looked at and experimented with using nanoscale structures such as nanofibers to control and encourage stem cell development. Through plasma treatment, Kumar et al. created polycaprolactone (PCL) nanofibers with increased hydrophilicity, showing good support for adhesion and Schwann cell growth [145]. Similar to this, Meiners et al. investigated nanofibers functionalized with neuroactive peptides to enhance adhesion, neurite production, and elongation in comparison to standard surface treatments [146]. These encouraging outcomes inspire additional research on how fiber-based scaffolds or aligned fibres can improve nerve tissue regeneration.

4.4 Drug delivery

Nanoparticles produced biologically are utilised in the diagnosis and treatment of several ailments, including cancer and cardiovascular disease. Additionally, bioconjugation of nanoparticles increases the potency of poorly soluble anticancer medications [147–149]. Nanomaterials can be functionalized for drug delivery and cell targeting, enabling them to convey their therapeutic payload directly to the needed location [8,150]. This improved nontoxic nanomaterial drug delivery approach can be utilised to diagnosis various cancer kinds [151]. Drug delivery systems can increase the effectiveness and safety of TE, so it is crucial to develop one that is efficient in order to direct in situ functional tissue regeneration without having negative effects on the remaining body [152]. In comparison to synthetic ones, biopolymers used in drug delivery systems ensure biocompatibility, biodegradability, and minimal immunogenicity. It is simple to create suspensions of biopolymers, such as silk fibroins, collagen, gelatin, starch, albumin, cellulose, and chitosan, that act as carriers for both large and small medicinal molecules [153,154]. Lipid-based

NPs as a delivery vehicle provide a number of benefits, including easy formulation, self-assembly, biocompatibility, high bioavailability, capacity for large payloads, and a variety of physicochemical properties that can be tuned to modify biological features [155,156]. These factors make lipid-based NPs the most prevalent class of nanomedicines that have received FDA approval [157]. Another material used for delivery of drugs are polymeric nanoparticles. Therapeutics may be chemically conjugated to the polymer, enclosed within the NP core, trapped in the polymer matrix, or attached to the NP surface. This makes polymeric NPs ideal for co-delivery applications because it permits administration of a variety of payloads, including hydrophobic and hydrophilic substances, as well as cargos with diverse molecular weights, like small molecules, biological macromolecules, proteins, and vaccines [158,159]. However, toxicity and a higher likelihood of particle aggregation are drawbacks of polymeric NPs.

5.   Toxicity of nanomaterials for tissue engineering

Ensuring the safety of synthesized, functionalized, and characterized nanomaterials is of utmost importance before their application in tissue engineering. Many of these nanostructures undergo surface modifications, making it essential to investigate changes in their properties, such as size, aggregation ability, chemical structure, and surface chemistry, prior to their use in biomedical applications. When these nanomaterials are administered via different routes, such as oral, inhalation, sub-cutaneous, or intraperitoneal, for tissue engineering purposes, they come into contact with the human body and may interact with other cells and proteins. This interaction could potentially lead to altered metabolism, accumulation within cells, or excretion from the body [160]. According to a study, the genotoxicity of biologically synthesized metallic nanomaterials may vary in different cases and is strongly influenced by factors such as synthesis parameters, the biological source used, and the specific assay employed for assessment [161]. Therefore, it is crucial to adhere to appropriate biosafety guidelines and checklists when selecting nanomaterials with good biocompatibility, suitable size, and effective surface modifications to improve their structure and solubility. By doing so, their potential for safer and more successful tissue engineering applications can be enhanced.

Conclusion

In this chapter, we have summarised the different nanomaterials, their synthesis, characterisation and surface modifications that allow for their application in tissue engineering. We have also discussed the different field of application of these nanomaterials. For the repair or regeneration of damaged tissue, the creation of nanomaterials and their use in tissue engineering are crucial. These materials have demonstrated excellent efficiency in tissue regeneration and good biocompatibility when compared to conventional dressings or scaffolds, showing their promising potential to be used in clinical practice. Although the development of nanomaterials for tissue engineering offers a promising method for repairing tissue defects, the majority of studies are still in the experimental stages and require much more work before they can be successfully used in clinical practice. More research must be done regarding the potential security problems with the use of these nanomaterials. According to current nanotechnology, more and more researchers are attempting to create new biomaterials by combining various nanomaterials. Moreover, new research reveals that directed stem cell growth and differentiation, the development of tissue and organs, implantation, and regenerative activities have substantial potential for medicinal use. The field of nanomedicine is at the frontier of research. However, with further investigation into their use in regenerative medicine and the resolution of stability and biosafety issues *in vivo,* the application of these nanomaterials will soon become a crucial strategy in medical research and clinical treatment, bringing good news to patients with a variety of tissue defects.

Acknowledgement

The authors are grateful to Science and Engineering Research Board (SERB), New Delhi (Grant Number-WEA/2020/000036 & CRG/2020/003014). The authors are grateful for the research endowment under the Intensification of Research in High Priority Area (IRHPA) program from Science and Engineering Research Board (SERB), New Delhi (Grant Number IPA/2020/000069). The authors are grateful for funding provided by Department of Biotechnology (DBT), New Delhi (Grant number BT/PR34216/AAQ/1/765/2019). D.P. would like to acknowledge DBT Fellowships (DBT/2021-22/NIAB/1706) provided by the Department of Biotechnology (DBT), New Delhi.

Reference

[1]   A. Hasan, A. Memic, N. Annabi, M. Hossain, A. Paul, M.R. Dokmeci, F. Dehghani, A. Khademhosseini, Electrospun scaffolds for tissue engineering of vascular grafts, Acta Biomater. 10 (2014) 11–25. https://doi.org/10.1016/J.ACTBIO.2013.08.022

[2]   A. Paul, V. Manoharan, D. Krafft, A. Assmann, J.A. Uquillas, S.R. Shin, A. Hasan, M.A. Hussain, A. Memic, A.K. Gaharwar, A. Khademhosseini, Nanoengineered biomimetic hydrogels for guiding human stem cell osteogenesis in three dimensional microenvironments, J Mater Chem B. 4 (2016) 3544–3554. https://doi.org/10.1039/C5TB02745D

[3]   M.E. Gomes, M.T. Rodrigues, R.M.A. Domingues, R.L. Reis, Tissue Engineering and Regenerative Medicine: New Trends and Directions-A Year in Review, Tissue Eng Part B Rev. 23 (2017) 211–224. https://doi.org/10.1089/TEN.TEB.2017.0081

[4]   Y. Zheng, X. Hong, J. Wang, L. Feng, T. Fan, R. Guo, H. Zhang, 2D Nanomaterials for Tissue Engineering and Regenerative Nanomedicines: Recent Advances and Future Challenges, Adv Healthc Mater. 10 (2021) 2001743. https://doi.org/10.1002/ADHM.202001743

[5]   A. Kaushik, R. Khan, P. Solanki, S. Gandhi, H. Gohel, Y.K. Mishra, From Nanosystems to a Biosensing Prototype for an Efficient Diagnostic: A Special Issue in Honor of Professor Bansi D. Malhotra, Biosensors 2021, Vol. 11, Page 359. 11 (2021) 359. https://doi.org/10.3390/BIOS11100359

[6]   P. Mishra, T. Munjal, S. Gandhi, Nanoparticles for detection, imaging, and diagnostic applications in animals, Nanoscience for Sustainable Agriculture. (2019) 437–477. https://doi.org/10.1007/978-3-319-97852-9_19/FIGURES/7

[7]   D. Shahdeo, A. Roberts, V. Kesarwani, M. Horvat, R.S. Chouhan, S. Gandhi, Polymeric biocompatible iron oxide nanoparticles labeled with peptides for imaging in ovarian cancer, Biosci Rep. 42 (2022). https://doi.org/10.1042/BSR20212622/230723

[8]   E. Yasun, S. Gandhi, S. Choudhury, R. Mohammadinejad, F. Benyettou, N. Gozubenli, H. Arami, Hollow micro and nanostructures for therapeutic and imaging applications, J Drug Deliv Sci Technol. 60 (2020) 102094. https://doi.org/10.1016/J.JDDST.2020.102094

[9]   D. Shahdeo, S. Gandhi, Next generation biosensors as a cancer diagnostic tool, Biosensor Based Advanced Cancer Diagnostics: From Lab to Clinics. (2022) 179–196. https://doi.org/10.1016/B978-0-12-823424-2.00016-8

[10]   M. Shah, P. Kolhe, S. Gandhi, Nano-assembly of multiwalled carbon nanotubes for sensitive voltammetric responses for the determination of residual levels of endosulfan, Chemosphere. 321 (2023) 138148. https://doi.org/10.1016/J.CHEMOSPHERE.2023.138148

[11]   A. Roberts, S. Mahari, D. Shahdeo, S. Gandhi, Label-free detection of SARS-CoV-2 Spike S1 antigen triggered by electroactive gold nanoparticles on antibody coated fluorine-doped tin oxide (FTO) electrode, Anal Chim Acta. 1188 (2021) 339207. https://doi.org/10.1016/J.ACA.2021.339207

[12]   J. Zhang, H. Chen, M. Zhao, G. Liu, J. Wu, 2D nanomaterials for tissue engineering application, Nano Res. 13 (2020) 2019–2034. https://doi.org/10.1007/S12274-020-2835-4/METRICS

[13]   D. Prakashan, A. Roberts, S. Gandhi, Recent advancement of nanotherapeutics in accelerating chronic wound healing process for surgical wounds and diabetic ulcers, Https://Doi.Org/10.1080/02648725.2023.2167432. (2023). https://doi.org/10.1080/02648725.2023.2167432

[14]   A. Hasan, M. Morshed, A. Memic, S. Hassan, T.J. Webster, H.E.S. Marei, Nanoparticles in tissue engineering: applications, challenges and prospects, Int J Nanomedicine. 13 (2018) 5637. https://doi.org/10.2147/IJN.S153758

[15]   R. Sensenig, Y. Sapir, C. MacDonald, S. Cohen, B. Polyak, Magnetic nanoparticle-based approaches to locally target therapy and enhance tissue regeneration in vivo, Https://Doi.Org/10.2217/Nnm.12.109. 7 (2012) 1425–1442. https://doi.org/10.2217/NNM.12.109

[16]   R.S. Chouhan, M. Horvat, J. Ahmed, N. Alhokbany, S.M. Alshehri, S. Gandhi, Magnetic Nanoparticles—A Multifunctional Potential Agent for Diagnosis and Therapy, Cancers 2021, Vol. 13, Page 2213. 13 (2021) 2213. https://doi.org/10.3390/CANCERS13092213

[17]   M. Borzenkov, G. Chirico, M. Collini, P. Pallavicini, Gold Nanoparticles for Tissue Engineering, in: 2018: pp. 343–390. https://doi.org/10.1007/978-3-319-76090-2_10

[18]   J.; Huang, F.; Liu, H.; Su, J.; Xiong, L.; Yang, J.; Xia, Y. Liang, J. Huang, F. Liu, H. Su, J. Xiong, L. Yang, J. Xia, Y. Liang, Advanced Nanocomposite Hydrogels for Cartilage Tissue Engineering, Gels 2022, Vol. 8, Page 138. 8 (2022) 138. https://doi.org/10.3390/GELS8020138

[19]   R. Eivazzadeh-Keihan, A. Maleki, M. de la Guardia, M.S. Bani, K.K. Chenab, P. Pashazadeh-Panahi, B. Baradaran, A. Mokhtarzadeh, M.R. Hamblin, Carbon based

nanomaterials for tissue engineering of bone: Building new bone on small black scaffolds: A review, J Adv Res. 18 (2019) 185–201. https://doi.org/10.1016/j.jare.2019.03.011

[20]   J.-O. You, M. Rafat, G.J.C. Ye, D.T. Auguste, Nanoengineering the Heart: Conductive Scaffolds Enhance Connexin 43 Expression, Nano Lett. 11 (2011) 3643–3648. https://doi.org/10.1021/nl201514a

[21]   G.N. Abdelrasoul, B. Farkas, I. Romano, A. Diaspro, S. Beke, Nanocomposite scaffold fabrication by incorporating gold nanoparticles into biodegradable polymer matrix: Synthesis, characterization, and photothermal effect, Materials Science and Engineering: C. 56 (2015) 305–310. https://doi.org/10.1016/j.msec.2015.06.037

[22]   O.-J. SUL, J.-C. KIM, T.-W. KYUNG, H.-J. KIM, Y.-Y. KIM, S.-H. KIM, J.-S. KIM, H.-S. CHOI, Gold Nanoparticles Inhibited the Receptor Activator of Nuclear Factor-κB Ligand (RANKL)-Induced Osteoclast Formation by Acting as an Antioxidant, Biosci Biotechnol Biochem. 74 (2010) 2209–2213. https://doi.org/10.1271/bbb.100375

[23]   D.N. Heo, W.-K. Ko, M.S. Bae, J.B. Lee, D.-W. Lee, W. Byun, C.H. Lee, E.-C. Kim, B.-Y. Jung, I.K. Kwon, Enhanced bone regeneration with a gold nanoparticle–hydrogel complex, J. Mater. Chem. B. 2 (2014) 1584–1593. https://doi.org/10.1039/C3TB21246G

[24]   X. Li, H. Wang, H. Rong, W. Li, Y. Luo, K. Tian, D. Quan, Y. Wang, L. Jiang, Effect of composite SiO 2 @AuNPs on wound healing: In vitro and in vivo studies, J Colloid Interface Sci. 445 (2015) 312–319. https://doi.org/10.1016/j.jcis.2014.12.084

[25]   S. Anees Ahmad, S. Sachi Das, A. Khatoon, M. Tahir Ansari, Mohd. Afzal, M. Saquib Hasnain, A. Kumar Nayak, Bactericidal activity of silver nanoparticles: A mechanistic review, Mater Sci Energy Technol. 3 (2020) 756–769. https://doi.org/10.1016/j.mset.2020.09.002

[26]   F.A. Sheikh, N.A.M. Barakat, M.A. Kanjwal, R. Nirmala, J.H. Lee, H. Kim, H.Y. Kim, Electrospun titanium dioxide nanofibers containing hydroxyapatite and silver nanoparticles as future implant materials, J Mater Sci Mater Med. 21 (2010) 2551–2559. https://doi.org/10.1007/s10856-010-4102-9

[27]   K. Madhumathi, P.T. Sudheesh Kumar, S. Abhilash, V. Sreeja, H. Tamura, K. Manzoor, S. V. Nair, R. Jayakumar, Development of novel chitin/nanosilver composite scaffolds for wound dressing applications, J Mater Sci Mater Med. 21 (2010) 807–813. https://doi.org/10.1007/s10856-009-3877-z

[28]   J.-J. Kim, R.K. Singh, S.-J. Seo, T.-H. Kim, J.-H. Kim, E.-J. Lee, H.-W. Kim, Magnetic scaffolds of polycaprolactone with functionalized magnetite nanoparticles: physicochemical, mechanical, and biological properties effective for bone regeneration, RSC Adv. 4 (2014) 17325–17336. https://doi.org/10.1039/C4RA00040D

[29]   R. De Santis, A. Russo, A. Gloria, U. D'Amora, T. Russo, S. Panseri, M. Sandri, A. Tampieri, M. Marcacci, V.A. Dediu, C.J. Wilde, L. Ambrosio, Towards the Design of 3D Fiber-Deposited Poly(-caprolactone)/Iron-Doped Hydroxyapatite Nanocomposite Magnetic Scaffolds for Bone Regeneration, J Biomed Nanotechnol. 11 (2015) 1236–1246. https://doi.org/10.1166/jbn.2015.2065

[30]   B. Anu Priya, K. Senthilguru, T. Agarwal, S.N. Gautham Hari Narayana, S. Giri, K. Pramanik, K. Pal, I. Banerjee, Nickel doped nanohydroxyapatite: vascular endothelial growth factor inducing biomaterial for bone tissue engineering, RSC Adv. 5 (2015) 72515–72528. https://doi.org/10.1039/C5RA09560C

[31]   K.A. Mosa, M. El-Naggar, K. Ramamoorthy, H. Alawadhi, A. Elnaggar, S. Wartanian, E. Ibrahim, H. Hani, Copper Nanoparticles Induced Genotoxicty, Oxidative Stress, and Changes in Superoxide Dismutase (SOD) Gene Expression in Cucumber (Cucumis sativus) Plants, Front Plant Sci. 9 (2018). https://doi.org/10.3389/fpls.2018.00872

[32]   E. Tomaszewska, S. Muszyński, K. Ognik, P. Dobrowolski, M. Kwiecień, J. Juśkiewicz, D. Chocyk, M. Świetlicki, T. Blicharski, B. Gładyszewska, Comparison of the effect of dietary copper nanoparticles with copper (II) salt on bone geometric and structural parameters as well as material characteristics in a rat model, Journal of Trace Elements in Medicine and Biology. 42 (2017) 103–110. https://doi.org/10.1016/j.jtemb.2017.05.002

[33]   J. Nicolas, S. Mura, D. Brambilla, N. Mackiewicz, P. Couvreur, Design, functionalization strategies and biomedical applications of targeted biodegradable/biocompatible polymer-based nanocarriers for drug delivery, Chem. Soc. Rev. 42 (2013) 1147–1235. https://doi.org/10.1039/C2CS35265F

[34]   M. Mehrasa, M.A. Asadollahi, K. Ghaedi, H. Salehi, A. Arpanaei, Electrospun aligned PLGA and PLGA/gelatin nanofibers embedded with silica nanoparticles for tissue engineering, Int J Biol Macromol. 79 (2015) 687–695. https://doi.org/10.1016/j.ijbiomac.2015.05.050

[35]   A. des Rieux, B. Ucakar, B.P.K. Mupendwa, D. Colau, O. Feron, P. Carmeliet, V. Préat, 3D systems delivering VEGF to promote angiogenesis for tissue engineering, Journal of Controlled Release. 150 (2011) 272–278. https://doi.org/10.1016/j.jconrel.2010.11.028

[36]   D. Shi, X. Xu, Y. Ye, K. Song, Y. Cheng, J. Di, Q. Hu, J. Li, H. Ju, Q. Jiang, Z. Gu, Photo-Cross-Linked Scaffold with Kartogenin-Encapsulated Nanoparticles for Cartilage Regeneration, ACS Nano. 10 (2016) 1292–1299. https://doi.org/10.1021/acsnano.5b06663

[37]   Q. Tang, T. Lim, L.Y. Shen, G. Zheng, X.J. Wei, C.Q. Zhang, Z.Z. Zhu, Well-dispersed platelet lysate entrapped nanoparticles incorporate with injectable PDLLA-PEG-PDLLA triblock for preferable cartilage engineering application, Biomaterials. 268 (2021) 120605. https://doi.org/10.1016/J.BIOMATERIALS.2020.120605

[38]   Y. XIE, W. LU, X. JIANG, Improvement of cationic albumin conjugated pegylated nanoparticles holding NC-1900, a vasopressin fragment analog, in memory deficits induced by scopolamine in mice, Behavioural Brain Research. 173 (2006) 76–84. https://doi.org/10.1016/j.bbr.2006.06.001

[39]   M.D. Chavanpatil, A. Khdair, J. Panyam, Surfactant-polymer Nanoparticles: A Novel Platform for Sustained and Enhanced Cellular Delivery of Water-soluble Molecules, Pharm Res. 24 (2007) 803–810. https://doi.org/10.1007/s11095-006-9203-2

[40]   Z. Sun, X. Wang, J. Liu, Z. Wang, W. Wang, D. Kong, X. Leng, ICG/ <scp>l</scp> - Arginine Encapsulated PLGA Nanoparticle-Thermosensitive Hydrogel Hybrid Delivery System

for Cascade Cancer Photodynamic-NO Therapy with Promoted Collagen Depletion in Tumor Tissues, Mol Pharm. 18 (2021) 928–939. https://doi.org/10.1021/acs.molpharmaceut.0c00937

[41]   J. Huang, F. Liu, H. Su, J. Xiong, L. Yang, J. Xia, Y. Liang, Advanced Nanocomposite Hydrogels for Cartilage Tissue Engineering, Gels. 8 (2022) 138. https://doi.org/10.3390/gels8020138

[42]   J. Prakash, D. Prema, K.S. Venkataprasanna, K. Balagangadharan, N. Selvamurugan, G.D. Venkatasubbu, Nanocomposite chitosan film containing graphene oxide/hydroxyapatite/gold for bone tissue engineering, Int J Biol Macromol. 154 (2020) 62–71. https://doi.org/10.1016/j.ijbiomac.2020.03.095

[43]   A. Liu, Z. Hong, X. Zhuang, X. Chen, Y. Cui, Y. Liu, X. Jing, Surface modification of bioactive glass nanoparticles and the mechanical and biological properties of poly(l-lactide) composites, Acta Biomater. 4 (2008) 1005–1015. https://doi.org/10.1016/j.actbio.2008.02.013

[44]   Inamuddin, J.N. Cruz, T. Altalhi, Green Sustainable Process for Chemical and Environmental Engineering and Science: Recent Advances in Nanocarriers, Elsevier, 2023. https://doi.org/10.1016/C2021-0-02836-4

[45]   S.K. Balu, V. Sampath, S. Andra, S. Alagar, S. Manisha Vidyavathy, Fabrication of carbon and silver nanomaterials incorporated hydroxyapatite nanocomposites: Enhanced biological and mechanical performances for biomedical applications, Materials Science and Engineering: C. 128 (2021) 112296. https://doi.org/10.1016/J.MSEC.2021.112296

[46]   D. Jiao, F. Lossada, J. Guo, O. Skarsetz, D. Hoenders, J. Liu, A. Walther, Electrical switching of high-performance bioinspired nanocellulose nanocomposites, Nature Communications 2021 12:1. 12 (2021) 1–10. https://doi.org/10.1038/s41467-021-21599-1

[47]   E. De Giglio, M.A. Bonifacio, A.M. Ferreira, S. Cometa, Z.Y. Ti, A. Stanzione, K. Dalgarno, P. Gentile, Multi-compartment scaffold fabricated via 3D-printing as in vitro co-culture osteogenic model, Scientific Reports 2018 8:1. 8 (2018) 1–13. https://doi.org/10.1038/s41598-018-33472-1

[48]   Inamuddin, T. Altalhi, J.N. Cruz, M.S.E.-D. Refat, Drug design using machine learning, 2022. https://doi.org/10.1002/9781394167258

[49]   C.J. Mortimer, C.J. Wright, The fabrication of iron oxide nanoparticle-nanofiber composites by electrospinning and their applications in tissue engineering, Biotechnol J. 12 (2017) 1600693. https://doi.org/10.1002/BIOT.201600693

[50]   M.L. Carmo Bastos, J.V. Silva-Silva, J. Neves Cruz, A.R. Palheta da Silva, A.A. Bentaberry-Rosa, G. da Costa Ramos, J.E. de Sousa Siqueira, M.R. Coelho-Ferreira, S. Percário, P. Santana Barbosa Marinho, A.M. do R. Marinho, M. de Oliveira Bahia, M.F. Dolabela, Alkaloid from Geissospermum sericeum Benth. & Hook.f. ex Miers (Apocynaceae) Induce Apoptosis by Caspase Pathway in Human Gastric Cancer Cells, Pharmaceuticals. 16 (2023) 765. https://doi.org/10.3390/ph16050765

Materials Research Forum LLC
https://doi.org/10.21741/9781644902998-2

[51] J. Cui, X. Yu, Y. Shen, B. Sun, W. Guo, M. Liu, Y. Chen, L. Wang, X. Zhou, M. Shafiq, X. Mo, Electrospinning Inorganic Nanomaterials to Fabricate Bionanocomposites for Soft and Hard Tissue Repair, Nanomaterials 2023, Vol. 13, Page 204. 13 (2023) 204. https://doi.org/10.3390/NANO13010204

[52] M. Dubský, Š. Kubinová, J. Širc, L. Voska, R. Zajíček, A. Zajícová, P. Lesný, A. Jirkovská, J. Michálek, M. Munzarová, V. Holáň, E. Syková, Nanofibers prepared by needleless electrospinning technology as scaffolds for wound healing, J Mater Sci Mater Med. 23 (2012) 931–941. https://doi.org/10.1007/S10856-012-4577-7/METRICS

[53] V. Singh, P. Yadav, V. Mishra, Recent Advances on Classification, Properties, Synthesis, and Characterization of Nanomaterials, Green Synthesis of Nanomaterials for Bioenergy Applications. (2020) 83–97. https://doi.org/10.1002/9781119576785.CH3

[54] K.A. Ozada N, Novel Microstructure Mechanical Activated Nano Composites for Tissue Engineering Applications, J Bioeng Biomed Sci. 05 (2015). https://doi.org/10.4172/2155-9538.1000143

[55] E. Ryan, S. Yin, Compressive strength of β-TCP scaffolds fabricated via lithography-based manufacturing for bone tissue engineering, Ceram Int. 48 (2022) 15516–15524. https://doi.org/10.1016/J.CERAMINT.2022.02.085

[56] A. Al-Kattan, V.P. Nirwan, A. Popov, Y. V. Ryabchikov, G. Tselikov, M. Sentis, A. Fahmi, A. V. Kabashin, Recent Advances in Laser-Ablative Synthesis of Bare Au and Si Nanoparticles and Assessment of Their Prospects for Tissue Engineering Applications, International Journal of Molecular Sciences 2018, Vol. 19, Page 1563. 19 (2018) 1563. https://doi.org/10.3390/IJMS19061563

[57] J.P. Fan, P. Kalia, L. Di Silvio, J. Huang, In vitro response of human osteoblasts to multi-step sol–gel derived bioactive glass nanoparticles for bone tissue engineering, Materials Science and Engineering: C. 36 (2014) 206–214. https://doi.org/10.1016/J.MSEC.2013.12.009

[58] S. Shokri, B. Movahedi, M. Rafieinia, H. Salehi, A new approach to fabrication of Cs/BG/CNT nanocomposite scaffold towards bone tissue engineering and evaluation of its properties, Appl Surf Sci. 357 (2015) 1758–1764. https://doi.org/10.1016/J.APSUSC.2015.10.048

[59] L.D.F.B. Torres, J.N. Cruz, Natural Products from the Amazon Used by the Cosmetic Industry, in: Drug Discovery and Design Using Natural Products, Springer, 2023: pp. 525–537. https://doi.org/10.1007/978-3-031-35205-8_19

[60] R. Jose Varghese, E.H.M. Sakho, S. Parani, S. Thomas, O.S. Oluwafemi, J. Wu, Introduction to nanomaterials: synthesis and applications, Nanomaterials for Solar Cell Applications. (2019) 75–95. https://doi.org/10.1016/B978-0-12-813337-8.00003-5

[61] A. Pal, R. Vel, S.H. Rahaman, S. Sengupta, S. Bodhak, Synthesis and characterizations of sugar-glass nanoparticles mediated protein delivery system for tissue engineering application, Nano Futures. 6 (2022) 025008. https://doi.org/10.1088/2399-1984/AC7832

[62]   M.S. Orellano, G.S. Longo, C. Porporatto, N.M. Correa, R.D. Falcone, Role of micellar interface in the synthesis of chitosan nanoparticles formulated by reverse micellar method, Colloids Surf A Physicochem Eng Asp. 599 (2020) 124876. https://doi.org/10.1016/J.COLSURFA.2020.124876

[63]   S. Moeini, M.R. Mohammadi, A. Simchi, In-situ solvothermal processing of polycaprolactone/hydroxyapatite nanocomposites with enhanced mechanical and biological performance for bone tissue engineering, Bioact Mater. 2 (2017) 146–155. https://doi.org/10.1016/J.BIOACTMAT.2017.04.004

[64]   Q. Zong, H. Chen, Y. Zhao, J. Wang, J. Wu, Bioactive carbon dots for tissue engineering applications, Smart Mater Med. 5 (2024) 1–14. https://doi.org/10.1016/j.smaim.2023.06.006

[65]   V. Dutta, R. Verma, C. Gopalkrishnan, M.H. Yuan, K.M. Batoo, R. Jayavel, A. Chauhan, K.Y.A. Lin, R. Balasubramani, S. Ghotekar, Bio-Inspired Synthesis of Carbon-Based Nanomaterials and Their Potential Environmental Applications: A State-of-the-Art Review, Inorganics 2022, Vol. 10, Page 169. 10 (2022) 169. https://doi.org/10.3390/INORGANICS10100169

[66]   A. Rónavári, N. Igaz, D.I. Adamecz, B. Szerencsés, C. Molnar, Z. Kónya, I. Pfeiffer, M. Kiricsi, Green Silver and Gold Nanoparticles: Biological Synthesis Approaches and Potentials for Biomedical Applications, Molecules 2021, Vol. 26, Page 844. 26 (2021) 844. https://doi.org/10.3390/MOLECULES26040844

[67]   S. Fahimirad, F. Ajalloueian, M. Ghorbanpour, Synthesis and therapeutic potential of silver nanomaterials derived from plant extracts, Ecotoxicol Environ Saf. 168 (2019) 260–278. https://doi.org/10.1016/J.ECOENV.2018.10.017

[68]   S.K. Sahoo, G.K. Panigrahi, M.K. Sahu, A. Arzoo, J.K. Sahoo, A. Sahoo, A.K. Pradhan, A. Dalbehera, Biological synthesis of GO-MgO nanomaterial using Azadirachta indica leaf extract: A potential bio-adsorbent for removing Cr(VI) ions from aqueous media, Biochem Eng J. 177 (2022) 108272. https://doi.org/10.1016/J.BEJ.2021.108272

[69]   H. Dabhane, S. Ghotekar, M. Zate, S. Kute, G. Jadhav, V. Medhane, Green synthesis of MgO nanoparticles using aqueous leaf extract of Ajwain (Trachyspermum ammi) and evaluation of their catalytic and biological activities, Inorg Chem Commun. 138 (2022) 109270. https://doi.org/10.1016/J.INOCHE.2022.109270

[70]   M.U. Zahid, E. Pervaiz, A. Hussain, M.I. Shahzad, M.B.K. Niazi, Synthesis of carbon nanomaterials from different pyrolysis techniques: a review, Mater Res Express. 5 (2018) 052002. https://doi.org/10.1088/2053-1591/AAC05B

[71]   Z. Li, Y. Sun, S. Ge, F. Zhu, F. Yin, L. Gu, F. Yang, P. Hu, G. Chen, K. Wang, A.A. Volinsky, An Overview of Synthesis and Structural Regulation of Magnetic Nanomaterials Prepared by Chemical Coprecipitation, Metals 2023, Vol. 13, Page 152. 13 (2023) 152. https://doi.org/10.3390/MET13010152

[72]   X. Hangxun, B.W. Zeiger, K.S. Suslick, Sonochemical synthesis of nanomaterials, Chem Soc Rev. 42 (2013) 2555–2567. https://doi.org/10.1039/C2CS35282F

[73]   G.-R. Li, H. Xu, X.-F. Lu, J.-X. Feng, Y.-X. Tong, C.-Y. Su, Electrochemical synthesis of nanostructured materials for electrochemical energy conversion and storage, Nanoscale. 5 (2013) 4056–4069. https://doi.org/10.1039/C3NR00607G

[74]   D. Zhang, K. Ye, Y. Yao, F. Liang, T. Qu, W. Ma, B. Yang, Y. Dai, T. Watanabe, Controllable synthesis of carbon nanomaterials by direct current arc discharge from the inner wall of the chamber, Carbon N Y. 142 (2019) 278–284. https://doi.org/10.1016/J.CARBON.2018.10.062

[75]   N. Sakono, Y. Ishida, K. Ogo, N. Tsumori, H. Murayama, M. Sakono, Molar-Fraction-Tunable Synthesis of Ag-Au Alloy Nanoparticles via a Dual Evaporation-Condensation Method as Supported Catalysts for CO Oxidation, ACS Appl Nano Mater. (2023). https://doi.org/10.1021/ACSANM.3C00089/SUPPL_FILE/AN3C00089_SI_001.PDF

[76]   N.M. Chu, N.D. Hieu, D.T.M. Do, R. Sarathi, T. Nakayama, H. Suematsu, Synthesis of molybdenum carbide nanoparticles using pulsed wire discharge in mixed atmosphere of kerosene and argon, Journal of the American Ceramic Society. 102 (2019) 7108–7115. https://doi.org/10.1111/JACE.16621

[77]   M. Behnke, P. Klemm, P. Dahlke, B. Shkodra, B. Beringer-Siemers, J.A. Czaplewska, S. Stumpf, P.M. Jordan, S. Schubert, S. Hoeppener, A. Vollrath, O. Werz, U.S. Schubert, Ethoxy acetalated dextran nanoparticles for drug delivery: A comparative study of formulation methods, Int J Pharm X. 5 (2023) 100173. https://doi.org/10.1016/J.IJPX.2023.100173

[78]   H. Shimoshige, H. Kobayashi, S. Shimamura, T. Mizuki, A. Inoue, T. Maekawa, Isolation and cultivation of a novel sulfate-reducing magnetotactic bacterium belonging to the genus Desulfovibrio, PLoS One. 16 (2021) e0248313. https://doi.org/10.1371/JOURNAL.PONE.0248313

[79]   A. Michael, A. Singh, A. Roy, M.R. Islam, Fungal- and Algal-Derived Synthesis of Various Nanoparticles and Their Applications, Bioinorg Chem Appl. 2022 (2022). https://doi.org/10.1155/2022/3142674

[80]   M. Nasrollahzadeh, S. Mahmoudi-Gom Yek, N. Motahharifar, M. Ghafori Gorab, Recent Developments in the Plant-Mediated Green Synthesis of Ag-Based Nanoparticles for Environmental and Catalytic Applications, The Chemical Record. 19 (2019) 2436–2479. https://doi.org/10.1002/TCR.201800202

[81]   N. Kumar, S. Sinha Ray, Synthesis and functionalization of nanomaterials, Springer Series in Materials Science. 277 (2018) 15–55. https://doi.org/10.1007/978-3-319-97779-9_2/COVER

[82]   Inamuddin, T. Altalhi, J.N. Cruz, M. Luqman, Nanomaterial-Supported Enzymes, Materials Research Forum LLC, 2022. https://doi.org/10.21741/9781644901977

[83]   X. Liu, M.N. George, S. Park, A.L. Miller II, B. Gaihre, L. Li, B.E. Waletzki, A. Terzic, M.J. Yaszemski, L. Lu, 3D-printed scaffolds with carbon nanotubes for bone tissue engineering: Fast and homogeneous one-step functionalization, Acta Biomater. 111 (2020) 129–140. https://doi.org/10.1016/J.ACTBIO.2020.04.047

[84] K. Elkhoury, C.S. Russell, L. Sanchez-Gonzalez, A. Mostafavi, T.J. Williams, C. Kahn, N.A. Peppas, E. Arab-Tehrany, A. Tamayol, Soft-Nanoparticle Functionalization of Natural Hydrogels for Tissue Engineering Applications, Adv Healthc Mater. 8 (2019). https://doi.org/10.1002/adhm.201900506

[85] F. Ahmad, M.M. Salem-Bekhit, F. Khan, S. Alshehri, A. Khan, M.M. Ghoneim, H.F. Wu, E.I. Taha, I. Elbagory, Unique Properties of Surface-Functionalized Nanoparticles for Bio-Application: Functionalization Mechanisms and Importance in Application, Nanomaterials. 12 (2022). https://doi.org/10.3390/nano12081333

[86] M.A.M. Tarkistani, V. Komalla, V. Kayser, Recent Advances in the Use of Iron–Gold Hybrid Nanoparticles for Biomedical Applications, Nanomaterials 2021, Vol. 11, Page 1227. 11 (2021) 1227. https://doi.org/10.3390/NANO11051227

[87] M. Pourmadadi, A. Tajiki, S.M. Hosseini, A. Samadi, M. Abdouss, S. Daneshnia, F. Yazdian, A comprehensive review of synthesis, structure, properties, and functionalization of MoS2; emphasis on drug delivery, photothermal therapy, and tissue engineering applications, J Drug Deliv Sci Technol. 76 (2022) 103767. https://doi.org/10.1016/J.JDDST.2022.103767

[88] Y. Wang, W. Zhang, C. Gong, B. Liu, Y. Li, L. Wang, Z. Su, G. Wei, Recent advances in the fabrication, functionalization, and bioapplications of peptide hydrogels, Soft Matter. 16 (2020) 10029–10045. https://doi.org/10.1039/D0SM00966K

[89] S.N. Mali, In silico Methods for Evaluating the Mode of Interaction of Nanoparticles with Molecular Target, Nanobiomaterials: Perspectives for Medical Applications in the Diagnosis and Treatment of Diseases. 145 (2023) 236–249. https://doi.org/10.21741/9781644902370-9

[90] L.A. Kolahalam, I. V. Kasi Viswanath, B.S. Diwakar, B. Govindh, V. Reddy, Y.L.N. Murthy, Review on nanomaterials: Synthesis and applications, Mater Today Proc. 18 (2019) 2182–2190. https://doi.org/10.1016/J.MATPR.2019.07.371

[91] S.R. Falsafi, H. Rostamabadi, E. Assadpour, S.M. Jafari, Morphology and microstructural analysis of bioactive-loaded micro/nanocarriers via microscopy techniques; CLSM/SEM/TEM/AFM, Adv Colloid Interface Sci. 280 (2020) 102166. https://doi.org/10.1016/J.CIS.2020.102166

[92] M. Kaliva, M. Vamvakaki, Nanomaterials characterization, Polymer Science and Nanotechnology: Fundamentals and Applications. (2020) 401–433. https://doi.org/10.1016/B978-0-12-816806-6.00017-0

[93] O.M. Lemine, Microstructural characterisation of α-Fe2O3 nanoparticles using, XRD line profiles analysis, FE-SEM and FT-IR, Superlattices Microstruct. 45 (2009) 576 582. https://doi.org/10.1016/J.SPMI.2009.02.004

[94] M. Šetka, R. Calavia, L. Vojkůvka, E. Llobet, J. Drbohlavová, S. Vallejos, Raman and XPS studies of ammonia sensitive polypyrrole nanorods and nanoparticles, Scientific Reports 2019 9:1. 9 (2019) 1–10. https://doi.org/10.1038/s41598-019-44900-1

[95]   J.N. Cruz, Nanobiomaterials: Perspectives for Medical Applications in the Diagnosis and Treatment of Diseases, 2023. https://doi.org/10.21741/9781644902370

[96]   M.M. Abutalib, A. Rajeh, Influence of Fe3O4 nanoparticles on the optical, magnetic and electrical properties of PMMA/PEO composites: Combined FT-IR/DFT for electrochemical applications, J Organomet Chem. 920 (2020) 121348.
https://doi.org/10.1016/J.JORGANCHEM.2020.121348

[97]   A. Naskar, K.S. Kim, Recent Advances in Nanomaterial-Based Wound-Healing Therapeutics, Pharmaceutics 2020, Vol. 12, Page 499. 12 (2020) 499.
https://doi.org/10.3390/PHARMACEUTICS12060499

[98]   S. Sharifi, M.J. Hajipour, L. Gould, M. Mahmoudi, Nanomedicine in Healing Chronic Wounds: Opportunities and Challenges, Mol Pharm. 18 (2021) 550–575.
https://doi.org/10.1021/ACS.MOLPHARMACEUT.0C00346/ASSET/IMAGES/MEDIUM/MP0 C00346_0013.GIF

[99]   M. Rodrigues, N. Kosaric, C.A. Bonham, G.C. Gurtner, Wound healing: A cellular perspective, Physiol Rev. 99 (2019) 665–706.
https://doi.org/10.1152/PHYSREV.00067.2017/ASSET/IMAGES/LARGE/Z9J0041828900006.J PEG

[100]   S. Chakrabarti, P. Chattopadhyay, J. Islam, S. Ray, P.S. Raju, B. Mazumder, Aspects of Nanomaterials in Wound Healing, Curr Drug Deliv. 16 (2018) 26–41.
https://doi.org/10.2174/1567201815666180918110134

[101]   I. Negut, V. Grumezescu, A.M. Grumezescu, Treatment Strategies for Infected Wounds, Molecules 2018, Vol. 23, Page 2392. 23 (2018) 2392.
https://doi.org/10.3390/MOLECULES23092392

[102]   R.A.F. Clark, K. Ghosh, M.G. Tonnesen, Tissue Engineering for Cutaneous Wounds, Journal of Investigative Dermatology. 127 (2007) 1018–1029.
https://doi.org/10.1038/SJ.JID.5700715

[103]   A. Barroso, H. Mestre, A. Ascenso, S. Simões, C. Reis, Nanomaterials in wound healing: From material sciences to wound healing applications, Nano Select. 1 (2020) 443–460.
https://doi.org/10.1002/NANO.202000055

[104]   R. Yu, H. Zhang, B. Guo, Conductive Biomaterials as Bioactive Wound Dressing for Wound Healing and Skin Tissue Engineering, Nano-Micro Letters 2021 14:1. 14 (2021) 1–46.
https://doi.org/10.1007/S40820-021-00751-Y

[105]   Z. Mbese, S. Alven, B.A. Aderibigbe, M. Meyer, I. Prade, E. Klüver, Collagen-Based Nanofibers for Skin Regeneration and Wound Dressing Applications, Polymers 2021, Vol. 13, Page 4368. 13 (2021) 4368. https://doi.org/10.3390/POLYM13244368

[106]   R.K. Thapa, K.L. Kiick, M.O. Sullivan, Encapsulation of collagen mimetic peptide-tethered vancomycin liposomes in collagen-based scaffolds for infection control in wounds, Acta Biomater. 103 (2020) 115–128. https://doi.org/10.1016/J.ACTBIO.2019.12.014

[107] S.R. Gomes, G. Rodrigues, G.G. Martins, M.A. Roberto, M. Mafra, C.M.R. Henriques, J.C. Silva, In vitro and in vivo evaluation of electrospun nanofibers of PCL, chitosan and gelatin: A comparative study, Materials Science and Engineering: C. 46 (2015) 348–358. https://doi.org/10.1016/J.MSEC.2014.10.051

[108] H. Bilgic, M. Demiriz, M. Ozler, T. Ide, N. Dogan, S. Gumus, A. Kiziltay, T. Endogan, V. Hasirci, N. Hasirci, Gelatin Based Scaffolds and Effect of EGF Dose on Wound Healing, J Biomater Tissue Eng. 3 (2013) 205–211. https://doi.org/10.1166/JBT.2013.1077

[109] T. Hakkarainen, R. Koivuniemi, M. Kosonen, C. Escobedo-Lucea, A. Sanz-Garcia, J. Vuola, J. Valtonen, P. Tammela, A. Mäkitie, K. Luukko, M. Yliperttula, H. Kavola, Nanofibrillar cellulose wound dressing in skin graft donor site treatment, Journal of Controlled Release. 244 (2016) 292–301. https://doi.org/10.1016/J.JCONREL.2016.07.053

[110] A.H. Tayeb, E. Amini, S. Ghasemi, M. Tajvidi, Cellulose Nanomaterials—Binding Properties and Applications: A Review, Molecules 2018, Vol. 23, Page 2684. 23 (2018) 2684. https://doi.org/10.3390/MOLECULES23102684

[111] R.J. Hickey, A.E. Pelling, Cellulose biomaterials for tissue engineering, Front Bioeng Biotechnol. 7 (2019) 45. https://doi.org/10.3389/FBIOE.2019.00045/BIBTEX

[112] L.M. Anaya-Esparza, J.M. Ruvalcaba-Gómez, C.I. Maytorena-Verdugo, N. González-Silva, R. Romero-Toledo, S. Aguilera-Aguirre, A. Pérez-Larios, E. Montalvo-González, Chitosan-TiO2: A Versatile Hybrid Composite, Materials 2020, Vol. 13, Page 811. 13 (2020) 811. https://doi.org/10.3390/MA13040811

[113] A. Chanda, J. Adhikari, A. Ghosh, S.R. Chowdhury, S. Thomas, P. Datta, P. Saha, Electrospun chitosan/polycaprolactone-hyaluronic acid bilayered scaffold for potential wound healing applications, Int J Biol Macromol. 116 (2018) 774–785. https://doi.org/10.1016/J.IJBIOMAC.2018.05.099

[114] M. Ovais, I. Ahmad, A.T. Khalil, S. Mukherjee, R. Javed, M. Ayaz, A. Raza, Z.K. Shinwari, Wound healing applications of biogenic colloidal silver and gold nanoparticles: recent trends and future prospects, Appl Microbiol Biotechnol. 102 (2018) 4305–4318. https://doi.org/10.1007/S00253-018-8939-Z/FIGURES/7

[115] S. Alizadeh, B. Seyedalipour, S. Shafieyan, A. Kheime, P. Mohammadi, N. Aghdami, Copper nanoparticles promote rapid wound healing in acute full thickness defect via acceleration of skin cell migration, proliferation, and neovascularization, Biochem Biophys Res Commun. 517 (2019) 684–690. https://doi.org/10.1016/J.BBRC.2019.07.110

[116] N. Asadi, H. Pazoki-Toroudi, A.R. Del Bakhshayesh, A. Akbarzadeh, S. Davaran, N. Annabi, Multifunctional hydrogels for wound healing: Special focus on biomacromolecular based hydrogels, Int J Biol Macromol. 170 (2021) 728–750. https://doi.org/10.1016/J.IJBIOMAC.2020.12.202

[117] S. Vieira, S. Vial, R.L. Reis, J.M. Oliveira, Nanoparticles for bone tissue engineering, Biotechnol Prog. 33 (2017) 590–611. https://doi.org/10.1002/BTPR.2469

[118] R.E. McMahon, L. Wang, R. Skoracki, A.B. Mathur, Development of nanomaterials for bone repair and regeneration, J Biomed Mater Res B Appl Biomater. 101B (2013) 387–397. https://doi.org/10.1002/JBM.B.32823

[119] M. J Hill, B. Qi, R. Bayaniahangar, V. Araban, Z. Bakhtiary, M.R. Doschak, B.C. Goh, M. Shokouhimehr, H. Vali, J.F. Presley, A.A. Zadpoor, M.B. Harris, P.P.S.S. Abadi, M. Mahmoudi, Nanomaterials for bone tissue regeneration: updates and future perspectives, Https://Doi.Org/10.2217/Nnm-2018-0445. 14 (2019) 2987–3006. https://doi.org/10.2217/NNM-2018-0445

[120] D. Zhang, D. Liu, J. Zhang, C. Fong, M. Yang, Gold nanoparticles stimulate differentiation and mineralization of primary osteoblasts through the ERK/MAPK signaling pathway, Materials Science and Engineering: C. 42 (2014) 70–77. https://doi.org/10.1016/J.MSEC.2014.04.042

[121] D.N. Heo, W.K. Ko, M.S. Bae, J.B. Lee, D.W. Lee, W. Byun, C.H. Lee, E.C. Kim, B.Y. Jung, I.K. Kwon, Enhanced bone regeneration with a gold nanoparticle–hydrogel complex, J Mater Chem B. 2 (2014) 1584–1593. https://doi.org/10.1039/C3TB21246G

[122] W.K. Ko, D.N. Heo, H.J. Moon, S.J. Lee, M.S. Bae, J.B. Lee, I.C. Sun, H.B. Jeon, H.K. Park, I.K. Kwon, The effect of gold nanoparticle size on osteogenic differentiation of adipose-derived stem cells, J Colloid Interface Sci. 438 (2015) 68–76. https://doi.org/10.1016/J.JCIS.2014.08.058

[123] J. Li, J. Zhang, Y. Chen, N. Kawazoe, G. Chen, TEMPO-Conjugated Gold Nanoparticles for Reactive Oxygen Species Scavenging and Regulation of Stem Cell Differentiation, ACS Appl Mater Interfaces. 9 (2017) 35683–35692. https://doi.org/10.1021/ACSAMI.7B12486/SUPPL_FILE/AM7B12486_SI_001.PDF

[124] S.Y. Choi, M.S. Song, P.D. Ryu, A.T.N. Lam, S.W. Joo, S.Y. Lee, Gold nanoparticles promote osteogenic differentiation in human adipose-derived mesenchymal stem cells through the Wnt/β-catenin signaling pathway, Int J Nanomedicine. 10 (2015) 4383–4392. https://doi.org/10.2147/IJN.S78775

[125] N.L. Rosi, D.A. Giljohann, C.S. Thaxton, A.K.R. Lytton-Jean, M.S. Han, C.A. Mirkin, Oligonucleotide-modified gold nanoparticles for infracellular gene regulation, Science (1979). 312 (2006) 1027–1030. https://doi.org/10.1126/SCIENCE.1125559/SUPPL_FILE/ROSI_SOM.PDF

[126] T. Qing, M. Mahmood, Y. Zheng, A.S. Biris, L. Shi, D.A. Casciano, A genomic characterization of the influence of silver nanoparticles on bone differentiation in MC3T3-E1 cells, Journal of Applied Toxicology. 38 (2018) 172–179. https://doi.org/10.1002/JAT.3528

[127] J. Radwan-Pragłowska, Ł. Janus, M. Piatkowski, D. Bogdał, D. Matysek, 3D Hierarchical, Nanostructured Chitosan/PLA/HA Scaffolds Doped with TiO2/Au/Pt NPs with Tunable Properties for Guided Bone Tissue Engineering, Polymers 2020, Vol. 12, Page 792. 12 (2020) 792. https://doi.org/10.3390/POLYM12040792

[128] M. Ragothaman, A. Kannan Villalan, A. Dhanasekaran, T. Palanisamy, Bio-hybrid hydrogel comprising collagen-capped silver nanoparticles and melatonin for accelerated tissue regeneration in skin defects, Materials Science and Engineering: C. 128 (2021) 112328. https://doi.org/10.1016/J.MSEC.2021.112328

[129] A. Halim, K.Y. Qu, X.F. Zhang, N.P. Huang, Recent Advances in the Application of Two-Dimensional Nanomaterials for Neural Tissue Engineering and Regeneration, ACS Biomater Sci Eng. 7 (2021) 3503–3529. https://doi.org/10.1021/ACSBIOMATERIALS.1C00490/ASSET/IMAGES/MEDIUM/AB1C004 90_0007.GIF

[130] M.E. Marti, A.D. Sharma, D.S. Sakaguchi, S.K. Mallapragada, Nanomaterials for neural tissue engineering, Nanomaterials in Tissue Engineering: Fabrication and Applications. (2013) 275–301. https://doi.org/10.1533/9780857097231.2.275

[131] R. Kumar, K.R. Aadil, S. Ranjan, V.B. Kumar, Advances in nanotechnology and nanomaterials based strategies for neural tissue engineering, J Drug Deliv Sci Technol. 57 (2020) 101617. https://doi.org/10.1016/J.JDDST.2020.101617

[132] E.R.R. Collazo, Repair of Stump Neuroma Using AxoGuardÂ® Nerve Protector and AvanceÂ® Nerve Graft in the Lower Extremity, Orthopedics and Rheumatology Open Access Journals. 1 (2015) 69–70. https://doi.org/10.19080/OROAJ.2015.01.555566

[133] C. Bibbo, E. Rodrigues-Colazzo, A.G. Finzen, Superficial Peroneal Nerve to Deep Peroneal Nerve Transfer With Allograft Conduit for Neuroma in Continuity, The Journal of Foot and Ankle Surgery. 57 (2018) 514–517. https://doi.org/10.1053/J.JFAS.2017.11.022

[134] W.H. Suh, K.S. Suslick, G.D. Stucky, Y.H. Suh, Nanotechnology, nanotoxicology, and neuroscience, Prog Neurobiol. 87 (2009) 133–170. https://doi.org/10.1016/J.PNEUROBIO.2008.09.009

[135] S.K. Seidlits, J.Y. Lee, C.E. Schmidt, Nanostructured scaffolds for neural applications, Https://Doi.Org/10.2217/17435889.3.2.183. 3 (2008) 183–199. https://doi.org/10.2217/17435889.3.2.183

[136] C.N.R. Rao, G.U. Kulkarni, P.J. Thomas, P.P. Edwards, Metal nanoparticles and their assemblies, Chem Soc Rev. 29 (2000) 27–35. https://doi.org/10.1039/A904518J

[137] M. Liong, J. Lu, M. Kovochich, T. Xia, S.G. Ruehm, A.E. Nel, F. Tamanoi, J.I. Zink, Multifunctional inorganic nanoparticles for imaging, targeting, and drug delivery, ACS Nano. 2 (2008) 889–896. https://doi.org/10.1021/NN800072T/SUPPL_FILE/NN800072T-FILE003.PDF

[138] Y.S. Lin, C.L. Haynes, Impacts of mesoporous silica nanoparticle size, pore ordering, and pore integrity on hemolytic activity, J Am Chem Soc. 132 (2010) 4834–4842. https://doi.org/10.1021/JA910846Q/SUPPL_FILE/JA910846Q_SI_001.PDF

[139] S. Shah, A. Solanki, P.K. Sasmal, K.B. Lee, Single vehicular delivery of siRNA and small molecules to control stem cell differentiation, J Am Chem Soc. 135 (2013) 15682–15685. https://doi.org/10.1021/JA4071738/SUPPL_FILE/JA4071738_SI_001.PDF

[140] C. Lois, A. Alvarez-Buylla, Proliferating subventricular zone cells in the adult mammalian forebrain can differentiate into neurons and glia., Proceedings of the National Academy of Sciences. 90 (1993) 2074–2077. https://doi.org/10.1073/PNAS.90.5.2074

[141] G. Orive, E. Anitua, J.L. Pedraz, D.F. Emerich, Biomaterials for promoting brain protection, repair and regeneration, Nature Reviews Neuroscience 2009 10:9. 10 (2009) 682–692. https://doi.org/10.1038/nrn2685

[142] S. Bhattacharya, K.M. Alkharfy, R. Janardhanan, D. Mukhopadhyay, Nanomedicine: Pharmacological perspectives, Nanotechnol Rev. 1 (2012) 235–253. https://doi.org/10.1515/NTREV-2011-0010/MACHINEREADABLECITATION/RIS

[143] D. Maysinger, A. Morinville, Drug delivery to the nervous system, Trends Biotechnol. 15 (1997) 410–418. https://doi.org/10.1016/S0167-7799(97)01095-0

[144] T. Yin, P. Wang, J. Li, R. Zheng, B. Zheng, D. Cheng, R. Li, J. Lai, X. Shuai, Ultrasound-sensitive siRNA-loaded nanobubbles formed by hetero-assembly of polymeric micelles and liposomes and their therapeutic effect in gliomas, Biomaterials. 34 (2013) 4532–4543. https://doi.org/10.1016/J.BIOMATERIALS.2013.02.067

[145] Z.M. Huang, Y.Z. Zhang, M. Kotaki, S. Ramakrishna, A review on polymer nanofibers by electrospinning and their applications in nanocomposites, Compos Sci Technol. 63 (2003) 2223–2253. https://doi.org/10.1016/S0266-3538(03)00178-7

[146] I. Ahmed, H.Y. Liu, P.C. Mamiya, A.S. Ponery, A.N. Babu, T. Weik, M. Schindler, S. Meiners, Three-dimensional nanofibrillar surfaces covalently modified with tenascin-C-derived peptides enhance neuronal growth in vitro, J Biomed Mater Res A. 76A (2006) 851–860. https://doi.org/10.1002/JBM.A.30587

[147] J.A. Hubbell, A. Chilkoti, Nanomaterials for Drug Delivery, Science (1979). 337 (2012) 303–305. https://doi.org/10.1126/SCIENCE.1219657

[148] J. Jacob, J.T. Haponiuk, S. Thomas, S. Gopi, Biopolymer based nanomaterials in drug delivery systems: A review, Mater Today Chem. 9 (2018) 43–55. https://doi.org/10.1016/J.MTCHEM.2018.05.002

[149] Inamuddin, T. Altalhi, J.N. Cruz, Nutraceuticals: Sources, Processing Methods, Properties, and Applications, Elsevier, 2023. https://doi.org/10.1016/C2021-0-03574-4

[150] H.K.S. Yadav, A.A. Almokdad, S.I.M. Shaluf, M.S. Debe, Polymer-Based Nanomaterials for Drug-Delivery Carriers, Nanocarriers for Drug Delivery: Nanoscience and Nanotechnology in Drug Delivery. (2019) 531–556. https://doi.org/10.1016/B978-0-12-814033-8.00017-5

[151] W. Wang, K.J. Lu, C.H. Yu, Q.L. Huang, Y.Z. Du, Nano-drug delivery systems in wound treatment and skin regeneration, Journal of Nanobiotechnology 2019 17:1. 17 (2019) 1–15. https://doi.org/10.1186/S12951-019-0514-Y

[152] M. Biondi, F. Ungaro, F. Quaglia, P.A. Netti, Controlled drug delivery in tissue engineering, Adv Drug Deliv Rev. 60 (2008) 229–242. https://doi.org/10.1016/J.ADDR.2007.08.038

[153] M. Xie, Y. Li, Z. Zhao, A. Chen, J. Li, Z. Li, G. Li, X. Lin, Development of silk fibroin-derived nanofibrous drug delivery system in supercritical CO2, Mater Lett. 167 (2016) 175–178. https://doi.org/10.1016/J.MATLET.2015.12.151

[154] S. Perteghella, B. Crivelli, L. Catenacci, M. Sorrenti, G. Bruni, V. Necchi, B. Vigani, M. Sorlini, M.L. Torre, T. Chlapanidas, Stem cell-extracellular vesicles as drug delivery systems: New frontiers for silk/curcumin nanoparticles, Int J Pharm. 520 (2017) 86–97. https://doi.org/10.1016/J.IJPHARM.2017.02.005

[155] L. Sercombe, T. Veerati, F. Moheimani, S.Y. Wu, A.K. Sood, S. Hua, Advances and challenges of liposome assisted drug delivery, Front Pharmacol. 6 (2015) 163819. https://doi.org/10.3389/FPHAR.2015.00286/BIBTEX

[156] M.H. Sarfraz, M. Zubair, B. Aslam, A. Ashraf, M.H. Siddique, S. Hayat, J.N. Cruz, S. Muzammil, M. Khurshid, M.F. Sarfraz, A. Hashem, T.M. Dawoud, G.D. Avila-Quezada, E.F. Abd_Allah, Comparative analysis of phyto-fabricated chitosan, copper oxide, and chitosan-based CuO nanoparticles: antibacterial potential against Acinetobacter baumannii isolates and anticancer activity against HepG2 cell lines, Front Microbiol. 14 (2023) 1188743. https://doi.org/10.3389/fmicb.2023.1188743

[157] O.S. Fenton, K.N. Olafson, P.S. Pillai, M.J. Mitchell, R. Langer, O.S. Fenton, K.N. Olafson, P.S. Pillai, R. Langer, M.J. Mitchell, Advances in Biomaterials for Drug Delivery, Advanced Materials. 30 (2018) 1705328. https://doi.org/10.1002/ADMA.201705328

[158] L. Zhang, A. Beatty, L. Lu, A. Abdalrahman, T.M. Makris, G. Wang, Q. Wang, Microfluidic-assisted polymer-protein assembly to fabricate homogeneous functionalnanoparticles, Materials Science and Engineering: C. 111 (2020) 110768. https://doi.org/10.1016/J.MSEC.2020.110768

[159] X. Liu, C. Li, J. Lv, F. Huang, Y. An, L. Shi, R. Ma, Glucose and H2O2 Dual-Responsive Polymeric Micelles for the Self-Regulated Release of Insulin, ACS Appl Bio Mater. 3 (2020) 1598–1606. https://doi.org/10.1021/ACSABM.9B01185/SUPPL_FILE/MT9B01185_SI_001.PDF

[160] X. Zheng, P. Zhang, Z. Fu, S. Meng, L. Dai, H. Yang, Applications of nanomaterials in tissue engineering, RSC Adv. 11 (2021) 19041–19058. https://doi.org/10.1039/d1ra01849c

[161] H. Barabadi, M. Najafi, H. Samadian, A. Azarnezhad, H. Vahidi, M.A. Mahjoub, M. Koohiyan, A. Ahmadi, A Systematic Review of the Genotoxicity and Antigenotoxicity of Biologically Synthesized Metallic Nanomaterials: Are Green Nanoparticles Safe Enough for Clinical Marketing?, Medicina 2019, Vol. 55, Page 439. 55 (2019) 439. https://doi.org/10.3390/MEDICINA55080439

Materials Research Forum LLC
https://doi.org/10.21741/9781644902998-3

Chapter 3

Interaction of Nanoparticles with Macromolecules: Biomedical and Food Science Applications

Deepa Sharma[1], Mohit Kumar[1], Nandlal Verma[2], Shivangi Sharma[3], Nisha Rathore[2], Sanjay Kataria[3] and Gautam Jaiswar[1,*]

[1]Department of Chemistry, Dr. Bhimrao Ambedkar University, Agra

[2]Department of Chemistry, K.R. College, Mathura

[3]Department of Chemistry, B.S.A. PG College, Mathura

gjaiswar@gmail.com

Abstract

In this chapter we focus on recent developments in nanotechnology, especially emphasizing the interaction between nanoparticles (NPs) and macromolecules. Nanoparticles and macromolecules depend on various parameters, including the sizes, shapes, charges, and chemical functionalities of these moieties. Macromolecules are nontoxic and extremely biocompatible but show strong affinity to bioactive compounds through various molecular interaction, such as π-π interactions, hydrophobic and hydrogen bonds. Nanoparticles and macromolecule carriers are capable of improving the efficacy of chemo therapeutics by their enhanced passive accumulation in tumor tissue as compared with low molecular weight drugs. Further we discussed the role of nanoparticles in foods science.

Keywords

Nanoparticles, Protein, Carbohydrate, Nucleic Acid and Lipids

Contents

Materials Research Forum LLC
https://doi.org/10.21741/9781644902998-3

1.   Introduction

Macromolecules are generally complex polymeric molecules made up of simple, orderly arrangement of repeated small chemical molecules or monomers linked covalently [1]. They can either be chemically synthesized or naturally occurring analogues of these macromolecules. The term "Macromolecules" is generalized for the large molecules that include protein, carbohydrates, nucleic acid, lipids, and polymers, arranged repeating units of covalently bonded atoms [2]. The macromolecules like nucleic acid (DNA and RNA) are created by the polymerization of monomer nucleotides and they store the genetic information of life. Whereas carbohydrates are created through the polymerization of a monomeric unit of sugar molecules. Carbohydrate includes simple monosaccharides like glucose and complex polysaccharides such as cellulose [3]. Lipids, which include fats/oils, hormones, waxes, and the energy-storing components of cell membranes, are nonpolar and hydrophobic macromolecules [4]. Proteins are the most abundant class of macromolecules; every living organism is made up of one or more long chains of amino acids that are folded to form complex structures [5].

In recent years, interest in the field of nanotechnology has increased significantly. Some fundamental and common terminology and ideas have been suggested in order to encourage the better development of nanotechnology. For example, with relation to nanoscale characteristics and features [6]. Nanoparticles are units of any shape with a size range of 1-100 nm in any one dimension. These particles have characteristics of both molecular structures and bulk materials due to their unusual size. As a result, nanoparticles are thought of as the "bridge" between macro and micro structures. One of their most interesting qualities is their compact size, which results in a high surface to volume ratio [7]. Nanoparticles can have a 0D, 1D, 2D, or 3D structure depending on the shape [8]. They can be divided into different categories depending on their source or type of material such as ceramic-based, semiconductor-based, carbon-based, polymer-based lipid-based, metal and metal oxide-based nanoparticles [9].

For the synthesis of nanomaterials, top-down and bottom-up techniques are used as the major methods. Top-down methods separate bulk materials to create nanostructured materials. Top-down techniques include electro-explosion, mechanical milling, laser ablation, chemical ablation and lithography. Whereas the bottom-up approach, which assembles small material to create a larger nanostructures. Self-assembly, microwave irradiation, hydrothermal solvothermal therapy, and other techniques are all part of the bottom-up approach [10]. The properties of the materials determine the many characterization techniques used to describe the nanoparticles. The morphology of the nanoparticles has been analyzed using scanning electron microscopy (SEM), transmission electron microscopy (TEM), atomic force microscopy (AFM), X-ray diffraction (XRD), photoelectron spectroscopy (XPS), Raman spectroscopy, polarized optical microscopy (POM), Brunauer-Emmett-Teller (BET) and Zeta size analyzer [11,12]. These properties provide unique applications of nanoparticles in a wide variety of research fields such as biomedical sciences, agriculture and food science, electronics optics and more.

2. Classification of macromolecules

Macromolecules are extensively investigated because of their significance as cellular building blocks, which play a crucial part in the functioning mechanisms required for each stage of growth in the living organism. The macromolecules are majorly classified into four groups based on the biological class: proteins, carbohydrates, lipids, and nucleic acids as shown in fig. 1. Without these macromolecules, cell processes cannot perform their important core functions, which are carried out by each type of macromolecule in a living cell [13].

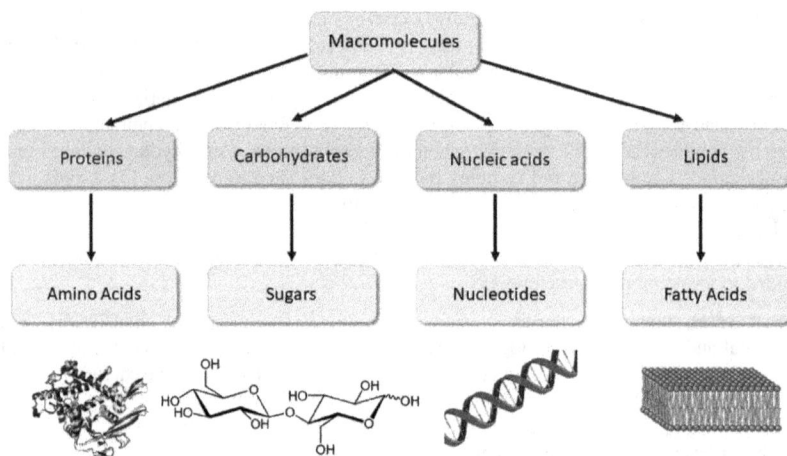

Figure 1. Classification of macromolecules.

Nanoparticle Toxicity and Compatibility   Materials Research Forum LLC
Materials Research Foundations 161 (2024) 64-82   https://doi.org/10.21741/9781644902998-3

2.1 Proteins

The basic building blocks of protein are amino acids (AAs), which are linked together by peptide bonds. Mostly, 20 natural AAs are the fundamental building blocks of proteins known as oligopeptides and polypeptides depending on the length of the peptides [14]. All 20 AAs, with the exception of glycine, are optically active. The primary structure of the protein is referred to as the linear polypeptide chain. The 20 naturally occurring amino acids are used to describe the primary structure. Each amino acid is assigned to the appropriate secondary structure element (such as alpha, beta, or gamma) in a sequence that specifies the secondary structure. The components of the secondary structure are further arranged to create the tertiary structure. The three-dimensional (3-D) tertiary structure of proteins is created via intramolecular interactions, hydrogen bonds, and disulfide linkages Multiple protein molecules are linking together using specific force to generate the protein are final quaternary structure [15]. Deoxyribonucleic acid (DNA), which codes for commands, directs biological systems to make protein molecules. All of the essential processes of the biological system are tightly regulated by protein molecules. Each protein is created for a specific purpose based on the codons that are present. According to the central dogma of the cell, a polypeptide chain is formed when certain AAs on ribonucleic acid (RNA) are linked one by one via peptides [16].

A protein's structure has a significant impact on its function during biological processes. As a result, a better understanding of protein structure is essential to understanding the mechanisms behind numerous biological activities. X-ray crystallography, nuclear magnetic resonance spectroscopy (NMR), and single-particle cryo-electron microscopy (cryo-EM) techniques are some of the techniques used to determine the protein structure [17]. The unfastened bonds are amorphous, but the secondary structure, such as α-helix and β-pleated sheets are commonly crystalline [18]. The influence of nanoparticles on living organisms at the protein level is a critical issue that is attracting more attention since proteins are significant biological molecules that are fundamental to the proper functioning of cells and organisms.

2.2 Carbohydrates

Carbohydrates, is the empirical formula $C_n(H_2O)_n$ and n=3 or higher with the chemical substance containing carbon, hydrogen, and oxygen. It is chemically described as hydrates of carbon because of the proportion of hydrogen and oxygen available with the similar to the water molecule. These carbohydrates are easily capable of forming the glycosidic bond (O and N) with other molecular atoms and each other [19]. The most basic kinds of carbohydrates are five- and six-carbon monosaccharides, such as xylose, glucose and fructose which are the basic building blocks of complex lignocellulosic biomass, a nonedible resource. A strong composite material made mostly of several high-molecular weight polymers with oxygen containing lignocellulosic biomass as follow; Lignocellulosic biomass comprises a robust composite material constructed primarily from different oxygen-containing high-molecular weight polymers as follows: (i) Cellulose is a polymer made up of C6-glucose units that are joined together by (1-4) glycosidic linkages, (ii) galactose, mannose, rhamnose and arabinose as well as C5-xylose units, composition the majority of the heteropolymer hemicellulose and (iii) lignin is an amorphous polymer made up of three phenyl propanolic monomers (monolignols), which are connected by ether and carbon-carbon linkages [20].

2.3   Nucleic acid

The three major components of nucleic acids are sugar molecules, bases that contain nitrogen, and phosphate molecules. All of the genetic information and the coding needed for protein production are stored in nucleic acids, which act as gene libraries. Except for red blood cells, all other types of cells and viruses have rather complicated molecules called nucleic acids [21]. The four nitrogenous bases adenine (A), thymine (T), cytosine (C), and Guanine(G) are present in each nucleic acid molecule. The five nitrogenous bases are A, G, C, T, and Uracil U. Of these, A, G, and C are present in all forms of nucleic acids, whereas T and U are only present in some types of nucleic acids [22]. Deoxyribonucleic acid (DNA) and ribonucleic acid (RNA) are the two different forms of nucleic acids found in a living organism. DNA is regarded as genetic material since it contains all of the hereditary information and serves as a blueprint for the complete living organism. In the past, there was disagreement on the statement of DNA as genetic material [23]. All four nitrogenous bases occur in complementary pairs in DNA, which is structurally a double-stranded helix that contains all of the nitrogenous bases and thymine (T) as the fourth nucleotide base pair. Thymine binds with adenine by two hydrogen bonds, while guanine binds with cytosine by three hydrogen bonds. A single strand of RNA has the same number of nitrogenous bases as DNA-three. Instead of thymine, uracil serves as the fourth nucleotide base pair, and it complements uracil instead of thymine. Ribonucleic acids are playing important role decoding and getting the information acquired from the DNA code and putting it into protein synthesis through transcription [24,25].

2.4   Lipids

Another significant macromolecule that is essential to the entire living organism is lipids. The word "lipid" is derived from the Greek word *lipos*, which means fat or greasy. A wide variety of organic compounds called lipids are present in microorganisms, plants, and animals. Lipids served as one of the three large classes of foods and vital components of living organisms and are significant class of macromolecules of all living cells. Some lipids are insoluble in water and are spread throughout the cell as small droplets. Sometimes, the minute lipid droplets can connect to other lipid chemical groups and bind to water molecules [26]. Certain lipids, such cephalin and lecithin, are highly soluble in both fats and water. As a result, these lipids are essential in binding other components, like proteins and lipid-soluble chemicals, that are soluble in water. The small lipid droplets act as energy storage molecules instead of carbohydrates, because of the decreased state of the carbons on the acyl chains, which increases energy. It does not provide a lot of energy in the case of partially oxidized carbons in carbohydrates[27].

3.   Interaction between nanoparticles and macromolecules

3.1   Nanoparticles and proteins

Proteins and nanoparticles are dependent on a number of factors, such as these moieties sizes, shapes, charges, and chemical functions. Proteins are highly biocompatible and nontoxic molecule, they have considerable attraction for bioactive compound via a variety of molecular interactions such as disulfide bonds, hydrophobic interactions, hydrogen bonding and π-π interactions [28]. Furthermore, the hydration layer may be significant influence by the adsorption, theoretically

Materials Research Foundations 161 (2024) 64-82
https://doi.org/10.21741/9781644902998-3

changing structure and function of the protein [29]. Apart from the impact on function, modifying the surface properties of nanoparticles is also help in reducing the toxicity of the final products.

Bovine serum albumin (BSA) conformational changes were studied in albumin gold nanoparticle bioconjugates. Albumin in bioconjugates prepare by the common adsorption method encountered substantial conformational changes at both the secondary and tertiary structure levels. As a result of the conformational changes in the bioconjugates, BSA was shown to adopt a more flexible conformational state on the boundary surface of gold nanoparticles [30]. Whereas the interaction between AuNPs and proteins (such as ubiquitin and fibrinogen) is significantly influenced by the structural and compositional variation on their surfaces. Because surface heterogeneity can provide force other van der Waals force and electrostatic interaction for proteins, it may be assumed that proteins stability in electrolytes is generally instigated by heterogeneity. The surface chemistry and structure of the AuNPs, in particular the quantity of surface heterogeneity, were observed to have an influence on the adsorption of ubiquitin and fibrinogen on the AuNPs [31]. The proteins at the surface of the particle adsorb/desorb, and rearrange, leading to the corona to harden and determines the particle final identity. For larger particles, the formation of a hard protein corona formed by many layers of proteins is a feasible possibility. Multiple protein adsorptions occur when proteins adsorb to flat surfaces because they are subjected to more significant conformational changes. The appearance of defects on the first protein layer docking other proteins, as well as an increase in the corona thickness of the larger particles respect to the smaller particles, may therefore be related to this occurrence shown in fig. 2 [32].

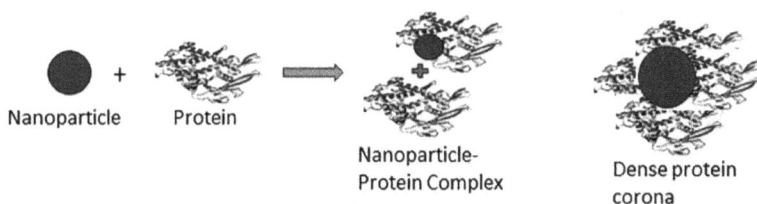

Nanoparticle   Protein

Nanoparticle-
Protein Complex

Dense protein
corona

Figure 2. Protein gold nanoparticle interaction

A hybrid nanocomposite was produced via *in-situ* fibrillation of plant proteins in the presence of superparamagnetic iron oxide nanoparticles. But the extent and specificity of the binding are determined by a complex interaction of components. By introducing surface-modified iron oxide (Fe_3O_4) NPs using oleate, we investigated the mechanism behind the self-assembly of plant protein fibrils. The length of the fibril-NP composite was concentration-dependent, and NP assisted in accelerating fibril formation within an ideal range. This may be caused by the efficient surface energy transfer between NPs and proteins during *in-situ* protein fibrillation, the formation of favorable hydrogen bonds, and electrostatic interactions [33]. The π-π interactions occur single walled carbon nanotubes (SWCNTs) and aromatic amino acids (tyrosine phenylalanine tryptophan) are found to play a critical role in their binding capabilities. Protein-coated SWCNTs can exhibit decreased cytotoxicity and altered biological interaction due to competitive protein binds on the surface, which is determined by the adsorption capacity of each protein [34].

Titanium dioxide (TiO_2) particles are commonly used as lightening agents in food and supplements. When TiO_2 particles get incorporated into food matrices containing plant proteins, a biocorona can form around them. The molecular interactions of two hydrophobic plant proteins (gliadin and zein) with four types of surface-modified TiO_2 nanoparticles have been studied. The plant proteins interacted with the surfaces of the TiO_2 nanoparticles, resulting in changes in their molecular conformations, as determined by ultraviolet-visible absorption, fluorescence quenching, and Fourier transform infrared spectroscopy. Quartz crystal microbalance studies indicated that the mass of the hard corona increased as the number of hydroxyl groups on the surfaces of the TiO_2 nanoparticles increased, showing that hydrogen bonding was important in the formation of the biocorona [35]. The longevity of the protein on the surface of nanoparticles interaction is based on the associative and dissociative rate of each protein which decide the hard binding or soft binding of the interaction.

3.2   Nanoparticles and carbohydrates

Saccharides are involved in many different biological processes, although they primarily serve as sources of structure and energy storage. Polysaccharides are composed of monosaccharides joined together in either a linear or branching structure by glycosidic linkages [36]. Saccharides provide mechanical rigidity and metabolic sustainability to living organisms in their most ubiquitous roles [37,38]. Whereas each polymer contains unique physicochemical features, it may also change in length, charge, monosaccharide sequence, and stereochemistry, making it challenging to properly recognize patterns in their adsorption on nanoparticles. Few interaction studies have focused on this macromolecular class, possibly because complex analytical techniques are required for identifying saccharide molecules in aqueous solutions [39,40]. Innovate and optimize the surface characteristics of the complex of active ingredients has mostly focused on the carbohydrate. The surface is functionalized by the properties of carbohydrates, which act as the primary drivers of plasma clearance, cell absorption, and biodegradation. The types of carbohydrates used in this capacity are highly heterogeneous (e.g., sucrose and carboxy-maltose), and the complex of the carbohydrate portion along with how it is bound to the iron core defines the overall surface characteristics, particle size, particle size distribution, and particle morphology. Despite the relatively apparent simplicity of producing iron-carbohydrate compounds. It has been clearly demonstrated that the process of manufacturing at all check point is critical for ensuring therapeutic drugs are safe and effective [41] TiO_2 nanoparticles interact with seven common carbohydrates, including monosaccharides, disaccharides, and polysaccharides. All of the carbohydrates investigated reacted with the surfaces of the TiO_2 nanoparticles and formed biocoronas. After interacting with the carbohydrates, the surface potential and size of the TiO_2 nanoparticles altered [42].

3.3   Nanoparticles and nucleic acids

Nucleic acids are polymers that direct the creation of proteins and encode a cell's genetic information. These nucleotide-based polyanionic molecules are made up of an anionic phosphate group, pentose, and a base (adenine, guanine, thymine, cytosine, or uracil) that may exist in both double-stranded (DNA) and single-stranded (RNA) forms [43]. Nucleosides and a phosphate backbone make up DNA, and it is known that DNA interacts with many materials via these chemical groups. DNA and RNA can also be adsorbed on surfaces by nucleobases in addition to the phosphate group, this has been indicated for metallic nanoparticles like gold nanoparticles and carbon nanotubes [44]. Single-stranded DNA (ss-DNA) exhibits high electrostatic attraction (van der Waals force) with citrate-coated Au-nanoparticles [45]. With producing hydrogen bonds between the DNA phosphate group and the hydroxyl groups, functionalized iron oxide nanoparticle surfaces interact with DNA. The hydroxyl groups on the nanoparticle surface provide a high degree of aqueous phase dispersion and enhance the formation of hydrogen bonds with DNA [46]. NA adsorption by graphene oxide (GO) is known to be significantly influenced by pi-pi stacking and hydrogen bonding. Molybdenum disulfide (MoS_2) and tungsten disulfide (WS_2) nanoparticles are smaller surface area and weaker binding forces (van der Waals interaction) than GO, possess a lower binding capacity. Single-layered MoS_2 and WS_2 have a similar structure, with the transition metal atoms coated on both sides by sulphur. GO has a wide range of oxygen-containing moieties, including hydroxyl, epoxy, and carboxyl groups, allowing hydrogen bonding as well as pi-pi stacking with DNA [47].

The importance of evaluating the kinetics of small molecule attachment with and dissociation from DNA therefore seems to be of greatest importance. Basically, there are two main methods that binding reactions between NPs and DNA might happen: either by employing electrostatic interactions or by using thiolated fragments. Although direct binding is favored, when possible, covalent attachment using basic reagent such as carbodiimides can inhibit DNA activity. Covalent attachment could also be thought of as a potential binding approach.

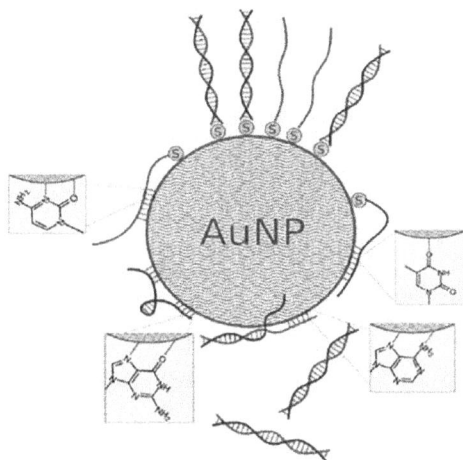

Figure 3. Interaction between DNA and gold nanoparticle

It is important to note that NPs are characterized not only by the properties of the metal cluster core, but also by the organic molecules, or capping agents, that prevent the colloid from aggregation and building a monolayer. Due to the possibility of controlling and modulating synthesis, the study of non-covalent metal-ligand interactions with various substrates is therefore of enormous interest shown in fig. 3 [48].

The essential interactions within the folded RNA appeared to be outcompeted by the stacking interactions between the graphene surface and nucleobases. At sufficiently high concentrations, the unfolded RNA molecules develop a film that traps nanobubbles; at sufficiently low concentrations, they organize into clusters of unfolded RNA oligonucleotides that are sparsely dispersed throughout the graphene surface. The natural folded hairpin structures of the RNA molecules, which are thermally stable in water, melt on the graphene surface [49,50]. The applications of DNAs nanoparticles interactions have been promoted by all of these advantages. An in-depth investigation of the interfacial characteristics is required for better utilization of the properties of both DNA and nanoparticles, which can lead to new insights to build platforms with improved performance.

Table:1. Interaction mechanism between Macromolecule and Nanoparticles.

Macromolecules	Type of nanoparticles	Interaction mechanism	References
Proteins			
L-arginine, L-lysine L-glutamine and glycine	IONPs	Electrostatic interactions	[56]
Plant proteins	TiO_2	Hydrogen bonding	[35]
lysozyme, BSA, papain and pepsin	SWCNT	Hydrophobic interaction	[57]
Carbohydrates			
Glucose, fructose, lactose, maltose	TiO_2	Hydrogen bonding	[42]
Nucleic acids			
pDNA	Silica coated IONPs	Electrostatic interactions: hydroxylic groups and phosphate	[58]
Calf thymus (CT)- DNA;	RuO_2	Electrostatic interactions	[59]
DNA	GO, MoS_2/WS_2	hydrogen bonding and pi-pi stacking	[47]
Lipids			
Oleic acid, palmitic acid, stearic acid, and linoleic acid	IONPs	Covalent binding	[60]

3.4 Nanoparticles and lipids

Lipids are organic molecules that are hydrophobic; they are soluble in non-polar solvents but insoluble or only partially soluble in water. These molecules include a wide range of substances, including terpenes, phospholipids, sterols, and fatty acids [51]. There have been several reports of fatty acids, including ricinoleic acid and oleic acid, being used as coating. Iron oxide nanoparticles (IONPs) to increase the nanoparticles' colloidal stability [52–54]. For fatty acids to bind to nanoparticles, carboxylate groups are one of the main mediators. Fatty acids have shown a bilayer formation around the IONPs surface, in which, the polar carboxylic head group binds to the surface in the inner layer, and thus the non-polar, hydrophobic tail is directed to the solvent. The carboxylic acid heads of this second layer are exposed to the aqueous media, and the non-polar tail groups that interact with the tail groups of the connected fatty acids produce the second layer [55].

4.　Application of interaction of nanoparticles with macromolecules

4.1　Biomedical

Based on this, a successful biomedical application requires the NPs to be colloidally stable in the biological environment. In an ideal world, the NPs would also be able to get past biological barriers and avoid immune system identification, which would cause an unintended build-up of the particles in the liver and spleen and compromise their intended application. Serum proteins known as opsonin's are absorbed by the body and end on their surface as the primary immune response to foreign substances such as NPs. According to opsonization, NPs that move freely within the body are recognized by phagocytes and cleared from the bloodstream [61]. Since standard imaging techniques rely on the greater permeability of the blood-brain barrier (BBB) in later stages of neurological diseases like multiple sclerosis and stroke, it is very difficult to recognize inflammatory processes associated with these disorders early. First example of a non-toxic, MRI-visible iron oxide glyconanoparticle serving as a very effective agent and enabling the in vivo presymptomatic recognition of brain disorders [62].

Most significantly, iron oxide particles are used as high contrast imaging agents for magnetic resonance imaging (MRI) [63]. As the least hazardous and most stable metal nanoparticle (NP) composition, gold nanoparticles (AuNPs) act as a safe and reliable diagnostic and therapeutic equipment [64]. With great potential as an oral drug delivery platform for a variety of therapeutics, including macromolecules (especially peptides and proteins), silica nanoparticles have emerged as a multifunctional, biocompatible, and biodegradable inorganic nanocarrier. Their specific structure facilitates the easier to load substantial therapeutic payloads at the appropriate loading capacities for a controlled and location-specific oral delivery [65]. The creation of mesoporous silica nanoparticles demonstrates their great potential as multipurpose drug delivery nanocarriers. These nanoparticles can have several functions for entering and targeting different types of cells, a high loading capacity, and a well-defined and adjustable porosity at the nanoscale scale[66]. Induced anti-inflammatory responses in macrophages have been shown by single-walled carbon nanotubes (SWCNTs) and 10 nm amorphous silica coated with albumin. This is demonstrated by the inhibition of cyclooxygenase-2 (Cox-2) activation by lipopolysaccharide in a serum free condition. The anti-inflammatory effects of the nanoparticles are also inhibited by precoating surfaces with a nonionic surfactant, which blocks albumin's adsorption. These results point to a significant function for adsorbed proteins in controlling the toxicity and absorption of nanoscale amorphous silica and SWCNTs [67].

Small interfering RNA (siRNA) was bound to gold nanoparticles via biodegradable disulfide bonds, with about 30 strands of siRNA per nanoparticle, after the gold nanoparticles had been coated with the hydrophilic polymer poly(ethylene glycol) (PEG). After that, a library of end-modified poly(amino ester)s (PBAEs), which have been shown to facilitate intracellular DNA delivery, were applied to the particles[68]. Since the DNA bound to AuNP is effectively protected from nuclease digestion, it is predicted that these AuNP conjugates will also provide a shield for the entrapped DNA during gene delivery. On the other hand, Au NPs in particular have been studied and utilized to enhance siRNA distribution with some effectiveness without showing any off-target or negative immunological effects [69,70]. The creation of iron-carbohydrate products improved the treatment of iron deficiency and iron deficiency anemia, but regulatory assessment of these nanomedicines remains challenging because of their very small particle size and complexity overall [41].

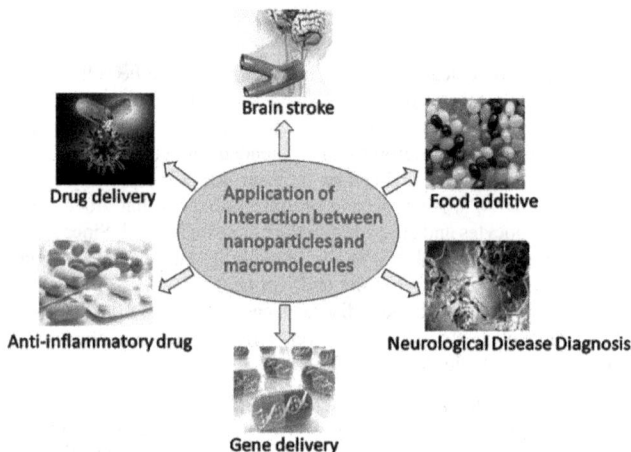

Figure 4. Application of interaction of nanoparticles with macromolecules.

4.2   Food science

Inorganic nanoparticles (NPs) are frequently utilized for nutritious (iron oxide) anticaking (silicon dioxide), and lighting (titanium dioxide) applications, as well as to impart desired physicochemical and qualitative qualities in foods and beverages [71]. In order to improve the lightness or brightness of food, food-grade titanium dioxide (TiO_2) is commonly used. TiO_2 is frequently added to food high in carbohydrates, such as chocolates, sweets, chewing gum, baked goods, and other sugary foods [72,73]. The U.S. Food and Drug Administration (FDA) has approved TiO_2 and MgO as food flavor carriers, food color additives, and antiseptics. Products that are coloring substance gums, cake icings, sweets, pies, desserts, and white sauces [74]. Important food processing organizations, including Nestlé, Kellogg's, and Coca-Cola Unilever, have commercially adopted TiO_2 nanoparticles as whitening and brightening agents. These companies have leveraged the interactions between TiO_2 nanoparticles and carbohydrates to understand and control the behavior of nanoparticles in complex food structures and the human gastrointestinal tract. TiO_2 nanoparticles can infiltrate body tissues and organs such as the liver, kidney, and spleen after being consumed and absorbed through the digestive tract [75]. Similarly, Food whitening products like chocolates, sweets, and cakes with significant amounts of carbohydrates frequently include titanium dioxide (TiO_2) [76]. Interaction between TiO_2 nanoparticles and carbohydrates, enter the body through food additives, and take part in a series of chemical interactions with the food matrix. The adsorption of food polysaccharides on the surface of nanoparticles is greater than that of both monosaccharides or disaccharides [42]

The consumption of plant-based foods is increasing in the human diet. To better understand their influence on the behavior of nanoparticles in food and inside the gastrointestinal system, we investigated the interactions between TiO_2 nanoparticles and plant proteins (glutenin, gliadin, zein, and soy) [35,77]. Similarly, TiO_2 nanoparticles and four heat-treated plant proteins-glutenin, soy protein isolate, gliadin, and zein-interacted with each other. Initially, the impact of heat treatment at various temperatures (37°C, 60°C, 90°C and 100°C) on the protein's structural characteristics was studied. The impact of heat treatment-induced protein structural modifications on the proteins' tendency for binding to TiO_2 nanoparticles was investigated [78]

Conclusion

Nanoparticles can interact with other molecules such as chemical compounds, carbohydrate, lipid, nucleic acid, and proteins attached to the surface through covalent bonds or by adsorption due to their considerable adsorption capacities due to their large surface area. Macromolecules being nontoxic and extremely biocompatible but show strong affinity to bioactive compounds through a various molecular interaction, such as π-π interactions, hydrophobic interaction, hydrogen bonding and van der Waals interaction. As compared to low molecular weight medicine, nanoparticles and macromolecule carriers have a higher capacity to passively accumulate in tumor tissue, which can increase the efficacy of chemotherapy. We also discussed about how nanoparticles are used in food science.

References

[1]　R. Aksakal, C. Mertens, M. Soete, N. Badi, F. Du Prez, Applications of Discrete Synthetic Macromolecules in Life and Materials Science: Recent and Future Trends, Advanced Science. 8 (2021) 2004038. https://doi.org/10.1002/advs.202004038

[2]　K. Matyjaszewski, N. V. Tsarevsky, Macromolecular Engineering by Atom Transfer Radical Polymerization, J Am Chem Soc. 136 (2014) 6513–6533. https://doi.org/10.1021/ja408069v

[3]　D. Pasini, D. Takeuchi, Cyclopolymerizations: Synthetic Tools for the Precision Synthesis of Macromolecular Architectures, Chem Rev. 118 (2018) 8983–9057. https://doi.org/10.1021/acs.chemrev.8b00286

[4]　S. Ramadurai, A. Kohut, N.K. Sarangi, O. Zholobko, V.A. Baulin, A. Voronov, T.E. Keyes, Macromolecular inversion-driven polymer insertion into model lipid bilayer membranes, J Colloid Interface Sci. 542 (2019) 483–494. https://doi.org/10.1016/j.jcis.2019.01.093

[5]　T. Fischer, J. Pietruszka, Key Building Blocks via Enzyme-Mediated Synthesis, in: 2010: pp. 1–43. https://doi.org/10.1007/128_2010_62

[6]   P.J. Borm, D. Robbins, S. Haubold, T. Kuhlbusch, H. Fissan, K. Donaldson, R. Schins, V. Stone, W. Kreyling, J. Lademann, J. Krutmann, D. Warheit, E. Oberdorster, The potential risks of nanomaterials: a review carried out for ECETOC, Part Fibre Toxicol. 3 (2006) 11. https://doi.org/10.1186/1743-8977-3-11

[7]   A. Radomska, J. Leszczyszyn, M. Radomski, The Nanopharmacology and Nanotoxicology of Nanomaterials: New Opportunities and Challenges, Advances in Clinical and Experimental Medicine. 25 (2016) 151–162. https://doi.org/10.17219/acem/60879

[8]   A. Mostofizadeh, Y. Li, B. Song, Y. Huang, Synthesis, Properties, and Applications of Low-Dimensional Carbon-Related Nanomaterials, J Nanomater. 2011 (2011) 1–21. https://doi.org/10.1155/2011/685081

[9]   Z. Alhalili, Metal Oxides Nanoparticles: General Structural Description, Chemical, Physical, and Biological Synthesis Methods, Role in Pesticides and Heavy Metal Removal through Wastewater Treatment, Molecules. 28 (2023) 3086. https://doi.org/10.3390/molecules28073086

[10]   N. Baig, I. Kammakakam, W. Falath, Nanomaterials: a review of synthesis methods, properties, recent progress, and challenges, Mater Adv. 2 (2021) 1821–1871. https://doi.org/10.1039/D0MA00807A

[11]   M.B. Gawande, A. Goswami, F.-X. Felpin, T. Asefa, X. Huang, R. Silva, X. Zou, R. Zboril, R.S. Varma, Cu and Cu-Based Nanoparticles: Synthesis and Applications in Catalysis, Chem Rev. 116 (2016) 3722–3811. https://doi.org/10.1021/acs.chemrev.5b00482

[12]   C. Liu, U. Burghaus, F. Besenbacher, Z.L. Wang, Preparation and Characterization of Nanomaterials for Sustainable Energy Production, ACS Nano. 4 (2010) 5517–5526. https://doi.org/10.1021/nn102420c

[13]   R.J. Ellis, Macromolecular crowding: an important but neglected aspect of the intracellular environment, Curr Opin Struct Biol. 11 (2001) 114–119. https://doi.org/10.1016/S0959-440X(00)00172-X

[14]   Y. Hou, Z. Wu, Z. Dai, G. Wang, G. Wu, Protein hydrolysates in animal nutrition: Industrial production, bioactive peptides, and functional significance, J Anim Sci Biotechnol. 8 (2017) 24. https://doi.org/10.1186/s40104-017-0153-9

[15]   Jianlin Cheng, A.N. Tegge, P. Baldi, Machine Learning Methods for Protein Structure Prediction, IEEE Rev Biomed Eng. 1 (2008) 41–49. https://doi.org/10.1109/RBME.2008.2008239

[16]   F. CRICK, Central Dogma of Molecular Biology, Nature. 227 (1970) 561–563. https://doi.org/10.1038/227561a0

[17]   E. Nwanochie, V.N. Uversky, Structure Determination by Single-Particle Cryo-Electron Microscopy: Only the Sky (and Intrinsic Disorder) is the Limit, Int J Mol Sci. 20 (2019) 4186. https://doi.org/10.3390/ijms20174186

[18] A. Bolje, S. Gobec, Analytical Techniques for Structural Characterization of Proteins in Solid Pharmaceutical Forms: An Overview, Pharmaceutics. 13 (2021) 534. https://doi.org/10.3390/pharmaceutics13040534

[19] G. Enkavi, M. Javanainen, W. Kulig, T. Róg, I. Vattulainen, Multiscale Simulations of Biological Membranes: The Challenge To Understand Biological Phenomena in a Living Substance, Chem Rev. 119 (2019) 5607–5774. https://doi.org/10.1021/acs.chemrev.8b00538

[20] C.-H. Zhou, X. Xia, C.-X. Lin, D.-S. Tong, J. Beltramini, Catalytic conversion of lignocellulosic biomass to fine chemicals and fuels, Chem Soc Rev. 40 (2011) 5588. https://doi.org/10.1039/c1cs15124j

[21] C. Manzoni, D.A. Kia, J. Vandrovcova, J. Hardy, N.W. Wood, P.A. Lewis, R. Ferrari, Genome, transcriptome and proteome: the rise of omics data and their integration in biomedical sciences, Brief Bioinform. 19 (2018) 286–302. https://doi.org/10.1093/bib/bbw114

[22] P. Rajput, H. Singh, A. Bandral, Richu, Q. Majid, A. Kumar, Explorations on thermophysical properties of nitrogenous bases (uracil/thymine) in aqueous l-histidine solutions at various temperatures, J Mol Liq. 367 (2022) 120548. https://doi.org/10.1016/j.molliq.2022.120548

[23] D.E. Marsh, The Origins of Diversity: Darwin's Conditions and Epigenetic Variations, Nutr Health. 19 (2007) 103–132. https://doi.org/10.1177/026010600701900213

[24] F. Javadi-Zarnaghi, C. Höbartner, Strategies for Characterization of Enzymatic Nucleic Acids, in: 2017: pp. 37–58. https://doi.org/10.1007/10_2016_59

[25] K. Shahane, M. Kshirsagar, S. Tambe, D. Jain, S. Rout, M.K.M. Ferreira, S. Mali, P. Amin, P.P. Srivastav, J. Cruz, R.R. Lima, An Updated Review on the Multifaceted Therapeutic Potential of Calendula officinalis L., Pharmaceuticals. 16 (2023) 611. https://doi.org/10.3390/ph16040611

[26] C. de Carvalho, M. Caramujo, The Various Roles of Fatty Acids, Molecules. 23 (2018) 2583. https://doi.org/10.3390/molecules23102583

[27] M.F. Ramadan, H.F. Oraby, Fatty Acids and Bioactive Lipids of Potato Cultivars: An Overview, J Oleo Sci. 65 (2016) 459–470. https://doi.org/10.5650/jos.ess16015

[28] Y. Luo, Q. Wang, Y. Zhang, Biopolymer-Based Nanotechnology Approaches To Deliver Bioactive Compounds for Food Applications: A Perspective on the Past, Present, and Future, J Agric Food Chem. 68 (2020) 12993–13000. https://doi.org/10.1021/acs.jafc.0c00277

[29] M. Mahmoudi, I. Lynch, M.R. Ejtehadi, M.P. Monopoli, F.B. Bombelli, S. Laurent, Protein–Nanoparticle Interactions: Opportunities and Challenges, Chem Rev. 111 (2011) 5610–5637. https://doi.org/10.1021/cr100440g

Materials Research Forum LLC
https://doi.org/10.21741/9781644902998-3

[30] L. Shang, Y. Wang, J. Jiang, S. Dong, pH-Dependent Protein Conformational Changes in Albumin:Gold Nanoparticle Bioconjugates: A Spectroscopic Study, Langmuir. 23 (2007) 2714–2721. https://doi.org/10.1021/la062064e

[31] R. Huang, R.P. Carney, F. Stellacci, B.L.T. Lau, Protein–nanoparticle interactions: the effects of surface compositional and structural heterogeneity are scale dependent, Nanoscale. 5 (2013) 6928. https://doi.org/10.1039/c3nr02117c

[32] J. Piella, N.G. Bastús, V. Puntes, Size-Dependent Protein–Nanoparticle Interactions in Citrate-Stabilized Gold Nanoparticles: The Emergence of the Protein Corona, Bioconjug Chem. 28 (2017) 88–97. https://doi.org/10.1021/acs.bioconjchem.6b00575

[33] J. Li, I. Pylypchuk, D.P. Johansson, V.G. Kessler, G.A. Seisenbaeva, M. Langton, Self-assembly of plant protein fibrils interacting with superparamagnetic iron oxide nanoparticles, Sci Rep. 9 (2019) 8939. https://doi.org/10.1038/s41598-019-45437-z

[34] C. Ge, J. Du, L. Zhao, L. Wang, Y. Liu, D. Li, Y. Yang, R. Zhou, Y. Zhao, Z. Chai, C. Chen, Binding of blood proteins to carbon nanotubes reduces cytotoxicity, Proceedings of the National Academy of Sciences. 108 (2011) 16968–16973. https://doi.org/10.1073/pnas.1105270108

[35] B. Yuan, B. Jiang, H. Li, X. Xu, F. Li, D.J. McClements, C. Cao, Interactions between TiO2 nanoparticles and plant proteins: Role of hydrogen bonding, Food Hydrocoll. 124 (2022) 107302. https://doi.org/10.1016/j.foodhyd.2021.107302

[36] A.M.G.C. Dias, A. Hussain, A.S. Marcos, A.C.A. Roque, A biotechnological perspective on the application of iron oxide magnetic colloids modified with polysaccharides, Biotechnol Adv. 29 (2011) 142–155. https://doi.org/10.1016/j.biotechadv.2010.10.003

[37] B. Chang, M. Zhang, G. Qing, T. Sun, Dynamic Biointerfaces: From Recognition to Function, Small. 11 (2015) 1097–1112. https://doi.org/10.1002/smll.201402038

[38] F.S. Alves, J.N. Cruz, I.N. de Farias Ramos, D.L. do Nascimento Brandão, R.N. Queiroz, G.V. da Silva, G.V. da Silva, M.F. Dolabela, M.L. da Costa, A.S. Khayat, J. de Arimatéia Rodrigues do Rego, D. do Socorro Barros Brasil, Evaluation of Antimicrobial Activity and Cytotoxicity Effects of Extracts of Piper nigrum L. and Piperine, Separations. 10 (2023) 21. https://doi.org/10.3390/separations10010021

[39] C. Wang, X. Gao, Z. Chen, Y. Chen, H. Chen, Preparation, Characterization and Application of Polysaccharide-Based Metallic Nanoparticles: A Review, Polymers (Basel). 9 (2017) 689. https://doi.org/10.3390/polym9120689

[40] S. Muzammil, J. Neves Cruz, R. Mumtaz, I. Rasul, S. Hayat, M.A. Khan, A.M. Khan, M.U. Ijaz, R.R. Lima, M. Zubair, Effects of Drying Temperature and Solvents on In Vitro Diabetic Wound Healing Potential of Moringa oleifera Leaf Extracts, Molecules. 28 (2023) 710. https://doi.org/10.3390/molecules28020710

[41] N. Nikravesh, G. Borchard, H. Hofmann, E. Philipp, B. Flühmann, P. Wick, Factors influencing safety and efficacy of intravenous iron-carbohydrate nanomedicines:

Materials Research Forum LLC
https://doi.org/10.21741/9781644902998-3

From production to clinical practice, Nanomedicine. 26 (2020) 102178.
https://doi.org/10.1016/j.nano.2020.102178

[42] Z. Qiaorun, S. Honghong, L. Yao, J. Bing, X. Xiao, D. Julian McClements, C. Chongjiang, Y. Biao, Investigation of the interactions between food plant carbohydrates and titanium dioxide nanoparticles, Food Research International. 159 (2022) 111574. https://doi.org/10.1016/j.foodres.2022.111574

[43] A. Brown, T. Brown, Curtailing their negativity, Nat Chem. 11 (2019) 501–503. https://doi.org/10.1038/s41557-019-0274-1

[44] B. Liu, J. Liu, Comprehensive Screen of Metal Oxide Nanoparticles for DNA Adsorption, Fluorescence Quenching, and Anion Discrimination, ACS Appl Mater Interfaces. 7 (2015) 24833–24838. https://doi.org/10.1021/acsami.5b08004

[45] Li, L.J. Rothberg, Label-Free Colorimetric Detection of Specific Sequences in Genomic DNA Amplified by the Polymerase Chain Reaction, J Am Chem Soc. 126 (2004) 10958–10961. https://doi.org/10.1021/ja048749n

[46] L. Abarca-Cabrera, P. Fraga-García, S. Berensmeier, Bio-nano interactions: binding proteins, polysaccharides, lipids and nucleic acids onto magnetic nanoparticles, Biomater Res. 25 (2021) 12. https://doi.org/10.1186/s40824-021-00212-y

[47] C. Lu, Y. Liu, Y. Ying, J. Liu, Comparison of MoS_2, WS_2, and Graphene Oxide for DNA Adsorption and Sensing, Langmuir. 33 (2017) 630–637. https://doi.org/10.1021/acs.langmuir.6b04502

[48] J.M. Carnerero, A. Jimenez-Ruiz, P.M. Castillo, R. Prado-Gotor, Covalent and Non-Covalent DNA–Gold-Nanoparticle Interactions: New Avenues of Research, ChemPhysChem. 18 (2017) 17–33. https://doi.org/10.1002/cphc.201601077

[49] Q. Li, J.P. Froning, M. Pykal, S. Zhang, Z. Wang, M. Vondrák, P. Banáš, K. Čépe, P. Jurečka, J. Šponer, R. Zbořil, M. Dong, M. Otyepka, RNA nanopatterning on graphene, 2d Mater. 5 (2018) 031006. https://doi.org/10.1088/2053-1583/aabdf7

[50] R.B.M. de Almeida, D.B. Barbosa, M.R. do Bomfim, J.A.O. Amparo, B.S. Andrade, S.L. Costa, J.M. Campos, J.N. Cruz, C.B.R. Santos, F.H.A. Leite, M.B. Botura, Identification of a Novel Dual Inhibitor of Acetylcholinesterase and Butyrylcholinesterase: In Vitro and In Silico Studies, Pharmaceuticals. 16 (2023) 95. https://doi.org/10.3390/ph16010095

[51] E. Fahy, D. Cotter, M. Sud, S. Subramaniam, Lipid classification, structures and tools, Biochimica et Biophysica Acta (BBA) - Molecular and Cell Biology of Lipids. 1811 (2011) 637–647. https://doi.org/10.1016/j.bbalip.2011.06.009

[52] S. Gyergyek, D. Makovec, M. Drofenik, Colloidal stability of oleic- and ricinoleic-acid-coated magnetic nanoparticles in organic solvents, J Colloid Interface Sci. 354 (2011) 498–505. https://doi.org/10.1016/j.jcis.2010.11.043

[53] M. Rudolph, J. Erler, U.A. Peuker, A TGA–FTIR perspective of fatty acid adsorbed on magnetite nanoparticles–Decomposition steps and magnetite reduction,

Colloids Surf A Physicochem Eng Asp. 397 (2012) 16–23.
https://doi.org/10.1016/j.colsurfa.2012.01.020

[54] M.H. Sarfraz, M. Zubair, B. Aslam, A. Ashraf, M.H. Siddique, S. Hayat, J.N. Cruz, S. Muzammil, M. Khurshid, M.F. Sarfraz, A. Hashem, T.M. Dawoud, G.D. Avila-Quezada, E.F. Abd_Allah, Comparative analysis of phyto-fabricated chitosan, copper oxide, and chitosan-based CuO nanoparticles: antibacterial potential against Acinetobacter baumannii isolates and anticancer activity against HepG2 cell lines, Front Microbiol. 14 (2023) 1188743. https://doi.org/10.3389/fmicb.2023.1188743

[55] H.-C. Roth, S. Schwaminger, P. Fraga García, J. Ritscher, S. Berensmeier, Oleate coating of iron oxide nanoparticles in aqueous systems: the role of temperature and surfactant concentration, Journal of Nanoparticle Research. 18 (2016) 99. https://doi.org/10.1007/s11051-016-3405-2

[56] J.L. Viota, F.J. Arroyo, A.V. Delgado, J. Horno, Electrokinetic characterization of magnetite nanoparticles functionalized with amino acids, J Colloid Interface Sci. 344 (2010) 144–149. https://doi.org/10.1016/j.jcis.2009.11.061

[57] K. Matsuura, T. Saito, T. Okazaki, S. Ohshima, M. Yumura, S. Iijima, Selectivity of water-soluble proteins in single-walled carbon nanotube dispersions, Chem Phys Lett. 429 (2006) 497–502. https://doi.org/10.1016/j.cplett.2006.08.044

[58] J.R. Sosa-Acosta, J.A. Silva, L. Fernández-Izquierdo, S. Díaz-Castañón, M. Ortiz, J.C. Zuaznabar-Gardona, A.M. Díaz-García, Iron Oxide Nanoparticles (IONPs) with potential applications in plasmid DNA isolation, Colloids Surf A Physicochem Eng Asp. 545 (2018) 167–178. https://doi.org/10.1016/j.colsurfa.2018.02.062

[59] J.-J. Xue, F. Bigdeli, J.-P. Liu, M.-L. Hu, A. Morsali, Ultrasonic-assisted synthesis and DNA interaction studies of two new Ru complexes; RuO $_2$ nanoparticles preparation, Nanomedicine. 13 (2018) 2691–2708. https://doi.org/10.2217/nnm-2018-0174

[60] M. Cano, K. Sbargoud, E. Allard, C. Larpent, Magnetic separation of fatty acids with iron oxide nanoparticles and application to extractive deacidification of vegetable oils, Green Chemistry. 14 (2012) 1786. https://doi.org/10.1039/c2gc35270b

[61] B. Pelaz, G. Charron, C. Pfeiffer, Y. Zhao, J.M. de la Fuente, X.-J. Liang, W.J. Parak, P. del Pino, Interfacing Engineered Nanoparticles with Biological Systems: Anticipating Adverse Nano-Bio Interactions, Small. 9 (2013) 1573–1584. https://doi.org/10.1002/smll.201201229

[62] S.I. van Kasteren, S.J. Campbell, S. Serres, D.C. Anthony, N.R. Sibson, B.G. Davis, Glyconanoparticles allow pre-symptomatic in vivo imaging of brain disease, Proceedings of the National Academy of Sciences. 106 (2009) 18–23. https://doi.org/10.1073/pnas.0806787106

[63] Z. Shen, A. Wu, X. Chen, Iron Oxide Nanoparticle Based Contrast Agents for Magnetic Resonance Imaging, Mol Pharm. 14 (2017) 1352–1364. https://doi.org/10.1021/acs.molpharmaceut.6b00839

[64] E.C. Dreaden, A.M. Alkilany, X. Huang, C.J. Murphy, M.A. El-Sayed, The golden age: gold nanoparticles for biomedicine, Chem. Soc. Rev. 41 (2012) 2740–2779. https://doi.org/10.1039/C1CS15237H

[65] M.M. Abeer, P. Rewatkar, Z. Qu, M. Talekar, F. Kleitz, R. Schmid, M. Lindén, T. Kumeria, A. Popat, Silica nanoparticles: A promising platform for enhanced oral delivery of macromolecules, Journal of Controlled Release. 326 (2020) 544–555. https://doi.org/10.1016/j.jconrel.2020.07.021

[66] C. Argyo, V. Weiss, C. Bräuchle, T. Bein, Multifunctional Mesoporous Silica Nanoparticles as a Universal Platform for Drug Delivery, Chemistry of Materials. 26 (2014) 435–451. https://doi.org/10.1021/cm402592t

[67] D. Dutta, S.K. Sundaram, J.G. Teeguarden, B.J. Riley, L.S. Fifield, J.M. Jacobs, S.R. Addleman, G.A. Kaysen, B.M. Moudgil, T.J. Weber, Adsorbed Proteins Influence the Biological Activity and Molecular Targeting of Nanomaterials, Toxicological Sciences. 100 (2007) 303–315. https://doi.org/10.1093/toxsci/kfm217

[68] J.-S. Lee, J.J. Green, K.T. Love, J. Sunshine, R. Langer, D.G. Anderson, Gold, Poly(β-amino ester) Nanoparticles for Small Interfering RNA Delivery, Nano Lett. 9 (2009) 2402–2406. https://doi.org/10.1021/nl9009793

[69] J. Conde, J.M. de la Fuente, P. V Baptista, In vitro transcription and translation inhibition via DNA functionalized gold nanoparticles, Nanotechnology. 21 (2010) 505101. https://doi.org/10.1088/0957-4484/21/50/505101

[70] J. Conde, J. Rosa, J.M. de la Fuente, P. V. Baptista, Gold-nanobeacons for simultaneous gene specific silencing and intracellular tracking of the silencing events, Biomaterials. 34 (2013) 2516–2523. https://doi.org/10.1016/j.biomaterials.2012.12.015

[71] A. Martirosyan, Y.-J. Schneider, Engineered Nanomaterials in Food: Implications for Food Safety and Consumer Health, Int J Environ Res Public Health. 11 (2014) 5720–5750. https://doi.org/10.3390/ijerph110605720

[72] Z. Chen, S. Han, S. Zhou, H. Feng, Y. Liu, G. Jia, Review of health safety aspects of titanium dioxide nanoparticles in food application, NanoImpact. 18 (2020) 100224. https://doi.org/10.1016/j.impact.2020.100224

[73] P. Talbot, J.M. Radziwill-Bienkowska, J.B.J. Kamphuis, K. Steenkeste, S. Bettini, V. Robert, M.-L. Noordine, C. Mayeur, E. Gaultier, P. Langella, C. Robbe-Masselot, E. Houdeau, M. Thomas, M. Mercier-Bonin, Food-grade TiO2 is trapped by intestinal mucus in vitro but does not impair mucin O-glycosylation and short-chain fatty acid synthesis in vivo: implications for gut barrier protection, J Nanobiotechnology. 16 (2018) 53. https://doi.org/10.1186/s12951-018-0379-5

[74] A. Weir, P. Westerhoff, L. Fabricius, K. Hristovski, N. von Goetz, Titanium Dioxide Nanoparticles in Food and Personal Care Products, Environ Sci Technol. 46 (2012) 2242–2250. https://doi.org/10.1021/es204168d

[75] X. Zhu, L. Zhao, Z. Liu, Q. Zhou, Y. Zhu, Y. Zhao, X. Yang, Long-term exposure to titanium dioxide nanoparticles promotes diet-induced obesity through exacerbating

intestinal mucus layer damage and microbiota dysbiosis, Nano Res. 14 (2021) 1512–1522. https://doi.org/10.1007/s12274-020-3210-1

[76]   C.E. Handford, M. Dean, M. Henchion, M. Spence, C.T. Elliott, K. Campbell, Implications of nanotechnology for the agri-food industry: Opportunities, benefits and risks, Trends Food Sci Technol. 40 (2014) 226–241. https://doi.org/10.1016/j.tifs.2014.09.007

[77]   J. Bing, X. Xiao, D.J. McClements, Y. Biao, C. Chongjiang, Protein corona formation around inorganic nanoparticles: Food plant proteins-TiO2 nanoparticle interactions, Food Hydrocoll. 115 (2021) 106594. https://doi.org/10.1016/j.foodhyd.2021.106594

[78]   B. Jiang, Q. Zhao, H. Shan, Y. Guo, X. Xu, D.J. McClements, C. Cao, B. Yuan, Impact of Heat Treatment on the Structure and Properties of the Plant Protein Corona Formed around TiO$_2$ Nanoparticles, J Agric Food Chem. 70 (2022) 6540–6551. https://doi.org/10.1021/acs.jafc.2c01650

Nanoparticle Toxicity and Compatibility
Materials Research Foundations 161 (2024) 83-121

Materials Research Forum LLC
https://doi.org/10.21741/9781644902998-4

Chapter 4

Cardiovascular Toxicity of Nanoparticles

Abhinay Thakur[1], Ashish Kumar[2]*

[1]Department of Chemistry, School of Chemical Engineering and Physical Sciences, Lovely Professional University, Phagwara, Punjab, 144411, India

[2]Department of Chemistry, Nalanda College of Engineering, Bihar Engineering University, Department of Science, Technology and Technical Education, Government of Bihar, Nalanda, Bihar, 803108, India

drashishchemlpu@gmail.com

Abstract

The integration of nanotechnology into various industries has raised concerns about potential health risks, particularly regarding cardiovascular toxicity. This book chapter comprehensively examines the impact of nanoparticles on the cardiovascular system, elucidating the underlying mechanisms and critical factors influencing toxicity. It reviews the current state of research on nanoparticle exposure via different routes, such as inhalation and ingestion, and their subsequent distribution in the cardiovascular system. Furthermore, the chapter delves into the cellular and molecular events triggering adverse cardiovascular effects, including oxidative stress, inflammation, and endothelial dysfunction. Understanding these toxicological aspects is crucial in developing effective safety guidelines and tailored nanomedical interventions for safer nanotechnology applications.

Keywords

Nanoparticles, Cardiovascular, Toxicity, Health Risks, Nanotechnology

Contents

1. Introduction

1.1 Nanoparticles: Definition and characteristics

Nanoparticles are minute particles with dimensions ranging from 1 to 100 nanometers, which is approximately 1000 times smaller than the width of human hair. These entities exhibit unique characteristics and behaviors compared to their bulk counterparts, primarily due to their reduced size and high surface area-to-volume ratio [1–3]. Their defining features make them fascinating and valuable in various fields of science and technology. The term "nanoparticles" encompasses a wide array of materials, offering versatility and adaptability in their applications. They can be made from metals like gold, silver, and iron; metal oxides like titanium dioxide and zinc oxide; carbon-based structures like carbon nanotubes and fullerenes; lipids such as liposomes; and polymers like polymeric nanoparticles. This diversity allows researchers and engineers to tailor nanoparticles for specific purposes, such as drug delivery, imaging, catalysis, electronics, and environmental remediation. The small size of nanoparticles grants them unique properties that set them apart from bulk materials. One of the most significant advantages is their high surface area-to-volume ratio, which facilitates enhanced reactivity. This property allows nanoparticles to be highly effective catalysts, making chemical reactions more efficient and precise. Additionally, the large surface area also enables more interactions with surrounding environments, making nanoparticles ideal candidates for sensing and biosensing applications. Nanoparticles' reactivity and stability also make them suitable for drug delivery systems. Their ability to encapsulate therapeutic agents and protect them from degradation ensures controlled and targeted drug release, maximizing efficacy while minimizing side effects [4,5]. This targeted drug delivery can significantly improve treatment outcomes for various diseases, including cancer and cardiovascular disorders.

Furthermore, the small size of nanoparticles grants them the ability to penetrate biological barriers, including cell membranes and the blood-brain barrier. This unique property opens up possibilities for targeted delivery of drugs and imaging agents to specific cells or tissues, thereby increasing their precision and reducing potential off-target effects. Despite their numerous beneficial applications, the nanoscale dimensions of nanoparticles raise concerns regarding their interactions with living organisms, especially the cardiovascular system. When nanoparticles come into contact with biological systems, they can trigger various responses that may be harmful. Some nanoparticles may induce oxidative stress, leading to cell damage and inflammation. Their ability to interact with blood components, such as platelets, can also potentially impact blood clotting and vascular function [6,7]. Hence, while nanoparticles offer promising potential in various industries, understanding their behavior and potential risks is vital, particularly when considering their use in medical applications. Careful evaluation of nanoparticle properties, biocompatibility, and biodistribution is essential to ensure their safe integration into consumer products, therapeutic agents, and medical devices. As researchers continue to explore the applications of nanoparticles, it is imperative to strike a balance between their advantages and potential risks, particularly in the context of cardiovascular health and overall safety.

1.2 Importance of studying cardiovascular toxicity

Studying cardiovascular toxicity is of paramount importance due to the critical role the cardiovascular system plays in maintaining overall health and wellbeing. The cardiovascular system comprises the heart, blood vessels, and blood, and it is responsible for delivering oxygen, nutrients, and hormones to various tissues and organs, as well as removing waste products and carbon dioxide from the body. Any adverse effects on this intricate system can have severe consequences for an individual's health [8,9]. One of the primary reasons for studying cardiovascular toxicity is the rapid advancement and widespread use of nanotechnology. Nanoparticles are increasingly integrated into consumer products, industrial processes, and medical applications. As a result, human exposure to nanoparticles has become more prevalent, raising concerns about potential health risks, including cardiovascular effects. Nanoparticles' unique properties, such as their small size and high reactivity, enable them to interact with biological systems in ways that larger particles cannot. When nanoparticles enter the bloodstream, they can be distributed throughout the body, including the cardiovascular system [10,11]. This raises questions about how nanoparticles may affect the heart, blood vessels, and other components of the cardiovascular system. Cardiovascular toxicity studies are also crucial for assessing the safety of nanomedicines. Nanoparticles are being actively explored as drug carriers for targeted drug delivery, allowing for more precise and efficient treatment of various diseases. However, their potential interactions with the cardiovascular system must be thoroughly understood to ensure patient safety and avoid unintended side effects.

Moreover, nanoparticles are not limited to medical applications. They are also widely used in industrial processes and consumer products, such as cosmetics, sunscreens, and food additives. This widespread exposure raises concerns about the potential long-term effects of nanoparticles on cardiovascular health, especially considering cumulative exposure from various sources as shown in Figure 1 [12]. Studying cardiovascular toxicity can help identify potential hazards associated with specific types of nanoparticles, their concentrations, and their exposure routes. This knowledge is essential for formulating safety guidelines and regulations to protect workers, consumers, and the general population from potential cardiovascular risks. Furthermore, understanding cardiovascular toxicity can aid in the design of safer nanoparticles and nanomedicines. By identifying the key factors that contribute to adverse effects, researchers can modify nanoparticle properties to reduce their potential for harm while retaining their beneficial applications.

1.3 Scope of the chapter

This chapter aims to provide a comprehensive overview of the cardiovascular toxicity associated with nanoparticles. It will delve into various aspects related to the subject matter, starting with a detailed explanation of nanoparticles, their definition, and key characteristics. The chapter will then justify the significance of studying cardiovascular toxicity specifically, emphasizing the potential health risks posed by nanoparticles [13,14]. By presenting evidence from existing research, this chapter will elucidate the mechanisms through which nanoparticles may exert harmful effects on the cardiovascular system. Furthermore, the discussion will explore factors influencing cardiovascular toxicity, including nanoparticle size, shape, and surface properties, as well as their clearance pathways. Case studies and experimental models will be reviewed to provide insights into the real-world impact of nanoparticles on cardiovascular health, drawing parallels between animal studies and human clinical trials. The chapter will also address the challenges in assessing cardiovascular toxicity and discuss the current regulatory framework for nanoparticles. Finally, the scope of this chapter will extend to the future prospects of nanomedicine and its potential benefits for cardiovascular applications. The potential for targeted drug delivery, theranostics, and other innovative approaches will be explored in the context of nanotechnology's role in advancing cardiovascular healthcare.

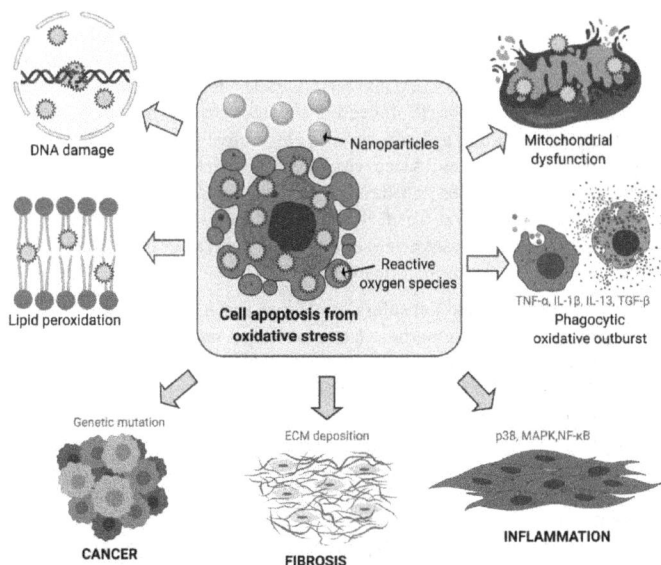

Figure 1: Toxic effects of nanoparticles in various vital organs and their metabolites. Adapted from Ref. [12] under CCBY 4.0.

2. Nanoparticle exposure routes

2.1 Inhalation exposure

Inhalation exposure to nanoparticles occurs when these tiny particles are present in the air and are breathed in by individuals. This exposure route is of significant concern because inhaled nanoparticles can easily reach the respiratory system, leading to potential adverse health effects. Nanoparticles, due to their small size, can remain suspended in the air for extended periods, allowing them to be inhaled deeply into the lungs [15–17]. The ability of nanoparticles to reach the lower respiratory tract is influenced by their size, shape, and surface properties. Once in the lungs, nanoparticles can interact with lung tissue and cells, potentially leading to inflammation and oxidative stress. The deposition of nanoparticles in different regions of the respiratory tract can vary based on their size. Ultrafine nanoparticles (less than 100 nanometers) can penetrate deep into the alveoli, the small air sacs in the lungs responsible for gas exchange, while larger nanoparticles may be trapped in the upper airways.

In occupational settings, workers involved in processes that generate or handle nanoparticles may face a higher risk of inhalation exposure. Industries such as manufacturing, construction, and nanotechnology research are among those with potential nanoparticle exposure risks. Workers in these settings may be exposed to airborne nanoparticles during activities like spray coating, grinding, or handling nanoparticle-containing materials. The health effects of inhalation exposure to nanoparticles are a subject of ongoing research. Studies have shown that certain nanoparticles, such as metal oxides like titanium dioxide and nanoparticles derived from combustion processes, can induce inflammation in the lungs. Prolonged or repeated exposure to such particles may lead to chronic lung diseases, including asthma, chronic obstructive pulmonary disease (COPD), and even lung cancer in extreme cases. Moreover, there is increasing evidence suggesting that inhalation exposure to nanoparticles can have systemic effects beyond the respiratory system. Nanoparticles that reach the bloodstream through the lungs can potentially be distributed throughout the body, including the cardiovascular system. This raises concerns about the potential impact of nanoparticles on cardiovascular health [18,19].

To address these concerns, occupational safety guidelines and regulations are being developed to protect workers from excessive nanoparticle exposure. Such measures may include engineering controls, personal protective equipment, and monitoring of airborne nanoparticle concentrations in workplaces. For the general population, inhalation exposure to nanoparticles from consumer products is also an area of interest. For instance, nanoparticles used in spray products like sunscreens and air fresheners can be inhaled when the products are used in confined spaces. To ensure the safe and responsible use of nanotechnology, it is essential to understand the potential risks associated with inhalation exposure to nanoparticles. Researchers continue to investigate the toxicity of different types of nanoparticles and the mechanisms through which they may impact respiratory and cardiovascular health [20,21]. Proactive risk assessment, along with appropriate safety measures, can help mitigate potential health hazards and support the sustainable development of nanotechnology for various applications.

2.2   Ingestion exposure

Ingestion exposure to nanoparticles occurs when these tiny particles are unintentionally or intentionally consumed through various routes, such as food, beverages, and other materials. This exposure route is relevant to nanoparticles used in food additives, food packaging, cosmetics, personal care products, and even some medications. Nanoparticles can be incorporated into food products as additives to improve texture, color, and shelf life. Additionally, they can be used in food packaging materials to enhance barrier properties and prevent spoilage. Consumers may also inadvertently ingest nanoparticles used in products like toothpaste, mouthwash, and lip balms. When nanoparticles are ingested, they encounter the gastrointestinal (GI) tract, which consists of the mouth, esophagus, stomach, and intestines. The behavior and potential impacts of nanoparticles in the GI tract depend on various factors, including their size, surface properties, and chemical composition [22]. The GI tract has natural defense mechanisms to prevent the absorption of harmful substances, including enzymes and mucus layers. However, the small size of nanoparticles enables them to interact with the gut lining and potentially cross the intestinal barrier. Some nanoparticles have been found to be absorbed into the bloodstream and distributed to various organs, including the liver, spleen, and kidneys. Ingested nanoparticles may also interact with the gut microbiota, which plays a crucial role in digestion and overall health. Studies have shown that

Materials Research Forum LLC
https://doi.org/10.21741/9781644902998-4

certain nanoparticles can influence the composition and function of the gut microbiota, which may have implications for gut health and immune responses [23–25].

One area of concern with ingestion exposure is the potential for nanoparticles to accumulate in organs over time. If nanoparticles are not efficiently cleared from the body, they may accumulate in tissues, leading to potential long-term health effects. The liver and spleen are among the organs that may accumulate nanoparticles after ingestion. Regulatory agencies worldwide are actively studying the potential risks of nanoparticles in food and consumer products. They are setting safety standards and guidelines for the use of nanoparticles in these applications to protect consumer health. Researchers are also investigating the potential health effects of nanoparticle ingestion, including short-term and long-term toxicity. Studies are being conducted to understand the behavior of nanoparticles in the GI tract, their potential for absorption, and their distribution in the body.

It is important to note that not all nanoparticles pose the same risks upon ingestion. The toxicity of nanoparticles can vary significantly based on their composition and surface properties. Additionally, the level of exposure plays a crucial role in determining potential health effects. To ensure the safe use of nanoparticles in food and consumer products, risk assessment studies are conducted, and safety evaluations are undertaken before these nanoparticles are approved for use. It is essential to strike a balance between the beneficial applications of nanoparticles and the potential risks associated with their ingestion [26,27]. Responsible use, proper risk assessment, and regulatory oversight are essential in harnessing the potential of nanotechnology while safeguarding consumer health.

2.3 Dermal and other routes

Dermal exposure to nanoparticles occurs when these tiny particles come into contact with the skin. Nanoparticles are used in a wide range of consumer products, including cosmetics, sunscreens, lotions, and topical medications [28,29]. When applied to the skin, these products may contain nanoparticles that can be absorbed through the skin and potentially enter the bloodstream. The skin is the body's largest organ and acts as a protective barrier against external substances. However, the small size and unique properties of nanoparticles allow them to interact with the skin's surface and potentially penetrate the skin's layers. Once nanoparticles enter the skin, they may reach underlying tissues and be distributed to other organs via the bloodstream. The potential health effects of dermal exposure to nanoparticles are an area of active research. Some studies have shown that certain nanoparticles can induce skin inflammation and oxidative stress. However, the toxicity of nanoparticles can vary based on factors such as their size, surface properties, and composition. Besides dermal exposure, nanoparticles can enter the body through other routes:

- Ocular Exposure: Nanoparticles in consumer products or workplace environments may come into contact with the eyes. This can happen through direct contact or from particles suspended in the air. Ocular exposure to nanoparticles raises concerns about potential eye irritation and other eye health effects.

- Parenteral Exposure: Parenteral exposure refers to the entry of nanoparticles into the body through routes other than the digestive or respiratory systems. This can occur through accidental punctures, injections (e.g., medical procedures), or wounds. In medical applications, nanoparticles may be directly injected into the bloodstream for targeted drug delivery or medical imaging purposes.

The potential risks of these exposure routes depend on factors such as the specific type of nanoparticles, their concentration, and the duration of exposure. Additionally, individual factors, such as pre-existing health conditions and skin conditions, can influence the body's response to nanoparticle exposure. To ensure the safe use of nanoparticles in consumer products and medical applications, regulatory agencies are evaluating the potential risks associated with dermal exposure and other routes of exposure. Guidelines and safety standards are being developed to minimize potential adverse effects and protect public health [30,31]. Furthermore, researchers are continuously investigating the behavior of nanoparticles in the skin and other tissues, as well as their potential long-term effects. Responsible use, proper risk assessment, and transparent labeling of products containing nanoparticles are essential to ensure consumer safety and build public trust in nanotechnology applications. As the field of nanotechnology continues to advance, ongoing research and regulatory oversight will play a crucial role in maximizing the benefits of nanoparticles while minimizing potential risks.

3.   Distribution of nanoparticles in the cardiovascular system

3.1   Nanoparticle transport and accumulation

Nanoparticle transport and accumulation are critical processes that determine the distribution of nanoparticles within the body, including the cardiovascular system. The movement and fate of nanoparticles in the body are influenced by various factors, such as their size, shape, surface properties, and the physiology of different organs and tissues.

- Transport in the Bloodstream: When nanoparticles enter the bloodstream through inhalation, ingestion, or dermal exposure, they become suspended in the blood. The circulation of nanoparticles in the blood is influenced by the flow dynamics and hemodynamics of the cardiovascular system.

Nanoparticles can move freely in the blood or adhere to blood components, such as red blood cells and platelets. The surface properties of nanoparticles play a crucial role in their interactions with blood components. Protein coronas, a layer of proteins that form around nanoparticles in the blood, can influence their behavior and potential interactions with cells [32,33]. The size of nanoparticles is a key determinant of their circulation time in the bloodstream. Smaller nanoparticles, particularly those in the nanometer range, have longer circulation times and may be able to evade clearance mechanisms for a more extended period.

- Accumulation in Organs and Tissues:

As nanoparticles circulate in the bloodstream, they can accumulate in various organs and tissues through different mechanisms:

a. Passive Accumulation: Nanoparticles may passively accumulate in certain organs based on their size and filtering capacity of the reticuloendothelial system. For instance, larger nanoparticles may be more likely to accumulate in organs like the liver and spleen, which have a high filtration capacity.

b. Active Targeting: In some cases, nanoparticles can be engineered to actively target specific organs or tissues. This is achieved by attaching ligands or molecules to the nanoparticle surface that recognize specific receptors or markers on the target cells. Active targeting enhances the nanoparticle's ability to accumulate in the desired site, such as tumor tissues in cancer therapy or inflamed endothelial cells in cardiovascular diseases.

c. Permeation: Nanoparticles can permeate tissues and cross cellular barriers, such as the blood-brain barrier (BBB). Some nanoparticles, particularly those with specific surface modifications, have shown the ability to cross the BBB and access the central nervous system.

- Biodistribution and Clearance:

The biodistribution of nanoparticles is the pattern of their accumulation in different organs and tissues over time. The accumulation of nanoparticles in specific organs can be monitored using imaging techniques such as positron emission tomography (PET) or magnetic resonance imaging (MRI). Clearance refers to the removal of nanoparticles from the body. Nanoparticles can be cleared through various routes, including excretion through the kidneys and liver, uptake by the spleen, or degradation and elimination by macrophages [34,35]. The kinetics of nanoparticle transport, accumulation, and clearance are complex and can vary depending on the nanoparticle properties and the physiological state of the individual. The distribution and accumulation of nanoparticles in the cardiovascular system are crucial considerations when evaluating the potential health effects of nanoparticles and designing safe and effective nanomedicines. Researchers are continually working to better understand these processes to ensure the responsible and beneficial use of nanotechnology for various applications.

3.2   Blood-brain barrier permeability

The blood-brain barrier (BBB) is a highly specialized and tightly regulated barrier that separates the bloodstream from the brain and central nervous system (CNS). Its primary function is to maintain the brain's microenvironment and protect it from potentially harmful substances circulating in the blood. The BBB is formed by a combination of endothelial cells, tight junctions, pericytes, astrocytes, and the basal lamina, creating a highly selective and semipermeable barrier. The unique structure of the BBB poses a challenge to the delivery of therapeutics to the brain, as it restricts the passage of many substances, including drugs and nanoparticles [36,37]. However, the BBB also plays a crucial role in preventing the entry of pathogens and toxins into the brain, safeguarding its delicate neuronal environment. Nanoparticles have emerged as promising candidates for drug delivery to the brain due to their ability to traverse biological barriers. However, their ability to cross the BBB is highly dependent on their physicochemical properties, such as size, surface charge, and surface functionalization.

- Size: Smaller nanoparticles, particularly those in the nanometer range, have an advantage in crossing the BBB. Particles with a size below 100 nm can potentially penetrate the tight junctions between endothelial cells that make up the BBB.

- Surface Charge: The surface charge of nanoparticles also influences their ability to cross the BBB. Positively charged nanoparticles may have higher chances of BBB penetration due to electrostatic interactions with negatively charged components on the BBB endothelial cell surface.

- Surface Functionalization: The attachment of specific ligands or molecules to the surface of nanoparticles can enhance their BBB permeability. These ligands can act as "Trojan horses" that bind to specific receptors on the BBB endothelial cells, facilitating transcytosis or active transport across the barrier.

- Nanoparticle Type: Different types of nanoparticles, such as liposomes, dendrimers, and polymeric nanoparticles, have shown varying degrees of BBB permeability based on their physicochemical properties.

Nanoparticles that successfully cross the BBB have the potential to deliver therapeutic agents to the brain, opening up new avenues for treating neurological disorders such as Alzheimer's disease, Parkinson's disease, and brain tumors. However, it is essential to ensure that these nanoparticles are safe and do not cause unintended side effects in the brain. The study of BBB permeability and the safe design of nanoparticle-based drug delivery systems require careful consideration. Researchers need to understand the potential risks associated with BBB penetration, including the possibility of inducing neurotoxicity or interfering with the brain's normal functions. Additionally, targeting nanoparticles to specific brain regions can be challenging due to the regional heterogeneity of the BBB [38,39].

Nonetheless, with continued research and advances in nanotechnology, the targeted delivery of therapeutics across the BBB holds tremendous promise for revolutionizing the treatment of various neurological disorders and improving patient outcomes. Ethical considerations and rigorous safety evaluations are critical to harnessing the full potential of nanoparticles for drug delivery to the brain while ensuring patient safety and well-being [40–47].

3.3 Nanoparticle interaction with endothelial cells

Nanoparticle interaction with endothelial cells is a crucial aspect of understanding the potential impacts of nanoparticles on the cardiovascular system. Endothelial cells form the inner lining of blood vessels, acting as a semi-permeable barrier between the blood and the surrounding tissues. These cells play a pivotal role in maintaining vascular health and regulating blood flow, as well as participating in immune responses and inflammation. When nanoparticles come into contact with endothelial cells, several important interactions can occur:

- Adhesion and Uptake: Nanoparticles can adhere to the surface of endothelial cells through various mechanisms, including electrostatic interactions and binding to specific receptors. Some nanoparticles may be internalized by endothelial cells through endocytosis, leading to their accumulation within these cells.

- Inflammation and Oxidative Stress: Nanoparticles have the potential to induce inflammation and oxidative stress in endothelial cells. This can trigger the release of inflammatory mediators, such as cytokines and chemokines, leading to an inflammatory response in the blood vessels. Chronic inflammation in the endothelium is a critical early step in the development of cardiovascular diseases, including atherosclerosis.

- Endothelial Dysfunction: Prolonged or excessive exposure to certain nanoparticles can lead to endothelial dysfunction, characterized by impaired vasodilation, increased permeability, and reduced nitric oxide production. Endothelial dysfunction disrupts the normal functioning of blood vessels, contributing to the development of cardiovascular diseases.

- Vascular Permeability: Some nanoparticles can increase vascular permeability by disrupting the tight junctions between endothelial cells. This enhanced permeability can lead to leakage of fluid and proteins from blood vessels into surrounding tissues, potentially contributing to edema and tissue damage.

- Reactive Oxygen Species (ROS) Generation: Nanoparticles with certain properties can induce the generation of reactive oxygen species (ROS) within endothelial cells. ROS are highly reactive molecules that can damage cellular structures and promote oxidative stress, potentially leading to endothelial cell dysfunction and cardiovascular pathology.

- Nitric Oxide Regulation: Nitric oxide (NO) is a crucial signaling molecule produced by endothelial cells, playing a key role in regulating vascular tone and blood flow. Some nanoparticles may modulate NO production and availability, affecting vascular function.

The impact of nanoparticle interaction with endothelial cells can vary based on nanoparticle properties, including size, shape, surface charge, and composition. Additionally, the duration and concentration of nanoparticle exposure are important factors that determine the extent of endothelial cell responses [48,49]. Studying nanoparticle interaction with endothelial cells is essential for understanding the potential risks associated with nanoparticle exposure and for developing safe and effective nanomedical interventions [50–57]. By gaining insights into the mechanisms underlying nanoparticle-endothelial cell interactions, researchers can design nanoparticles with reduced toxicity and targeted therapeutic effects for cardiovascular applications. Responsible nanotechnology development, along with rigorous safety assessments, is crucial to harness the potential of nanoparticles while ensuring cardiovascular health and overall well-being.

4.   Mechanisms of cardiovascular toxicity

4.1   Oxidative stress and reactive oxygen species (ROS) generation

Oxidative stress and the generation of reactive oxygen species (ROS) are central mechanisms through which nanoparticles can induce cellular and tissue damage in the cardiovascular system. Oxidative stress occurs when there is an imbalance between the production of ROS and the body's ability to neutralize them with antioxidants. This imbalance results in an excess of ROS, which can lead to oxidative damage to various cellular components, including lipids, proteins, and DNA. Oxidative stress is implicated in the development and progression of various cardiovascular diseases. When nanoparticles come into contact with cells in the cardiovascular system, they can initiate ROS generation through different pathways:

- Electron Transfer: Some nanoparticles, especially those composed of metals or metal oxides, can act as catalysts for electron transfer reactions. These reactions lead to the formation of ROS, such as superoxide radicals ($O_2^{\cdot-}$), hydrogen peroxide (H_2O_2), and hydroxyl radicals ($\cdot OH$).

- Cellular Enzyme Activation: Nanoparticles can activate certain cellular enzymes, such as NADPH oxidases, which play a significant role in ROS generation. NADPH oxidases transfer electrons across cellular membranes, leading to the production of superoxide radicals.

- Mitochondrial Dysfunction: Nanoparticles may interfere with mitochondrial function, leading to electron leakage from the electron transport chain. This leakage can result in the formation of ROS within the mitochondria.

The endothelium, which lines the inner surface of blood vessels, is particularly susceptible to oxidative stress induced by nanoparticles. Endothelial cells are exposed to circulating nanoparticles and are crucial in regulating vascular function, maintaining blood flow, and preventing platelet aggregation. Oxidative stress in endothelial cells can lead to endothelial dysfunction, a critical early step in the development of cardiovascular diseases. The excessive production of ROS impairs the bioavailability of nitric oxide (NO), a signaling molecule produced by endothelial cells that promotes vasodilation and inhibits platelet aggregation and adhesion. Reduced NO bioavailability results in impaired vasodilation, increased vascular permeability, and a pro-inflammatory environment in the blood vessels. This promotes the adhesion of immune cells and platelets to the endothelium, initiating the formation of atherosclerotic plaques.

Furthermore, oxidative stress-induced damage to lipids and proteins in the endothelium can result in inflammation and activation of the immune response, further contributing to endothelial dysfunction and vascular impairment. The interplay between oxidative stress, inflammation, and endothelial dysfunction creates a vicious cycle that perpetuates cardiovascular toxicity induced by nanoparticles. Prolonged or chronic exposure to nanoparticles with the ability to generate ROS can lead to the progression of cardiovascular diseases, including atherosclerosis, hypertension, and heart failure [58–60]. To address the cardiovascular toxicity of nanoparticles, researchers are exploring strategies to mitigate ROS generation and oxidative stress. These include surface modifications of nanoparticles to reduce ROS production, the development of antioxidant-loaded nanoparticles to counteract ROS, and the design of nanoparticles with reduced toxic potential. By understanding the mechanisms of oxidative stress induced by nanoparticles and their implications for cardiovascular health, researchers aim to harness the potential of nanotechnology safely for medical applications and other fields while minimizing potential risks.

4.2   Inflammation and immune response activation

Inflammation and immune response activation are crucial mechanisms through which nanoparticles can induce cardiovascular toxicity. When nanoparticles come into contact with cells in the cardiovascular system, they can initiate a cascade of events leading to inflammatory responses and immune activation. Chronic inflammation and immune response activation play significant roles in the development and progression of various cardiovascular diseases.

- Inflammatory Responses: Nanoparticles can trigger inflammatory responses in the cardiovascular system by interacting with endothelial cells, immune cells, and other vascular components. The presence of nanoparticles can lead to the activation of signaling pathways that promote the release of pro-inflammatory cytokines, chemokines, and other inflammatory mediators. Pro-inflammatory cytokines, such as interleukin-1 (IL-1), interleukin-6 (IL-6), and tumor necrosis factor-alpha (TNF-α), recruit immune cells to the site of nanoparticle exposure. Immune cells, including monocytes, macrophages, and neutrophils, are then recruited to the site of injury or inflammation [61,62,71,63–70]. These immune cells release additional inflammatory mediators, amplify the inflammatory response, and contribute to the clearance of nanoparticles. However, prolonged or excessive inflammation can lead to tissue damage, endothelial dysfunction, and the progression of cardiovascular diseases, such as atherosclerosis.

- Activation of Immune Responses: Nanoparticles can activate both innate and adaptive immune responses in the cardiovascular system. The innate immune system is the body's first line of defense against foreign invaders and pathogens. When nanoparticles interact with immune cells, such as macrophages and dendritic cells, they can trigger the release of pro-inflammatory cytokines and activate immune responses. In addition to the innate immune response, nanoparticles can also stimulate adaptive immune responses. Adaptive immunity involves the activation of T cells and B cells, leading to the production of specific antibodies against the nanoparticles. These immune responses can further exacerbate inflammation upon subsequent exposures to nanoparticles [72,73]. Immune activation can lead to the formation of immune complexes and the activation of complement cascades, which can contribute to tissue damage and inflammation.

- Role of Particle Surface Properties: The surface properties of nanoparticles play a critical role in determining their ability to induce inflammation and immune response activation. For example, nanoparticles with specific surface charges or surface coatings can interact differently with immune cells and endothelial cells, leading to varying degrees of immune activation and inflammatory responses.

- Immunomodulatory Effects: On the other hand, certain nanoparticles have been engineered for immunomodulatory purposes. These nanoparticles can be designed to dampen inflammatory responses or induce immunosuppression to treat certain inflammatory conditions or autoimmune diseases.

However, it is crucial to strike a balance between the potential benefits of immunomodulation and the risks of immune suppression, as both are essential for maintaining cardiovascular health and overall immune function. To address the cardiovascular toxicity associated with inflammation and immune response activation, researchers are actively investigating strategies to design nanoparticles with reduced immunogenicity and pro-inflammatory potential. Additionally, understanding the factors that contribute to immune activation and inflammatory responses is critical for the safe and effective use of nanoparticles in medical applications and other fields. Responsible nanotechnology development and rigorous safety evaluations are essential to harnessing the full potential of nanoparticles while ensuring cardiovascular health and overall well-being.

4.3   Endothelial dysfunction and vascular impairment

Endothelial dysfunction is a critical mechanism through which nanoparticles can induce vascular impairment and contribute to the development of cardiovascular diseases [73,74]. The endothelium, a single layer of endothelial cells lining the inner surface of blood vessels, plays a crucial role in maintaining vascular homeostasis and regulating various vascular functions. When nanoparticles interact with endothelial cells, they can trigger a series of events that lead to endothelial dysfunction:

- Impaired Nitric Oxide (NO) Production: One of the key functions of the endothelium is to produce and release nitric oxide (NO), a potent vasodilator. NO promotes relaxation of the smooth muscle cells in blood vessel walls, leading to vasodilation and increased blood flow. Nanoparticles can disrupt NO production in endothelial cells, leading to impaired vasodilation and altered blood flow.

- Increased Oxidative Stress: Nanoparticles can induce oxidative stress in endothelial cells, leading to an imbalance between the production of reactive oxygen species (ROS) and the body's antioxidant defenses. Excessive ROS generation can lead to oxidative damage to endothelial cells, proteins, and lipids, further contributing to endothelial dysfunction.

- Inflammation and Immune Responses: Nanoparticles can trigger inflammatory responses in endothelial cells, leading to the release of pro-inflammatory cytokines and chemokines. Chronic inflammation can contribute to endothelial dysfunction and promote the adhesion of immune cells to the endothelial surface.

- Vascular Permeability: Some nanoparticles can increase vascular permeability by disrupting the tight junctions between endothelial cells[75]. This enhanced permeability can lead to leakage of fluid and proteins from blood vessels into surrounding tissues, contributing to tissue edema and inflammation.

Endothelial dysfunction can have profound effects on vascular health and contribute to the development of various cardiovascular diseases:

- Atherosclerosis: Endothelial dysfunction is a critical early step in the development of atherosclerosis, a condition characterized by the accumulation of fatty deposits (plaques) in arterial walls. Impaired NO production and increased oxidative stress promotes the adhesion of immune cells and lipids to the endothelium, initiating the formation of atherosclerotic plaques.

- Hypertension: Endothelial dysfunction can lead to increased vascular resistance and impaired vasodilation, contributing to elevated blood pressure and hypertension. Reduced NO bioavailability and increased ROS generation are key factors in the development of hypertension.

- Thrombosis: Disrupted endothelial function can lead to altered blood flow and the activation of platelets and the coagulation system. These factors increase the risk of thrombus formation and can lead to blockages in blood vessels, potentially causing heart attacks or strokes.

- Vascular Inflammation: Chronic endothelial dysfunction and inflammation can lead to a pro-inflammatory environment in blood vessels, perpetuating cardiovascular pathology and promoting the progression of cardiovascular diseases.

Understanding the mechanisms of endothelial dysfunction induced by nanoparticles is crucial for evaluating the potential cardiovascular risks associated with nanoparticle exposure. Researchers are actively exploring strategies to mitigate endothelial dysfunction, including nanoparticle surface modifications to reduce oxidative stress and inflammation, as well as the development of nanoparticles with reduced toxicity potential. Overall, responsible nanotechnology development and comprehensive risk assessments are essential to harnessing the potential of nanoparticles while ensuring cardiovascular health and minimizing potential adverse effects on the vasculature.

5.   Factors influencing cardiovascular toxicity

5.1   Particle size, shape, and surface properties

Particle size, shape, and surface properties are crucial factors that profoundly influence the behavior and interactions of nanoparticles with the cardiovascular system. These parameters play a significant role in determining the biodistribution, cellular uptake, and potential toxicity of nanoparticles [76,77]. Here's a detailed elaboration of each factor:

- Particle Size: The size of nanoparticles is a critical determinant of their biological fate and toxicity. Nanoparticles are generally defined as particles with dimensions in the nanometer range (1-100 nm). Their small size imparts unique physical properties compared to bulk materials.

a. Cellular Uptake: Smaller nanoparticles have a higher surface area-to-volume ratio, making them more reactive and facilitating greater cellular uptake. They can be internalized by cells more efficiently, including endothelial cells and immune cells in the cardiovascular system. Cellular uptake can lead to intracellular accumulation and potential toxicity.

b. Biodistribution: The size of nanoparticles affects their biodistribution in the body. Smaller nanoparticles can pass through fenestrations in blood vessels and accumulate in various tissues and organs, including the heart and blood vessels.

c. Clearance: Larger nanoparticles tend to be cleared more slowly from the body, as they are subject to the filtering capacity of the reticuloendothelial system in the liver and spleen. Smaller nanoparticles may be cleared through renal excretion more readily [78,79].

- Particle Shape: The shape of nanoparticles also influences their cellular interactions and biodistribution. Various nanoparticle shapes, such as spheres, rods, and tubes, have been explored for different applications.

a. Cellular Uptake: Some nanoparticle shapes, such as nanorods and nanowires, may have a higher aspect ratio, leading to enhanced cellular uptake compared to spherical nanoparticles.

b. Biodistribution: Particle shape can influence how nanoparticles interact with the vascular system. For instance, elongated nanoparticles may be more prone to accumulate at vascular sites with disturbed flow, where atherosclerotic plaques tend to develop.

- Surface Properties: The surface properties of nanoparticles are critical in determining their interactions with biological entities, including cells and biomolecules.

a. Surface Charge: The surface charge of nanoparticles, often expressed as zeta potential, can affect their stability in biological fluids and their interactions with cell membranes. Positively charged nanoparticles may have enhanced cellular uptake due to electrostatic interactions with negatively charged cell surfaces.

b. Surface Coatings and Functionalization: Nanoparticles can be coated with various materials or functionalized with specific ligands or molecules. Surface coatings can influence the nanoparticles' stability, biocompatibility, and interactions with the immune system.

c. Protein Corona Formation: Upon exposure to biological fluids, proteins can adsorb onto the nanoparticle surface, forming a protein corona. The protein corona can alter the nanoparticles' biological identity and affect their cellular uptake and clearance [80,81]. The combination of particle size, shape, and surface properties dictates the nanoparticles' overall behavior in the cardiovascular system. These factors collectively influence the nanoparticles' ability to cross biological barriers, interact with endothelial cells, trigger inflammatory responses, and induce oxidative stress.

5.2   Biodegradation and clearance pathways

Biodegradation and clearance pathways are crucial aspects that significantly influence the behavior and potential impact of nanoparticles on the cardiovascular system. Biodegradation refers to the process by which nanoparticles break down into smaller, biocompatible components over time. This property is particularly important for nanoparticles intended for drug delivery applications. Biodegradable nanoparticles can release their encapsulated therapeutic cargo gradually, allowing for sustained and controlled drug release. This feature is beneficial in treating various cardiovascular conditions, where precise drug dosing and prolonged therapeutic effects are desirable. The rate of biodegradation can vary depending on the nanoparticle's composition, surface properties, and the surrounding physiological environment. Biodegradable nanoparticles are often composed of polymers or natural materials that can undergo enzymatic degradation or hydrolysis, making them less likely to accumulate in the body and reducing potential long-term toxicity [82–84]. On the other hand, non-biodegradable nanoparticles may persist in the body for longer durations, increasing the likelihood of accumulation in tissues and organs. For instance, some metallic nanoparticles, such as gold or silver nanoparticles, may not undergo significant biodegradation and could potentially lead to toxicity if they accumulate in high concentrations over time.

Clearance pathways are equally important in determining nanoparticle fate and potential toxicity. Clearance refers to the process of removing nanoparticles from the body. Different organs and systems play essential roles in nanoparticle clearance:

- Renal Clearance: Nanoparticles that are small enough to pass through the renal filtration threshold (typically around 5-6 nm) can be cleared through the kidneys and excreted in the urine. This pathway is particularly relevant for smaller nanoparticles, facilitating their safe elimination from systemic circulation.

- Hepatic Clearance: Some nanoparticles, especially those larger than the renal filtration threshold, may be taken up by hepatocytes in the liver. These nanoparticles can be subsequently cleared through the biliary excretion pathway, contributing to their elimination from the body.

- Reticuloendothelial System (RES) Clearance: The RES, including macrophages in the liver and spleen, plays a vital role in recognizing and clearing nanoparticles from the bloodstream. This pathway is particularly relevant for larger and non-biodegradable nanoparticles.

- Lymphatic Clearance: Nanoparticles that are too large to be cleared by renal filtration can be transported through the lymphatic system and eventually reach the lymph nodes for clearance.

- Pulmonary Clearance: Inhalable nanoparticles can be cleared from the respiratory system through mechanisms such as mucociliary clearance or uptake by alveolar macrophages.

Understanding the biodegradation and clearance pathways of nanoparticles is essential for the responsible design and use of nanoparticles in various applications, especially in nanomedicine. Biodegradable nanoparticles with appropriate surface properties can minimize long-term accumulation and potential toxicity. Furthermore, nanoparticles designed for cardiovascular drug delivery should take into account the specific clearance mechanisms relevant to the intended route of administration and dosage. By carefully considering these factors, researchers can develop nanoparticles with improved biocompatibility and reduced cardiovascular risks while harnessing the full potential of nanotechnology for medical applications.

5.3   Dose-response relationships

Dose-response relationships play a crucial role in understanding the cardiovascular toxicity of nanoparticles and assessing their potential risks. A dose-response relationship describes the relationship between the dose (amount) of a substance administered and the magnitude of the biological response it elicits. In the context of nanoparticles, it refers to how the cardiovascular system responds to varying doses of nanoparticle exposure.

- Dose-Response Curve: The dose-response relationship is typically represented graphically using a dose-response curve. The curve illustrates the relationship between different doses of nanoparticles and the corresponding cardiovascular response. The response can be measured in terms of toxicity, adverse effects, or therapeutic efficacy, depending on the context of the study.

- Threshold and Non-Threshold Effects: Dose-response relationships can exhibit two main patterns: threshold and non-threshold effects. In a threshold effect, there is a minimum dose below which no response is observed, but once the threshold is surpassed, the response increases in a dose-dependent manner [85,86]. Non-threshold effects, on the other hand, show a continuous relationship without a specific threshold, meaning that any dose, no matter how small, can elicit some level of response.

- High Dose vs. Low Dose Effects: The dose-response relationship can also reveal important information about high-dose and low-dose effects. At high doses, nanoparticles may overwhelm the body's defense mechanisms and lead to toxicity. At low doses, nanoparticles may have minimal or no observable effects, and the response may not be significant.

- Determining Safe Exposure Levels: Dose-response relationships are essential for determining safe exposure levels to nanoparticles. Regulatory agencies use this information to establish acceptable exposure limits and guidelines for nanoparticle usage in various applications. The goal is to identify doses that provide therapeutic benefits without causing adverse effects or toxicity.

- Individual Variability: Individual variability in responses to nanoparticle exposure can also be observed in dose-response relationships. Different people may respond differently to the same dose due to variations in genetic factors, pre-existing health conditions, and other individual characteristics.

- Assessing Therapeutic Efficacy: In nanomedicine, dose-response relationships are crucial for evaluating the therapeutic efficacy of nanoparticle-based treatments [87]. Understanding how different doses of therapeutic nanoparticles affect the cardiovascular system helps researchers optimize drug dosages for maximum effectiveness while minimizing potential side effects.

- Interpreting Animal Studies: Dose-response relationships obtained from animal studies are essential for extrapolating findings to humans. However, the relevance of animal data to human responses must be carefully considered due to potential species-specific differences in nanoparticle metabolism and toxicity.

Researchers use dose-response relationships to identify the appropriate dosage range for safe and effective nanoparticle applications. By understanding the potential risks associated with varying nanoparticle doses, researchers and clinicians can make informed decisions about nanoparticle-based therapies and develop guidelines to ensure the responsible use of nanoparticles in nanomedicine and other fields.

5.4 Species-specific variability

Species-specific variability refers to the differences in how different species respond to nanoparticle exposure. In the context of cardiovascular toxicity, this variability becomes particularly relevant when studying the effects of nanoparticles on various animal models and extrapolating findings to humans. Understanding species-specific differences is essential for assessing the relevance and safety of nanoparticle-based therapies and applications in humans. Here's a detailed elaboration of species-specific variability:

- Metabolism and Biodistribution: Different species have varying metabolic pathways and clearance mechanisms, which can influence how nanoparticles are processed and distributed in the body. For example, nanoparticles cleared primarily through the liver in one species may be predominantly cleared through the kidneys in another. These differences can lead to variations in nanoparticle accumulation in specific organs or tissues, affecting the overall toxicity profile.

- Immune Responses: The immune systems of different species can respond differently to nanoparticle exposure. Immune responses, including inflammatory reactions and immune cell interactions, may vary among species, influencing the overall immunotoxicity of nanoparticles.

- Organ Function: Variations in organ function, such as the functioning of the liver, kidneys, and cardiovascular system, can affect the body's ability to handle nanoparticle exposure. Differences in organ physiology and function may alter the biodistribution, clearance, and potential toxicity of nanoparticles.

- Cellular Receptors and Targets: Receptors and targets on cells that interact with nanoparticles may differ among species, influencing the cellular responses to nanoparticle exposure. For example, nanoparticle uptake by specific cell types or cellular internalization pathways may vary among species.

- Nanoparticle Interactions with Plasma Proteins: Nanoparticles can interact with various plasma proteins in the bloodstream, forming a "protein corona" around the nanoparticle surface. The composition of the protein corona can vary among species, affecting how nanoparticles are recognized and processed by the immune system.

- Sensitivity to Nanoparticle-Induced Toxicity: Species can differ in their sensitivity to nanoparticle-induced toxicity. Some species may be more susceptible to specific types of nanoparticle toxicity, while others may show resistance or reduced sensitivity. This can make it challenging to extrapolate findings from animal studies to humans accurately.

- Pharmacokinetics and Pharmacodynamics: Pharmacokinetics refers to how the body processes a substance, such as nanoparticle absorption, distribution, metabolism, and excretion. Pharmacodynamics refers to how the substance affects the body. Both factors can vary significantly among species, affecting nanoparticle behavior and toxicity.

- Pre-existing Health Conditions: Pre-existing health conditions or genetic factors in different species may modify the responses to nanoparticle exposure. For example, a species with a specific genetic predisposition to cardiovascular diseases may be more susceptible to nanoparticle-induced cardiovascular toxicity.

Due to these species-specific differences, the extrapolation of findings from animal studies to humans must be approached cautiously. While animal studies provide valuable insights into nanoparticle behavior and toxicity, it is essential to validate and corroborate these findings in human studies whenever possible. To enhance the predictive value of animal studies, researchers are developing advanced in vitro models, organ-on-a-chip systems, and other human-relevant platforms. These approaches aim to bridge the gap between animal data and human responses, ultimately improving the safety and effectiveness of nanoparticle-based therapies and applications in clinical settings.

6. Case studies and experimental models

6.1 Cardiovascular toxicity in animal studies

Cardiovascular toxicity studies in animal models are essential for assessing the potential risks associated with nanoparticle exposure. These studies involve administering nanoparticles to animals and observing their effects on the cardiovascular system. By doing so, researchers gain valuable insights into nanoparticle behavior, biodistribution, and toxicity, helping identify potential cardiovascular risks and mechanisms of toxicity. Researchers carefully select animal models based on their relevance to human physiology and the specific objectives of the study. Commonly used animals include rodents, such as mice and rats, which are readily available and easy to handle. Larger animals like rabbits or non-human primates may also be used to better mimic certain human responses [88–90]. Nanoparticles are administered to the animals using different routes, such as intravenous injection, oral gavage, inhalation, or dermal application, depending on the intended exposure scenario in humans. The chosen route of administration affects the biodistribution and accumulation of nanoparticles in various organs and tissues, including the cardiovascular system. Animal studies allow researchers to assess cardiovascular function through techniques like echocardiography, electrocardiography (ECG), blood pressure monitoring, and histological analysis. These assessments help researchers understand how nanoparticles may impact cardiac structure, contractility, and vascular function.

One important aspect of cardiovascular toxicity studies is the examination of endothelial function. Endothelial dysfunction is a significant factor in cardiovascular diseases, and studying how nanoparticles affect endothelial cells lining blood vessels provides critical insights into potential cardiovascular risks [91–93]. Inflammatory and oxidative stress responses induced by nanoparticles in the cardiovascular system are also explored. Increased oxidative stress and inflammation are associated with cardiovascular pathology, and understanding how nanoparticles influence these processes is crucial. Dose-response relationships are established in animal studies to understand how different nanoparticle doses impact cardiovascular outcomes. This information helps set safe exposure limits and determine appropriate dosages for nanoparticle-based therapies. Some studies involve prolonged exposure to nanoparticles to assess chronic effects and potential cumulative toxicity over time. This long-term evaluation is essential to understand the safety profile of nanoparticles used in long-lasting medical interventions. Animal studies also reveal species-specific variability in nanoparticle responses. Comparing results across species helps identify potential similarities and differences in nanoparticle behavior, aiding in the extrapolation of findings to humans.

While animal studies provide valuable data, they have limitations. Extrapolating animal data to humans requires considering potential differences in metabolism, clearance pathways, and cardiovascular physiology between species. Researchers often complement their findings with in vitro experiments using human cells or tissues and conduct human clinical trials and observational studies for more direct relevance to human health outcomes. By integrating findings from animal studies with other approaches, researchers gain a comprehensive understanding of nanoparticle cardiovascular toxicity [94,95]. This knowledge informs the responsible development and application of nanoparticle-based interventions in cardiovascular medicine, ultimately enhancing patient safety and healthcare outcomes.

6.2   Human clinical trials and observational studies

Human clinical trials and observational studies are crucial components in the assessment of nanoparticle cardiovascular toxicity [96–98]. These studies provide valuable insights into how nanoparticles behave in the human body and their potential impact on the cardiovascular system. Here's a detailed elaboration of human clinical trials and observational studies:

- Human Clinical Trials: Human clinical trials involve controlled experiments conducted with human participants to evaluate the safety, efficacy, and tolerability of nanoparticle-based interventions. These trials are typically conducted in several phases, starting with small-scale Phase I trials to establish the safety profile of the nanoparticles in healthy volunteers or patients with specific cardiovascular conditions.

a. Phase I Trials: In Phase I trials, a small group of participants receives a low dose of nanoparticles to assess their safety, pharmacokinetics (how the body processes the nanoparticles), and biodistribution.

b. Phase II Trials: If Phase I trials indicate safety, Phase II trials involve a larger group of participants to assess the nanoparticle's effectiveness in treating specific cardiovascular conditions.

c. Phase III Trials: Phase III trials involve a more extensive and diverse patient population to further evaluate the nanoparticles' efficacy, safety, and side effects on a larger scale.

- Observational Studies: Observational studies involve the analysis of existing data from human populations exposed to nanoparticles, either intentionally or unintentionally. These studies are retrospective and do not involve experimental interventions. Instead, researchers analyze data from individuals who have already been exposed to nanoparticles through occupational exposure, environmental exposure, or medical treatments.

a. Cohort Studies: In cohort studies, researchers follow a group of individuals with known exposure to nanoparticles and track their cardiovascular outcomes over time. This approach helps identify associations between nanoparticle exposure and specific cardiovascular effects.

b. Case-Control Studies: Case-control studies compare individuals with a particular cardiovascular outcome (cases) to those without the outcome (controls) and retrospectively analyze their nanoparticle exposure history. This study design helps establish potential links between nanoparticle exposure and cardiovascular outcomes.

c. Cross-Sectional Studies: Cross-sectional studies assess both nanoparticle exposure and cardiovascular outcomes simultaneously in a single population at a specific point in time. These studies provide a snapshot of the relationship between exposure and outcomes.

d. Nested Case-Control Studies: Nested case-control studies are conducted within existing cohort studies. Researchers select cases and controls from the cohort and investigate their nanoparticle exposure history to study specific cardiovascular outcomes.

- Advantages and Limitations: Human clinical trials and observational studies offer several advantages in understanding nanoparticle cardiovascular toxicity. They provide direct insights into human responses and help validate findings from animal studies and in vitro experiments. They also aid in determining appropriate dosages, understanding long-term effects, and identifying potential cardiovascular risks associated with nanoparticle-based interventions. However, these studies also have limitations[99–101]. Human clinical trials can be time-consuming, and costly, and may require extensive ethical considerations. Observational studies may be subject to confounding factors and biases due to the lack of experimental control. By integrating findings from human clinical trials and observational studies with data from animal models and in vitro experiments, researchers can obtain a comprehensive understanding of nanoparticle cardiovascular toxicity. This multidisciplinary approach informs the responsible development and safe implementation of nanoparticle-based therapies and applications in cardiovascular medicine, ultimately improving patient outcomes and safety.

6.3　Comparative analysis of different nanoparticles

Comparative analysis of different nanoparticles is a critical aspect of studying cardiovascular toxicity and assessing the potential risks and benefits associated with their use. This analysis involves systematically comparing various types of nanoparticles, considering their size, shape, surface properties, composition, and intended applications. Here's a detailed elaboration of the comparative analysis of different nanoparticles:

- Size and Shape: Nanoparticles come in various sizes and shapes, such as spheres, rods, tubes, and disks. Comparative analysis allows researchers to evaluate how these different geometries influence nanoparticle behavior, interactions with biological entities, and cardiovascular responses. For example, certain shapes may enhance cellular uptake or alter biodistribution patterns, affecting their potential cardiovascular toxicity.

- Surface Properties: The surface of nanoparticles can be modified with coatings or functional groups to improve biocompatibility, stability, and targeting capabilities. Comparative analysis helps assess how surface modifications influence the nanoparticle's interactions with the cardiovascular system and whether they reduce or exacerbate potential adverse effects [102–104].

- Composition: Nanoparticles can be made from various materials, such as metals, metal oxides, polymers, lipids, and carbon-based materials. Each material type has unique physicochemical properties that can impact its cardiovascular behavior and toxicity. Comparative analysis allows researchers to identify materials with more favorable safety profiles for specific applications.

- Biodegradability: Biodegradable nanoparticles can be broken down into harmless by-products over time, reducing the risk of long-term accumulation and toxicity. Comparative analysis helps identify biodegradable nanoparticles that may be safer for drug delivery and other medical applications.

- Targeting Capabilities: Some nanoparticles can be designed to target specific cells or tissues in the cardiovascular system, offering potential therapeutic benefits with reduced off-target effects. The comparative analysis evaluates the targeting efficiency and potential benefits of targeted nanoparticles compared to non-targeted ones.

- Cellular and Tissue Interactions: By comparing the interactions of different nanoparticles with cardiovascular cells and tissues, researchers can gain insights into their cellular uptake mechanisms, intracellular trafficking, and potential effects on cell function and tissue integrity.

- Pharmacokinetics: Comparative analysis helps assess the pharmacokinetics of different nanoparticles, including their absorption, distribution, metabolism, and excretion. Understanding how nanoparticles are processed in the body provides critical information about potential cardiovascular risks.

- Toxicity Profiles: Evaluating the toxicity profiles of different nanoparticles allows researchers to identify those with lower cardiovascular toxicity while retaining their desired functionalities for medical applications.

- Structure-Activity Relationships: Comparative analysis aids in establishing structure-activity relationships, linking specific nanoparticle characteristics to their biological effects. This knowledge can guide the rational design of nanoparticles with improved biocompatibility and reduced cardiovascular risks.

- Application-Specific Considerations: Comparative analysis considers the specific applications of nanoparticles, such as drug delivery, imaging, or biosensing. For each application, researchers weigh the benefits against potential cardiovascular risks to determine the overall suitability of different nanoparticles.

By conducting a comparative analysis, researchers can make informed decisions about which nanoparticles are more suitable for specific applications and how to optimize their design to minimize potential cardiovascular toxicity. This knowledge enhances the responsible development and safe implementation of nanoparticle-based therapies and technologies in cardiovascular medicine, contributing to improved patient outcomes and healthcare practices.

7.   Safety guidelines and regulation

7.1   Current regulatory framework for nanoparticles

The current regulatory framework for nanoparticles is a complex and evolving landscape that varies by country and region. As of my last update in September 2021, there were no specific regulations exclusively dedicated to nanoparticles [105–107]. Instead, nanoparticles are generally regulated under existing frameworks that govern chemicals, drugs, medical devices, and consumer products. Here's a detailed elaboration of the regulatory landscape for nanoparticles:

- Medical Applications: For nanoparticles used in medical applications, such as drug delivery systems or imaging agents, regulatory approval typically follows a stepwise process. Preclinical studies are conducted in animal models to assess the safety and efficacy of nanoparticle-based therapy. Data from these studies are submitted to regulatory agencies, such as the US Food and Drug Administration (FDA) in the United States or the European Medicines Agency (EMA) in Europe, for evaluation. If the preclinical data show promising results and the therapy is deemed safe and effective, the nanoparticle-based therapy can proceed to human clinical trials. These trials are conducted in phases (Phase I, II, and III) involving human participants to further assess safety, efficacy, and optimal dosage. Regulatory approval for medical use is granted if the clinical trial data demonstrate that the benefits outweigh the risks.

- Consumer Products and Industrial Applications: For nanoparticles used in consumer products or industrial applications, the regulatory landscape is less specific. In many countries, including the US and EU, the safety of consumer products containing nanoparticles is assessed through risk-based evaluations. Manufacturers may be required to provide safety data and conduct risk assessments before commercializing nanoparticle-containing products. Regulatory agencies, such as the US Environmental Protection Agency (EPA) and the European Chemicals Agency (ECHA), play a role in assessing the environmental and health risks associated with nanoparticles used in industrial applications. These agencies may require manufacturers to submit data on nanoparticle properties, potential exposure scenarios, and toxicity information to ensure safe handling and disposal.

- Challenges and Future Developments: The regulatory framework for nanoparticles faces several challenges and opportunities for improvement:

a. Lack of Standardization: Currently, there is no standardized testing protocol for evaluating nanoparticle safety across different applications, leading to variability in study designs and outcomes.

b. Definition and Classification: The lack of a universally accepted definition and classification of nanoparticles can complicate regulatory efforts to cover all relevant materials.

c. Long-Term Effects: The long-term health effects of nanoparticle exposure are still being studied, and more research is needed to understand the potential risks over extended periods.

d. Rapid Technological Advancements: The rapidly evolving field of nanotechnology necessitates continuous updates to regulations to keep up with new materials and applications.

e. Global Harmonization: Achieving global harmonization in nanoparticle regulation is challenging due to differing approaches in different regions and countries [108–110].

As nanotechnology continues to advance, regulatory authorities worldwide are working to update and adapt their frameworks to address the unique challenges posed by nanoparticles. There is a growing recognition of the importance of international collaboration and harmonization of standards to ensure the safe and responsible use of nanoparticles in various applications.

7.2   Challenges in assessing cardiovascular toxicity

Assessing cardiovascular toxicity associated with nanoparticles presents several challenges due to the complex interactions between nanoparticles and the cardiovascular system. These challenges arise from the unique characteristics of nanoparticles and the intricate physiology of the cardiovascular system [111,112]. Here's a detailed elaboration of the challenges in assessing cardiovascular toxicity:

- Nanoparticle Properties: Nanoparticles exhibit unique physicochemical properties, such as size, shape, surface charge, and composition, which can influence their behavior in biological systems. Understanding how these properties affect nanoparticle interactions with cardiovascular cells and tissues is essential but challenging.

- Biodistribution and Clearance: Nanoparticles can rapidly distribute throughout the body and accumulate in various organs, including the cardiovascular system. Assessing the biodistribution and clearance pathways of nanoparticles requires sophisticated techniques and may vary depending on the nanoparticle's characteristics.

- Nanoparticle-Cell Interactions: The interactions between nanoparticles and cardiovascular cells are complex and can lead to a range of responses, including cellular uptake, inflammation, oxidative stress, and apoptosis. Elucidating these interactions and their consequences demands advanced cellular and molecular techniques.

- Long-Term Effects: Cardiovascular toxicity assessments need to consider the potential long-term effects of nanoparticle exposure. Prolonged exposure or chronic toxicity may manifest differently than acute effects and may require extended study periods, making long-term evaluations challenging.

- Species-Specific Variability: Nanoparticle behavior and toxicity can vary among different animal species and humans due to differences in metabolism, immune responses, and physiology. Extrapolating findings from animal studies to humans requires careful consideration of these species-specific differences [113,114].

- Lack of Standardized Testing Protocols: There is a lack of standardized testing protocols specifically designed for assessing nanoparticle cardiovascular toxicity. This can lead to variability in study designs and outcomes, making it difficult to compare results across different studies.

- Complex Cardiovascular System: The cardiovascular system is highly intricate, consisting of various cell types and tissues with diverse functions. Understanding how nanoparticles interact with different components of the cardiovascular system presents challenges in experimental design and data interpretation.

- Mechanistic Understanding: While progress has been made in understanding some mechanisms of nanoparticle-induced cardiovascular toxicity, much remains to be elucidated. The intricate cellular and molecular pathways involved in nanoparticle toxicity demand further investigation.

- In Vitro vs. In Vivo Models: Researchers use both in vitro cell culture models and in vivo animal models to study nanoparticle toxicity. Bridging the gap between these two approaches and accurately translating findings from in vitro studies to in vivo responses is complex and requires careful validation.

- Real-World Exposure Scenarios: Assessing cardiovascular toxicity in real-world exposure scenarios, such as occupational settings or environmental exposure, introduces additional challenges due to the variability of exposure levels and exposure routes.

Addressing these challenges requires interdisciplinary collaboration between researchers, clinicians, toxicologists, and regulatory authorities. Advances in technology, standardization of testing protocols, and improvements in in vitro and in vivo models will play pivotal roles in overcoming these obstacles. A comprehensive understanding of the cardiovascular toxicity of nanoparticles is essential to guide the safe and responsible use of nanotechnology in various applications, including medicine, consumer products, and environmental remediation.

7.3   Recommendations for safer nanotechnology applications

To promote safer nanotechnology applications, several recommendations have been put forward to address the potential risks associated with nanoparticles. These recommendations aim to ensure the responsible development and use of nanotechnology while maximizing its benefits and minimizing potential harm [115–118]. Here's a detailed elaboration of the recommendations for safer nanotechnology applications:

- Standardized Testing Protocols: Developing standardized protocols for assessing nanoparticle safety is crucial. These protocols should consider various nanoparticle characteristics, exposure scenarios, and endpoints for toxicity assessment. Standardization will facilitate more consistent and comparable results across different studies, allowing for more reliable risk assessments.

- Improved In Vitro Models: Advancements in in vitro models using human-derived cells and tissues can provide more relevant data for predicting human responses to nanoparticles. Robust and validated in vitro models will reduce the reliance on animal studies and enable better assessment of nanoparticle toxicity, especially in the context of the cardiovascular system.

- Better Understanding of Mechanisms: In-depth research on the mechanisms of nanoparticle-induced toxicity is essential [119,120]. Understanding the cellular and molecular pathways involved in cardiovascular toxicity will help identify potential biomarkers and therapeutic targets for intervention.

- Multidisciplinary Collaboration: Collaboration between researchers, regulators, industries, and healthcare professionals is crucial to ensure the exchange of knowledge and data. Multidisciplinary approaches foster informed decision-making and a more comprehensive understanding of nanoparticle safety.

- Risk-Based Assessments: Implementing risk-based assessments for nanoparticles in various applications will help identify potential hazards and prioritize safety measures accordingly. Assessing risk throughout the lifecycle of nanoparticles, from production to disposal, ensures comprehensive safety evaluations.

- Long-Term Monitoring: Long-term monitoring of individuals exposed to nanoparticles, especially in medical applications, will provide valuable data on chronic effects and long-term safety profiles. This information is essential for understanding potential risks associated with prolonged nanoparticle exposure.

- Public Awareness and Education: Public awareness and education about the benefits and potential risks of nanotechnology are crucial. Educating the public, healthcare professionals, and stakeholders will foster responsible use and support evidence-based regulations.

- Ethical Considerations: Ethical considerations in nanotechnology research and applications are paramount. Ensuring transparency, informed consent, and ethical practices in research involving nanoparticles will enhance trust in the field and protect human rights.

- Global Harmonization: Achieving global harmonization in nanoparticle regulations is challenging but important. Collaboration among different countries and regions can lead to consistent and internationally accepted safety guidelines for nanoparticles.

- Continuous Monitoring and Updating: As nanotechnology continues to advance, safety guidelines and regulations must keep pace with new developments. Continuous monitoring of nanotechnology applications and updating of regulations based on emerging data and knowledge is essential to ensure ongoing safety.

By integrating these recommendations into the development, production, and regulation of nanoparticles, we can promote safer nanotechnology applications across various industries. Emphasizing safety and responsible use will enable the realization of nanotechnology's potential benefits while mitigating potential risks to human health, the environment, and society as a whole.

8. Nanomedicine and future perspectives

8.1 Targeted drug delivery using nanoparticles

Targeted drug delivery using nanoparticles is a cutting-edge approach in nanomedicine that holds tremendous promise for revolutionizing drug therapies. Traditional drug delivery methods often suffer from limitations, such as off-target effects and inadequate drug concentrations at the intended site of action [121–123]. Nanoparticles offer a solution to these challenges by acting as carriers for therapeutic agents, precisely delivering drugs to specific cells or tissues in the body. By engineering nanoparticles with suitable surface modifications, such as ligands or antibodies, they can be made to recognize and bind to specific receptors or biomarkers on the target cells. This active targeting strategy allows nanoparticles to bypass biological barriers, penetrate tissues, and accumulate at the site of disease, increasing drug concentration at the desired location. Furthermore, nanoparticles can also passively accumulate in certain diseased tissues due to the enhanced permeability and retention (EPR) effect, further enhancing their targeting capabilities. Once at the target site, nanoparticles can release their payload in a controlled and sustained manner, ensuring a continuous supply of drugs to the affected area. This approach not only enhances the therapeutic efficacy of drugs but also minimizes systemic toxicity, reducing side effects in healthy tissues. Targeted drug delivery using nanoparticles has shown promising results in various medical fields, including cancer therapy, infectious diseases, and inflammatory disorders. As research in nanotechnology advances, the design of more sophisticated and versatile nanoparticles with improved targeting strategies continues to evolve [124–126]. The future of targeted drug delivery lies in personalized medicine, where nanoparticles can be tailored to individual patients based on their unique biological profiles, leading to more precise and effective treatments. Overall, the development of targeted drug delivery using nanoparticles represents a significant milestone in healthcare, bringing us closer to the vision of personalized and precision medicine.

8.2 Nanoparticle-based theranostics

Nanoparticle-based theranostics is an innovative and multifaceted approach in the field of nanomedicine that combines therapy and diagnostics into a single integrated platform [127,128]. Theranostic nanoparticles serve a dual purpose by not only delivering therapeutic agents but also acting as imaging agents to monitor treatment responses in real time. These nanoparticles are engineered with precision to carry drugs and imaging agents concurrently, allowing clinicians to simultaneously visualize the target site and administer the therapeutic payload. The ability to monitor the therapeutic efficacy in real-time enables clinicians to tailor treatment regimens to individual patients, ensuring a more personalized and precise approach to healthcare [129,130]. This theranostic approach holds great promise in various medical applications, especially in oncology, where it allows for early detection of tumors, accurate tumor localization, and effective delivery of chemotherapeutic agents directly to cancerous cells. Theranostic nanoparticles can also be utilized in other areas, such as infectious diseases, cardiovascular disorders, and neurological conditions. The future of nanoparticle-based theranostics lies in the integration of emerging technologies, such as nanosensors and artificial intelligence, to enhance treatment monitoring and prediction of patient responses. By advancing this field, nanomedicine is paving the way for improved patient outcomes and a paradigm shift towards personalized and precision medicine.

8.3 Cardiovascular applications of nanomedicine

Cardiovascular applications of nanomedicine have emerged as a promising frontier in the battle against cardiovascular diseases, which remain a leading cause of morbidity and mortality worldwide. Nanomedicine harnesses the unique properties of nanoparticles to address various aspects of cardiovascular health, ranging from diagnostics to therapeutics [131–134]. Nanoparticle-based imaging agents offer enhanced visualization of cardiovascular structures and functions, enabling early detection and precise diagnosis of cardiovascular pathologies. Moreover, nanotechnology has revolutionized drug delivery by allowing targeted delivery of therapeutic agents to specific sites in the cardiovascular system. Targeted drug delivery enhances therapeutic efficacy, reduces off-target effects, and enables sustained drug release, all critical factors in treating cardiovascular diseases like atherosclerosis, hypertension, and myocardial infarction. Additionally, nanotechnology has facilitated the development of nanoscale devices and sensors that can monitor cardiovascular health in real-time, providing valuable insights into physiological parameters and allowing for timely interventions. Nanoparticles also play a pivotal role in enhancing stem cell therapy, supporting tissue repair and regeneration in damaged cardiovascular tissues. By promoting stem cell survival and homing, nanoparticles improve the potential of stem cells to contribute to cardiovascular repair. As nanomedicine continues to evolve, there is optimism that novel therapies and diagnostic tools will emerge, contributing to improved patient outcomes and a more comprehensive approach to managing cardiovascular diseases. The integration of nanotechnology with regenerative medicine and tissue engineering holds the promise of restoring cardiac function and ameliorating cardiovascular conditions [135,136]. Thus, cardiovascular applications of nanomedicine have the potential to transform the landscape of cardiovascular medicine, advancing patient care towards more personalized, precise, and effective strategies.

Conclusion

In conclusion, nanomedicine represents a transformative field that leverages the unique properties of nanoparticles to revolutionize various aspects of healthcare. Targeted drug delivery using nanoparticles offers a sophisticated approach to improving the efficacy of drug therapies while minimizing side effects, thus bringing us closer to the vision of personalized medicine. Nanoparticle-based theranostics, with its ability to integrate diagnostics and therapy, opens up new possibilities for early detection, precise treatment, and real-time monitoring of treatment responses. This approach has the potential to significantly impact patient outcomes, particularly in diseases like cancer and other complex conditions. Moreover, the cardiovascular applications of nanomedicine hold great promise for combating cardiovascular diseases, which remain a significant global health burden. Nanoparticle-based imaging agents enable advanced visualization of the cardiovascular system, leading to early disease detection and precise diagnosis. The development of nanoscale devices and sensors for real-time monitoring of cardiovascular health offers the potential for more proactive and personalized patient care. In the realm of cardiovascular therapeutics, targeted drug delivery using nanoparticles provides hope for more effective and safer treatment options. By delivering drugs directly to the affected tissues, nanoparticles enhance therapeutic outcomes and reduce systemic toxicity, advancing the management of cardiovascular diseases like atherosclerosis, hypertension, and myocardial infarction. Additionally, nanotechnology's ability to support stem cell therapy opens new avenues for cardiac repair and regeneration, providing potential breakthroughs in treating heart diseases and restoring cardiac function.

As nanomedicine progresses, continuous research, innovation, and collaboration among researchers, clinicians, regulatory authorities, and industries are essential to address challenges and optimize the potential of nanotechnology in healthcare. Standardized testing protocols, improved in vitro models, and a deeper understanding of nanoparticle interactions with biological systems are critical for safety assessments and responsible development of nanomedicine. While nanomedicine holds significant promise, it also calls for responsible and ethical use to ensure that potential risks and environmental impacts are carefully considered. Public awareness and education about nanotechnology's benefits and potential risks are essential to foster informed decision-making and acceptance among patients and the public. Looking to the future, the integration of nanotechnology with emerging technologies, such as artificial intelligence and nanosensors, offers exciting opportunities for advancing personalized and precision medicine further. These synergistic approaches have the potential to optimize treatment regimens, predict patient responses, and drive more tailored healthcare strategies. In conclusion, nanomedicine has the power to transform healthcare and improve patient outcomes across a wide range of diseases and conditions. The convergence of nanotechnology with medicine brings us closer to a future where personalized and precise therapies are the norm, leading to a better quality of life for patients worldwide. By harnessing the remarkable potential of nanomedicine while maintaining a commitment to safety and responsible practices, we can unlock a new era of healthcare advancements that will shape the future of medicine. The journey of nanomedicine continues, filled with possibilities and hope, as we work together to unlock its full potential for the betterment of human health and well-being.

Reference

[1] S. Rizal, H.P.S. Abdul Khalil, A.A. Oyekanmi, N.G. Olaiya, C.K. Abdullah, E.B. Yahya, T. Alfatah, F.A. Sabaruddin, A.A. Rahman, Cotton wastes functionalized biomaterials from micro to nano: A cleaner approach for a sustainable environmental application, Polymers (Basel). 13 (2021) 1–36. https://doi.org/10.3390/polym13071006

[2] I. Safarik, K. Horska, K. Pospiskova, M. Safarikova, Magnetically Responsive Activated Carbons for Bio - and Environmental Applications, Int. Rev. Chem. Eng. 4 (2012) 346–352

[3] M.. Tadda, A. Ahsan, A. Shifu, M. ElSergany, T. Arunkumar, B. Jose, M. Razzaque, Abdur, N.. Daud, Nik, A Review on Activated Carbon from Biowaste : Process , Application and Prospects, J. Adv. Civ. Eng. Pract. Res. 5 (2018) 82–83

[4] Y. Wang, J. Wang, X. Zhang, D. Bhattacharyya, E.M. Sabolsky, Quantifying Environmental and Economic Impacts of Highly Porous Activated Carbon from Lignocellulosic Biomass for High-Performance Supercapacitors, Energies. 15 (2022). https://doi.org/10.3390/en15010351

[5] M. Bernardo, N. Lapa, I. Matos, I. Fonseca, Critical discussion on activated carbons from bio-wastes - environmental risk assessment, Bol. Del Grup. Español Del Carbón. 40 (2011) 18–21. http://www.gecarbon.org/Boletines/Boletin/BoletinGEC_040.pdf

[6] S. Lahreche, I. Moulefera, A. El Kebir, L. Sabantina, M. Kaid, A. Benyoucef, Application of Activated Carbon Adsorbents Prepared from Prickly Pear Fruit Seeds and a Conductive Polymer Matrix to Remove Congo Red from Aqueous Solutions, Fibers. 10 (2022) 1–19. https://doi.org/10.3390/fib10010007

[7] N.D. Mu'azu, N. Jarrah, M. Zubair, O. Alagha, Removal of phenolic compounds from water using sewage sludge-based activated carbon adsorption: A review, Int. J. Environ. Res. Public Health. 14 (2017) 1–34. https://doi.org/10.3390/ijerph14101094

[8] L. Pan, Y. Cao, J. Zang, Q. Huang, L. Wang, Y. Zhang, S. Fan, J. Tang, Z. Xie, Preparation of iron-loaded granular activated carbon catalyst and its application in tetracycline antibiotic removal from aqueous solution, Int. J. Environ. Res. Public Health. 16 (2019) 5–7. https://doi.org/10.3390/ijerph16132270

[9] K. Shahane, M. Kshirsagar, S. Tambe, D. Jain, S. Rout, M.K.M. Ferreira, S. Mali, P. Amin, P.P. Srivastav, J. Cruz, R.R. Lima, An Updated Review on the Multifaceted Therapeutic Potential of Calendula officinalis L., Pharmaceuticals. 16 (2023) 611. https://doi.org/10.3390/ph16040611

[10] N.J. Riungu, M. Hesampour, A. Pihlajamaki, M. Manttari, H. Sirén, P.G. Home, G.M. Ndegwa, Investigating Removal of Pesticides From Water By Nanofiltration Membrane Technology, J. Eng. Comput. Appl. Sci. 1 (2012) 50–60

[11] H. Yu, Y.R. Son, H. Yoo, H.G. Cha, H. Lee, II.S. Kim, Chitosan-derived porous activated carbon for the removal of the chemical warfare agent simulant dimethyl methylphosphonate, Nanomaterials. 9 (2019). https://doi.org/10.3390/nano9121703

[12] J.A. Damasco, S. Ravi, J.D. Perez, D.E. Hagaman, M.P. Melancon, Understanding nanoparticle toxicity to direct a safe-by-design approach in cancer nanomedicine, Nanomaterials. 10 (2020) 1–41. https://doi.org/10.3390/nano10112186

[13] H. Soonmin, N.A. Kabbashi, Review on activated carbon: Synthesis, properties and applications, Int. J. Eng. Trends Technol. 69 (2021) 124–139.

https://doi.org/10.14445/22315381/IJETT-V69I9P216

[14] S. Muzammil, J. Neves Cruz, R. Mumtaz, I. Rasul, S. Hayat, M.A. Khan, A.M. Khan, M.U. Ijaz, R.R. Lima, M. Zubair, Effects of Drying Temperature and Solvents on In Vitro Diabetic Wound Healing Potential of Moringa oleifera Leaf Extracts, Molecules. 28 (2023) 710. https://doi.org/10.3390/molecules28020710

[15] S. Revel-Vilk, G. Chodick, V. Shalev, N. Gadir, Study Design: Development of an Advanced Machine Learning Algorithm for the Early Diagnosis of Gaucher Disease Using Real-World Data , Blood. 136 (2020) 13–14. https://doi.org/10.1182/blood-2020-134414

[16] A. Souri, M.Y. Ghafour, A.M. Ahmed, F. Safara, A. Yamini, M. Hoseyninezhad, A new machine learning-based healthcare monitoring model for student's condition diagnosis in Internet of Things environment, Soft Comput. 24 (2020) 17111–17121. https://doi.org/10.1007/s00500-020-05003-6

[17] P. Gard, L. Lalanne, A. Ambourg, D. Rousseau, F. Lesueur, C. Frindel, A Secured Smartphone-Based Architecture for Prolonged Monitoring of Neurological Gait, 2018. https://doi.org/10.1007/978-3-319-76213-5_1

[18] B. Al-Salemi, S.A. Mohd Noah, M.J. Ab Aziz, RFBoost: An improved multi-label boosting algorithm and its application to text categorisation, Knowledge-Based Syst. 103 (2016) 104–117. https://doi.org/10.1016/j.knosys.2016.03.029

[19] T. Ha, Y. Jung, J.Y. Kim, S.Y. Park, D.K. Kang, T.H. Kim, Comparison of the diagnostic performance of abbreviated MRI and full diagnostic MRI using a computer-aided diagnosis (CAD) system in patients with a personal history of breast cancer: the effect of CAD-generated kinetic features on reader performance, Clin. Radiol. 74 (2019) 817.e15-817.e21. https://doi.org/10.1016/j.crad.2019.06.025

[20] P. Chavan, A.K. Singh, G. Kaur, Recent progress in the utilization of industrial waste and by-products of citrus fruits: A review, J. Food Process Eng. 41 (2018). https://doi.org/10.1111/jfpe.12895

[21] H.S. Ng, P.E. Kee, H.S. Yim, P.T. Chen, Y.H. Wei, J. Chi-Wei Lan, Recent advances on the sustainable approaches for conversion and reutilization of food wastes to valuable bioproducts, Bioresour. Technol. 302 (2020) 122889. https://doi.org/10.1016/j.biortech.2020.122889

[22] S.K. Pramanik, F.B. Suja, S.M. Zain, B.K. Pramanik, The anaerobic digestion process of biogas production from food waste: Prospects and constraints, Bioresour. Technol. Reports. 8 (2019) 100310. https://doi.org/10.1016/j.biteb.2019.100310

[23] R.I. Kosheleva, A.C. Mitropoulos, G.Z. Kyzas, Synthesis of activated carbon from food waste, Environ. Chem. Lett. 17 (2019) 429–438. https://doi.org/10.1007/s10311-018-0817-5

[24] G. Rekleitis, K.J. Haralambous, M. Loizidou, K. Aravossis, Utilization of agricultural and livestock waste in anaerobic digestion (A.D): Applying the biorefinery concept in a circular economy, Energies. 13 (2020). https://doi.org/10.3390/en13174428

[25] S. Dahiya, A.N. Kumar, J. Shanthi Sravan, S. Chatterjee, O. Sarkar, S.V. Mohan, Food waste biorefinery: Sustainable strategy for circular bioeconomy, Elsevier Ltd, 2018. https://doi.org/10.1016/j.biortech.2017.07.176

[26] W. Peng, A. Pivato, Sustainable Management of Digestate from the Organic Fraction of Municipal Solid Waste and Food Waste Under the Concepts of Back to Earth Alternatives and

Circular Economy, Waste and Biomass Valorization. 10 (2019) 465–481.
https://doi.org/10.1007/s12649-017-0071-2

[27]   M.L. Carmo Bastos, J.V. Silva-Silva, J. Neves Cruz, A.R. Palheta da Silva, A.A.
Bentaberry-Rosa, G. da Costa Ramos, J.E. de Sousa Siqueira, M.R. Coelho-Ferreira, S. Percário,
P. Santana Barbosa Marinho, A.M. do R. Marinho, M. de Oliveira Bahia, M.F. Dolabela,
Alkaloid from Geissospermum sericeum Benth. & Hook.f. ex Miers (Apocynaceae) Induce
Apoptosis by Caspase Pathway in Human Gastric Cancer Cells, Pharmaceuticals. 16 (2023) 765.
https://doi.org/10.3390/ph16050765

[28]   E.M. Meemken, M. Qaim, Organic Agriculture, Food Security, and the Environment,
Annu. Rev. Resour. Econ. 10 (2018) 39–63. https://doi.org/10.1146/annurev-resource-100517-
023252

[29]   K. Spalvins, K. Ivanovs, D. Blumberga, Single cell protein production from waste
biomass: Review of various agricultural by-products, Agron. Res. 16 (2018) 1493–1508.
https://doi.org/10.15159/AR.18.129

[30]   J. Kim, S. Rundle-Thiele, K. Knox, Systematic literature review of best practice in food
waste reduction programs, J. Soc. Mark. 9 (2019) 447–466. https://doi.org/10.1108/JSOCM-05-
2019-0074

[31]   Y. Li, Y. Jin, A. Borrion, H. Li, Current status of food waste generation and management
in China, Bioresour. Technol. 273 (2019) 654–665.
https://doi.org/10.1016/j.biortech.2018.10.083

[32]   M.B. Baig, K.H. Al-Zahrani, F. Schneider, G.S. Straquadine, M. Mourad, Food waste
posing a serious threat to sustainability in the Kingdom of Saudi Arabia – A systematic review,
Saudi J. Biol. Sci. 26 (2019) 1743–1752. https://doi.org/10.1016/j.sjbs.2018.06.004

[33]   O. Obsa, M. Tadesse, D.G. Kim, Z. Asaye, F. Yimer, M. Gebrehiwot, N. Brüggemann, K.
Prost, Organic Waste Generation and Its Valorization Potential through Composting in
Shashemene, Southern Ethiopia, Sustain. 14 (2022) 1–19. https://doi.org/10.3390/su14063660

[34]   N. Bhargava, V.S. Sharanagat, R.S. Mor, K. Kumar, Active and intelligent biodegradable
packaging films using food and food waste-derived bioactive compounds: A review, Trends
Food Sci. Technol. 105 (2020) 385–401. https://doi.org/10.1016/j.tifs.2020.09.015

[35]   C.M. Galanakis, M. Rizou, T.M.S. Aldawoud, I. Ucak, N.J. Rowan, Innovations and
technology disruptions in the food sector within the COVID-19 pandemic and post-lockdown
era, Trends Food Sci. Technol. 110 (2021) 193–200. https://doi.org/10.1016/j.tifs.2021.02.002

[36]   M. Bilal, H.M.N. Iqbal, Sustainable bioconversion of food waste into high-value products
by immobilized enzymes to meet bio-economy challenges and opportunities – A review, Food
Res. Int. 123 (2019) 226–240. https://doi.org/10.1016/j.foodres.2019.04.066

[37]   S.K. Awasthi, S. Sarsaiya, M.K. Awasthi, T. Liu, J. Zhao, S. Kumar, Z. Zhang, Changes
in global trends in food waste composting: Research challenges and opportunities, Bioresour.
Technol. 299 (2020) 122555. https://doi.org/10.1016/j.biortech.2019.122555

[38]   M. Kumar, X. Xiong, M. He, D.C.W. Tsang, J. Gupta, E. Khan, S. Harrad, D. Hou, Y.S.
Ok, N.S. Bolan, Microplastics as pollutants in agricultural soils, Environ. Pollut. 265 (2020).
https://doi.org/10.1016/j.envpol.2020.114980

[39]   B. Sharma, B. Vaish, Monika, U.K. Singh, P. Singh, R.P. Singh, Recycling of Organic
Wastes in Agriculture: An Environmental Perspective, Int. J. Environ. Res. 13 (2019) 409–429.

https://doi.org/10.1007/s41742-019-00175-y

[40]　A. Thakur, S. Kaya, A.S. Abousalem, S. Sharma, R. Ganjoo, H. Assad, A. Kumar, Computational and experimental studies on the corrosion inhibition performance of an aerial extract of Cnicus Benedictus weed on the acidic corrosion of mild steel, Process Saf. Environ. Prot. 161 (2022) 801–818. https://doi.org/10.1016/j.psep.2022.03.082

[41]　A. Thakur, S. Kaya, A.S. Abousalem, A. Kumar, Experimental, DFT and MC simulation analysis of Vicia Sativa weed aerial extract as sustainable and eco-benign corrosion inhibitor for mild steel in acidic environment, Sustain. Chem. Pharm. 29 (2022) 100785. https://doi.org/10.1016/j.scp.2022.100785

[42]　C. Dhonchak, N. Agnihotri, A. Kumar, R. Kamal, A. Thakur, A. Kumar, Spectrophotometric Investigation and Computational Studies of Zirconium(IV)-3-hydroxy-2-[1'-phenyl-3'-(p-methoxyphenyl)-4'-pyrazolyl]-4H-chromen-4-one Complex, J. Anal. Chem. 78 (2023) 856–865. https://doi.org/10.1134/S1061934823070055

[43]　D. Sharma, A. Thakur, M.K. Sharma, R. Sharma, S. Kumar, A. Sihmar, H. Dahiya, G. Jhaa, A. Kumar, A.K. Sharma, H. Om, Effective corrosion inhibition of mild steel using novel 1,3,4-oxadiazole-pyridine hybrids: Synthesis, electrochemical, morphological, and computational insights, Environ. Res. 234 (2023) 116555. https://doi.org/10.1016/j.envres.2023.116555

[44]　A. Thakur, S. Kaya, A. Kumar, Recent Innovations in Nano Container-Based Self-Healing Coatings in the Construction Industry, Curr. Nanosci. 18 (2021) 203–216. https://doi.org/10.2174/1573413717666210216120741

[45]　S. Kaya, A. Thakur, A. Kumar, The role of in Silico/DFT investigations in analyzing dye molecules for enhanced solar cell efficiency and reduced toxicity, J. Mol. Graph. Model. 124 (2023) 108536. https://doi.org/10.1016/j.jmgm.2023.108536

[46]　A. Thakur, A. Kumar, Sustainable Inhibitors for Corrosion Mitigation in Aggressive Corrosive Media: A Comprehensive Study, J. Bio- Tribo-Corrosion. 7 (2021) 1–48. https://doi.org/10.1007/s40735-021-00501-y

[47]　S. Bashir, A. Thakur, H. Lgaz, I.M. Chung, A. Kumar, Computational and experimental studies on Phenylephrine as anti-corrosion substance of mild steel in acidic medium, J. Mol. Liq. 293 (2019) 111539. https://doi.org/10.1016/j.molliq.2019.111539

[48]　Y. Zhou, N. Engler, M. Nelles, Symbiotic relationship between hydrothermal carbonization technology and anaerobic digestion for food waste in China, Bioresour. Technol. 260 (2018) 404–412. https://doi.org/10.1016/j.biortech.2018.03.102

[49]　C. Zarbà, G. Chinnici, G. La Via, S. Bracco, B. Pecorino, M. D'amico, Regulatory elements on the circular economy: Driving into the agri-food system, Sustain. 13 (2021). https://doi.org/10.3390/su13158350

[50]　A. Thakur, A. Kumar, S. Sharma, R. Ganjoo, H. Assad, Computational and experimental studies on the efficiency of Sonchus arvensis as green corrosion inhibitor for mild steel in 0.5 M HCl solution, Mater. Today Proc. 66 (2022) 609–621. https://doi.org/10.1016/j.matpr.2022.06.479

[51]　A. Thakur, A. Kumar, Recent trends in nanostructured carbon-based electrochemical sensors for the detection and remediation of persistent toxic substances in real-time analysis, Mater. Res. Express. 10 (2023) 034001. https://doi.org/10.1088/2053-1591/acbd1a

[52]   C. Dhonchak, N. Agnihotri, A. Kumar, A. Thakur, A. Kumar, Computational Insights in the Spectrophotometrically Analyzed Niobium (V)-3-Hydroxy-2-(4-methylphenyl)-4H-chromen-4-one Complex using DFT Method, Biointerface Res. Appl. Chem. 13 (2023) 357. https://doi.org/10.33263/BRIAC134.357

[53]   A. Thakur, A. Kumar, R. Zhang, Alcoholic Beverage Purification Applications of Activated Carbon, in: C. Verma, M.A. Quraishi (Eds.), Act. Carbon, The Royal Society of Chemistry, 2023: pp. 152–178. https://doi.org/10.1039/bk9781839169861-00152

[54]   A. Thakur, S. Sharma, R. Ganjoo, H. Assad, A. Kumar, Anti-Corrosive Potential of the Sustainable Corrosion Inhibitors Based on Biomass Waste: A Review on Preceding and Perspective Research, J. Phys. Conf. Ser. 2267 (2022) 012079. https://doi.org/10.1088/1742-6596/2267/1/012079

[55]   A. Kumar, A. Thakur, Overview of the properties, applicability, and recent advancements of some natural products used as potential inhibitors in various corrosive systems, in: Handb. Res. Corros. Sci. Eng., 2023: pp. 275–310. https://doi.org/10.4018/978-1-6684-7689-5.ch010

[56]   A. Thakur, A. Kumar, Recent advances on rapid detection and remediation of environmental pollutants utilizing nanomaterials-based (bio)sensors, Sci. Total Environ. 834 (2022) 155219. https://doi.org/10.1016/j.scitotenv.2022.155219

[57]   G. Parveen, S. Bashir, A. Thakur, S.K. Saha, P. Banerjee, A. Kumar, Experimental and computational studies of imidazolium based ionic liquid 1-methyl- 3-propylimidazolium iodide on mild steel corrosion in acidic solution, Mater. Res. Express. 7 (2019) 016510. https://doi.org/10.1088/2053-1591/ab5c6a

[58]   E. Saborowski, A. Dittes, T. Lindner, T. Lampke, Nickel-aluminum thermal spray coatings as adhesion promoter and susceptor for inductively joined polymer-metal hybrids, Polymers (Basel). 13 (2021). https://doi.org/10.3390/polym13081320

[59]   M. Nur-e-alam, M.K. Basher, M. Vasiliev, N. Das, Physical vapor-deposited silver (Ag)-based metal-dielectric nanocomposites for thin-film and coating applications, Appl. Sci. 11 (2021). https://doi.org/10.3390/app11156746

[60]   S. Bagherifard, S. Slawik, I. Fernández-Pariente, C. Pauly, F. Mücklich, M. Guagliano, Nanoscale surface modification of AISI 316L stainless steel by severe shot peening, Mater. Des. 102 (2016) 68–77. https://doi.org/10.1016/j.matdes.2016.03.162

[61]   A. Thakur, S. Kaya, A. Kumar, Recent Trends in the Characterization and Application Progress of Nano-Modified Coatings in Corrosion Mitigation of Metals and Alloys, Appl. Sci. 13 (2023) 730. https://doi.org/10.3390/app13020730

[62]   S. Bashir, H. Lgaz, I.M. Chung, A. Kumar, Effective green corrosion inhibition of aluminium using analgin in acidic medium: an experimental and theoretical study, Chem. Eng. Commun. 208 (2021) 1121–1130. https://doi.org/10.1080/00986445.2020.1752680

[63]   S. Bashir, A. Thakur, H. Lgaz, I.M. Chung, A. Kumar, Corrosion Inhibition Performance of Acarbose on Mild Steel Corrosion in Acidic Medium: An Experimental and Computational Study, Arab. J. Sci. Eng. 45 (2020) 4773–4783. https://doi.org/10.1007/s13369-020-04514-6

[64]   A. Thakur, A. Kumar, Computational insights into the corrosion inhibition potential of some pyridine derivatives: A DFT approach, Eur. J. Chem. 14 (2023) 246–253. https://doi.org/10.5155/eurjchem.14.2.246-253.2408

[65]   A. Thakur, A. Kumar, S. Kaya, R. Marzouki, F. Zhang, L. Guo, Recent Advancements in

Surface Modification, Characterization and Functionalization for Enhancing the Biocompatibility and Corrosion Resistance of Biomedical Implants, Coatings. 12 (2022) 1459. https://doi.org/10.3390/coatings12101459

[66]   S. Bashir, A. Thakur, H. Lgaz, I.M. Chung, A. Kumar, Corrosion inhibition efficiency of bronopol on aluminium in 0.5 M HCl solution: Insights from experimental and quantum chemical studies, Surfaces and Interfaces. 20 (2020) 100542. https://doi.org/10.1016/j.surfin.2020.100542

[67]   D. Sharma, A. Thakur, M.K. Sharma, K. Jakhar, A. Kumar, A.K. Sharma, O.M. Hari, Synthesis, Electrochemical, Morphological, Computational and Corrosion Inhibition Studies of 3-(5-Naphthalen-2-yl-[1,3,4]oxadiazol-2-yl)-pyridine against Mild Steel in 1 M HCl, Asian J. Chem. 35 (2023) 1079–1088. https://doi.org/10.14233/ajchem.2023.27711

[68]   C. Verma, A. Thakur, R. Ganjoo, S. Sharma, H. Assad, A. Kumar, M.A. Quraishi, A. Alfantazi, Coordination bonding and corrosion inhibition potential of nitrogen-rich heterocycles: Azoles and triazines as specific examples, Coord. Chem. Rev. 488 (2023) 215177. https://doi.org/10.1016/j.ccr.2023.215177

[69]   A. Kumar, A. Thakur, Encapsulated nanoparticles in organic polymers for corrosion inhibition, Elsevier Inc., 2020. https://doi.org/10.1016/B978-0-12-819359-4.00018-0

[70]   S. Sharma, R. Ganjoo, S. Kr. Saha, N. Kang, A. Thakur, H. Assad, V. Sharma, A. Kumar, Experimental and theoretical analysis of baclofen as a potential corrosion inhibitor for mild steel surface in HCl medium, J. Adhes. Sci. Technol. 36 (2022) 2067–2092. https://doi.org/10.1080/01694243.2021.2000230

[71]   A. Thakur, A. Kumar, S. Kaya, D.V.N. Vo, A. Sharma, Suppressing inhibitory compounds by nanomaterials for highly efficient biofuel production: A review, Fuel. 312 (2022) 122934. https://doi.org/10.1016/j.fuel.2021.122934

[72]   S.K. Jaganathan, E. Supriyanto, S. Murugesan, A. Balaji, M.K. Asokan, Biomaterials in cardiovascular research: Applications and clinical implications, Biomed Res. Int. 2014 (2014). https://doi.org/10.1155/2014/459465

[73]   A. Kilic, Artificial Intelligence and Machine Learning in Cardiovascular Health Care, Ann. Thorac. Surg. 109 (2020) 1323–1329. https://doi.org/10.1016/j.athoracsur.2019.09.042

[74]   K. Zhang, T. Liu, J.A. Li, J.Y. Chen, J. Wang, N. Huang, Surface modification of implanted cardiovascular metal stents: From antithrombosis and antirestenosis to endothelialization, 2014. https://doi.org/10.1002/jbm.a.34714

[75]   J. lei Wang, B. chao Li, Z. jun Li, K. feng Ren, L. jiang Jin, S. miao Zhang, H. Chang, Y. xin Sun, J. Ji, Electropolymerization of dopamine for surface modification of complex-shaped cardiovascular stents, Biomaterials. 35 (2014) 7679–7689. https://doi.org/10.1016/j.biomaterials.2014.05.047

[76]   S. Aryal, A. Alimadadi, I. Manandhar, B. Joe, X. Cheng, Machine Learning Strategy for Gut Microbiome-Based Diagnostic Screening of Cardiovascular Disease, Hypertension. 76 (2020) 1555–1562. https://doi.org/10.1161/HYPERTENSIONAHA.120.15885

[77]   M.S. Islam, H. Muhamed Umran, S.M. Umran, M. Karim, Intelligent Healthcare Platform: Cardiovascular Disease Risk Factors Prediction Using Attention Module Based LSTM, 2019 2nd Int. Conf. Artif. Intell. Big Data, ICAIBD 2019. (2019) 167–175. https://doi.org/10.1109/ICAIBD.2019.8836998

[78]   N. Louridi, M. Amar, B. El Ouahidi, Identification of Cardiovascular Diseases Using Machine Learning, 7th Mediterr. Congr. Telecommun. 2019, C. 2019. (2019). https://doi.org/10.1109/CMT.2019.8931411

[79]   J. Wu, M. Dong, S. Santos, C. Rigatto, Y. Liu, F. Lin, Lab-on-a-chip platforms for detection of cardiovascular disease and cancer biomarkers, Sensors (Switzerland). 17 (2017). https://doi.org/10.3390/s17122934

[80]   M. Moravej, D. Mantovani, Biodegradable metals for cardiovascular stent application: Interests and new opportunities, Int. J. Mol. Sci. 12 (2011) 4250–4270. https://doi.org/10.3390/ijms12074250

[81]   P. Shyamkumar, P. Rai, S. Oh, M. Ramasamy, R.E. Harbaugh, V. Varadan, Wearable wireless cardiovascular monitoring using textile-based nanosensor and nanomaterial systems, Electron. 3 (2014) 504–520. https://doi.org/10.3390/electronics3030504

[82]   L. Panzella, F. Moccia, R. Nasti, S. Marzorati, L. Verotta, A. Napolitano, Bioactive Phenolic Compounds From Agri-Food Wastes: An Update on Green and Sustainable Extraction Methodologies, Front. Nutr. 7 (2020) 1–27. https://doi.org/10.3389/fnut.2020.00060

[83]   K. Arumugam, M. Naved, P.P. Shinde, O. Leiva-Chauca, A. Huaman-Osorio, T. Gonzales-Yanac, Multiple disease prediction using Machine learning algorithms, Mater. Today Proc. 80 (2023) 3682–3685. https://doi.org/10.1016/j.matpr.2021.07.361

[84]   M. Alshamrani, Broad-Spectrum Theranostics and Biomedical Application of Functionalized Nanomaterials, Polymers (Basel). 14 (2022) 1221. https://doi.org/10.3390/polym14061221

[85]   D. Chandna, Diagnosis of Heart Disease Using Data Mining Algorithm, 5 (2014) 1678–1680

[86]   K. Munir, A. Biesiekierski, C. Wen, Y. Li, Surface modifications of metallic biomaterials, LTD, 2020. https://doi.org/10.1016/B978-0-08-102965-7.00012-6

[87]   S.K. Metkar, K. Girigoswami, Diagnostic biosensors in medicine – A review, Biocatal. Agric. Biotechnol. 17 (2019) 271–283. https://doi.org/10.1016/j.bcab.2018.11.029

[88]   C. Li, C. Guo, V. Fitzpatrick, A. Ibrahim, M.J. Zwierstra, P. Hanna, A. Lechtig, A. Nazarian, S.J. Lin, D.L. Kaplan, Design of biodegradable, implantable devices towards clinical translation, Nat. Rev. Mater. 5 (2020) 61–81. https://doi.org/10.1038/s41578-019-0150-z

[89]   V. R, S. G, R. Velusamy, S. Ramakrishna, An in-vitro evaluation study on the effects of surface modification via physical vapor deposition on the degradation rates of magnesium-based biomaterials, Surf. Coatings Technol. 411 (2021) 126972. https://doi.org/10.1016/j.surfcoat.2021.126972

[90]   I. Cicha, R. Priefer, P. Severino, E.B. Souto, S. Jain, Biosensor-Integrated Drug Delivery Systems as New Materials for Biomedical Applications, Biomolecules. 12 (2022). https://doi.org/10.3390/biom12091198

[91]   R. Molinaro, A. Pasto, F. Taraballi, F. Giordano, J.A. Azzi, E. Tasciotti, C. Corbo, Biomimetic nanoparticles potentiate the anti-inflammatory properties of dexamethasone and reduce the cytokine storm syndrome: An additional weapon against COVID-19?, Nanomaterials. 10 (2020) 1–12. https://doi.org/10.3390/nano10112301

[92]   C. Buzea, I.I. Pacheco, K. Robbie, Nanomaterials and nanoparticles: Sources and

Materials Research Forum LLC
https://doi.org/10.21741/9781644902998-4

toxicity, Biointerphases. 2 (2007) MR17–MR71. https://doi.org/10.1116/1.2815690

[93] M.H. Sarfraz, M. Zubair, B. Aslam, A. Ashraf, M.H. Siddique, S. Hayat, J.N. Cruz, S. Muzammil, M. Khurshid, M.F. Sarfraz, A. Hashem, T.M. Dawoud, G.D. Avila-Quezada, E.F. Abd_Allah, Comparative analysis of phyto-fabricated chitosan, copper oxide, and chitosan-based CuO nanoparticles: antibacterial potential against Acinetobacter baumannii isolates and anticancer activity against HepG2 cell lines, Front. Microbiol. 14 (2023) 1188743. https://doi.org/10.3389/fmicb.2023.1188743

[94] S. Barua, S. Gogoi, R. Khan, Fluorescence biosensor based on gold-carbon dot probe for efficient detection of cholesterol, Synth. Met. 244 (2018) 92–98. https://doi.org/10.1016/j.synthmet.2018.07.010

[95] R. Alizadehsani, M. Abdar, M. Roshanzamir, A. Khosravi, P.M. Kebria, F. Khozeimeh, S. Nahavandi, N. Sarrafzadegan, U.R. Acharya, Machine learning-based coronary artery disease diagnosis: A comprehensive review, Comput. Biol. Med. 111 (2019) 103346. https://doi.org/10.1016/j.compbiomed.2019.103346

[96] O.M.L. Alharbi, A.A. Basheer, R.A. Khattab, I. Ali, Health and environmental effects of persistent organic pollutants, J. Mol. Liq. 263 (2018) 442–453. https://doi.org/10.1016/j.molliq.2018.05.029

[97] J. Levita, S. Megantara, Mutakin, S.A. Sumiwi, CHRONOTROPIC, INOTROPIC, AND RADICAL SCAVENGING ACTIVITY OF Curcuma heyneana EXTRACT BY MOLECULAR DOCKING SIMULATION AND In-vitro STUDY ON THE HEART OF Fejervarya cancrivora FROG, Rasayan J. Chem. 2022 (2022) 1–10. https://doi.org/10.31788/RJC.2022.1556978

[98] S. Nagaraja, A.R. Pelton, Corrosion resistance of a Nitinol ocular microstent: Implications on biocompatibility, J. Biomed. Mater. Res. - Part B Appl. Biomater. 108 (2020) 2681–2690. https://doi.org/10.1002/jbm.b.34599

[99] A.E. Segneanu, G. Vlase, A.T. Lukinich-Gruia, D.D. Herea, I. Grozescu, Untargeted Metabolomic Approach of Curcuma longa to Neurodegenerative Phytocarrier System Based on Silver Nanoparticles, Antioxidants. 11 (2022). https://doi.org/10.3390/antiox11112261

[100] B.D. Mansuriya, Z. Altintas, Graphene Quantum Dot-Based Electrochemical, Materials (Basel). 13 (2020) 1–30

[101] E. Adarkwa, R. Kotoka, S. Desai, 3D printing of polymeric Coatings on AZ31 Mg alloy Substrate for Corrosion Protection of biomedical implants, Med. Devices Sensors. 4 (2021) 1–15. https://doi.org/10.1002/mds3.10167

[102] M. Balamurugan, P. Santharaman, T. Madasamy, S. Rajesh, N.K. Sethy, K. Bhargava, S. Kotamraju, C. Karunakaran, Recent trends in electrochemical biosensors of superoxide dismutases, Biosens. Bioelectron. 116 (2018) 89–99. https://doi.org/10.1016/j.bios.2018.05.040

[103] R. Buettner, M. Schunter, Efficient machine learning based detection of heart disease, 2019 IEEE Int. Conf. E-Health Networking, Appl. Serv. Heal. 2019. (2019). https://doi.org/10.1109/HealthCom46333.2019.9009429

[104] A. Baldassarre, N. Mucci, M. Padovan, A. Pellitteri, S. Viscera, L.I. Lecca, R.P. Galea, G. Arcangeli, The role of electrocardiography in occupational medicine, from einthoven's invention to the digital era of wearable devices, Int. J. Environ. Res. Public Health. 17 (2020) 1–28. https://doi.org/10.3390/ijerph17144975

[105] N.R. Shanmugam, S. Muthukumar, S. Prasad, A review on ZnO-based electrical

biosensors for cardiac biomarker detection, Futur. Sci. OA. 3 (2017). https://doi.org/10.4155/fsoa-2017-0006

[106] R.D. Crapnell, N.C. Dempsey, E. Sigley, A. Tridente, C.E. Banks, Electroanalytical point-of-care detection of gold standard and emerging cardiac biomarkers for stratification and monitoring in intensive care medicine - a review, Springer Vienna, 2022. https://doi.org/10.1007/s00604-022-05186-9

[107] A. Rescalli, E.M. Varoni, F. Cellesi, P. Cerveri, Analytical Challenges in Diabetes Management: Towards Glycated Albumin Point-of-Care Detection, Biosensors. 12 (2022) 1–22. https://doi.org/10.3390/bios12090687

[108] K. Kannan, A. Menaga, Risk Factor Prediction by Naive Bayes Classifier, Logistic Regression Models, Various Classification and Regression Machine Learning Techniques, Proc. Natl. Acad. Sci. India Sect. B - Biol. Sci. 92 (2022) 63–79. https://doi.org/10.1007/s40011-021-01278-3

[109] B. Regan, R. O'Kennedy, D. Collins, Point-of-care compatibility of ultra-sensitive detection techniques for the cardiac biomarker troponin I—challenges and potential value, Biosensors. 8 (2018) 1–32. https://doi.org/10.3390/bios8040114

[110] A. Mehta, A. Mishra, S. Kainth, S. Basu, Carbon quantum dots/TiO2 nanocomposite for sensing of toxic metals and photodetoxification of dyes with kill waste by waste concept, Mater. Des. 155 (2018) 485–493. https://doi.org/10.1016/j.matdes.2018.06.015

[111] P.H. Quan, I. Antoniac, F. Miculescu, A. Antoniac, V. Manescu, A. Robu, A.I. Bița, M. Miculescu, A. Saceleanu, A.D. Bodog, V. Saceleanu, Fluoride Treatment and In Vitro Corrosion Behavior of Mg-Nd-Y-Zn-Zr Alloys Type, Materials (Basel). 15 (2022) 1–17. https://doi.org/10.3390/ma15020566

[112] N. Radha Shanmugam, S. Muthukumar, S. Chaudhry, J. Anguiano, S. Prasad, Ultrasensitive nanostructure sensor arrays on flexible substrates for multiplexed and simultaneous electrochemical detection of a panel of cardiac biomarkers, Biosens. Bioelectron. 89 (2017) 764–772. https://doi.org/10.1016/j.bios.2016.10.046

[113] C. Maraveas, I.S. Bayer, T. Bartzanas, Recent advances in antioxidant polymers: From sustainable and natural monomers to synthesis and applications, Polymers (Basel). 13 (2021) 1–31. https://doi.org/10.3390/polym13152465

[114] J. Yang, Y. Zheng, X. Gou, K. Pu, Z. Chen, Q. Guo, R. Ji, H. Wang, Y. Wang, Y. Zhou, Prevalence of comorbidities and its effects in coronavirus disease 2019 patients: A systematic review and meta-analysis, Int. J. Infect. Dis. 94 (2020) 91–95. https://doi.org/10.1016/j.ijid.2020.03.017

[115] M.M. Baig, H. GholamHosseini, A.A. Moqeem, F. Mirza, M. Lindén, A Systematic Review of Wearable Patient Monitoring Systems – Current Challenges and Opportunities for Clinical Adoption, J. Med. Syst. 41 (2017). https://doi.org/10.1007/s10916-017-0760-1

[116] F. Gao, Y. Hu, G. Li, S. Liu, L. Quan, Z. Yang, Y. Wei, C. Pan, Layer-by-layer deposition of bioactive layers on magnesium alloy stent materials to improve corrosion resistance and biocompatibility, Bioact. Mater. 5 (2020) 611–623. https://doi.org/10.1016/j.bioactmat.2020.04.016

[117] S.D. Deshmukh, S.N. Shilaskar, Wearable sensors and patient monitoring system: A Review, 2015 Int. Conf. Pervasive Comput. Adv. Commun. Technol. Appl. Soc. ICPC 2015. 00

Materials Research Forum LLC
https://doi.org/10.21741/9781644902998-4

(2015) 31–33. https://doi.org/10.1109/PERVASIVE.2015.7086982

[118] S. Dolati, J. Soleymani, S. Kazem Shakouri, A. Mobed, The trends in nanomaterial-based biosensors for detecting critical biomarkers in stroke, Clin. Chim. Acta. 514 (2021) 107–121. https://doi.org/10.1016/j.cca.2020.12.034

[119] M. Jorfi, E.J. Foster, Recent advances in nanocellulose for biomedical applications, J. Appl. Polym. Sci. 132 (2015) 1–19. https://doi.org/10.1002/app.41719

[120] J.C. Miller, D. Skoll, L.A. Saxon, Home Monitoring of Cardiac Devices in the Era of COVID-19, Curr. Cardiol. Rep. 23 (2021). https://doi.org/10.1007/s11886-020-01431-w

[121] A. Jain, S. Ranjan, N. Dasgupta, C. Ramalingam, Nanomaterials in food and agriculture: An overview on their safety concerns and regulatory issues, Crit. Rev. Food Sci. Nutr. 58 (2018) 297–317. https://doi.org/10.1080/10408398.2016.1160363

[122] L.H. Lin, H.P. Lee, M.L. Yeh, Characterization of a sandwich plga-gallic acid-plga coating on mg alloy zk60 for bioresorbable coronary artery stents, Materials (Basel). 13 (2020) 1–16. https://doi.org/10.3390/ma13235538

[123] Y. Qi, H. Qi, Y. He, W. Lin, P. Li, L. Qin, Y. Hu, L. Chen, Q. Liu, H. Sun, Q. Liu, G. Zhang, S. Cui, J. Hu, L. Yu, D. Zhang, J. Ding, Strategy of Metal-Polymer Composite Stent to Accelerate Biodegradation of Iron-Based Biomaterials, ACS Appl. Mater. Interfaces. 10 (2018) 182–192. https://doi.org/10.1021/acsami.7b15206

[124] R.D. Crapnell, N.C. Dempsey-Hibbert, M. Peeters, A. Tridente, C.E. Banks, Molecularly imprinted polymer based electrochemical biosensors: Overcoming the challenges of detecting vital biomarkers and speeding up diagnosis, Talanta Open. 2 (2020) 100018. https://doi.org/10.1016/j.talo.2020.100018

[125] C.O. Rolim, F.L. Koch, C.B. Westphall, J. Werner, A. Fracalossi, G.S. Salvador, A cloud computing solution for patient's data collection in health care institutions, 2nd Int. Conf. EHealth, Telemedicine, Soc. Med. ETELEMED 2010, Incl. MLMB 2010; BUSMMed 2010. (2010) 95–99. https://doi.org/10.1109/eTELEMED.2010.19

[126] S. Veličković, B. Stojanović, L. Ivanović, S. Miladinović, S. Milojević, Application of Nanocomposites in the Automotive Industry, Mobil. Veh. Mech. 45 (2019) 51–64. https://doi.org/10.24874/mvm.2019.45.03.05

[127] A. Mpupa, A. Nqombolo, B. Mizaikoff, P.N. Nomngongo, Beta-Cyclodextrin-Decorated Magnetic Activated Carbon as a Sorbent for Extraction and Enrichment of Steroid Hormones (Estrone, β-Estradiol, Hydrocortisone and Progesterone) for Liquid Chromatographic Analysis, Molecules. 27 (2022). https://doi.org/10.3390/molecules27010248

[128] D. Iannazzo, C. Espro, C. Celesti, A. Ferlazzo, G. Neri, Smart biosensors for cancer diagnosis based on graphene quantum dots, Cancers (Basel). 13 (2021). https://doi.org/10.3390/cancers13133194

[129] I.M. Maafa, Biodiesel Synthesis from High Free-Fatty-Acid Chicken Fat Using a Scrap-Tire Derived Solid Acid Catalyst and KOH, Polymers (Basel). 14 (2022). https://doi.org/10.3390/polym14030643

[130] S. Agarwal, J. Curtin, B. Duffy, S. Jaiswal, Biodegradable magnesium alloys for orthopaedic applications: A review on corrosion, biocompatibility and surface modifications, Mater. Sci. Eng. C. 68 (2016) 948–963. https://doi.org/10.1016/j.msec.2016.06.020

[131] Q. Wang, P. Zhou, S. Liu, S. Attarilar, R.L.W. Ma, Y. Zhong, L. Wang, Multi-scale surface treatments of titanium implants for rapid osseointegration: A review, Nanomaterials. 10 (2020) 1–27. https://doi.org/10.3390/nano10061244

[132] N. Badolati, R. Masselli, M. Maisto, A. Di Minno, G.C. Tenore, M. Stornaiuolo, E. Novellino, Genotoxicity assessment of three nutraceuticals containing natural antioxidants extracted from agri-food waste biomasses, Foods. 9 (2020). https://doi.org/10.3390/foods9101461

[133] S. Höhn, S. Virtanen, A.R. Boccaccini, Protein adsorption on magnesium and its alloys: A review, Appl. Surf. Sci. 464 (2019) 212–219. https://doi.org/10.1016/j.apsusc.2018.08.173

[134] A. Abraham, Lecture Notes in Networks and Systems 167 3rd International Conference on Computing Informatics and Networks, 2020

[135] E.M. Miramontes-Gutierrez, J.M. Ochoa-Rivero, H.O. Rubio-Arias, L. Ballinas-Casarrubias, B.A. Rocha-Gutiérrez, Impact of cations ($Na+$, $K+$, $Mg+2$) and anions ($F-$, $Cl-$, $SO42-$) leaching from filters packed with natural zeolite and ferric nanoparticles for wastewater treatment, Int. J. Environ. Res. Public Health. 18 (2021). https://doi.org/10.3390/ijerph18168525

[136] M. Raj, R.N. Goyal, Silver nanoparticles and electrochemically reduced graphene oxide nanocomposite based biosensor for determining the effect of caffeine on Estradiol release in women of child-bearing age, Sensors Actuators, B Chem. 284 (2019) 759–767. https://doi.org/10.1016/j.snb.2019.01.018

Materials Research Forum LLC
https://doi.org/10.21741/9781644902998-5

Chapter 5

Colon Targeted Nano Drug Delivery Systems

H. Tarannum[1], B.S. Nayak[2*], A. Ojha[3*], and S. Nandi[4*]

[1]College of Pharmacy, Surajmal University, Kichha, Uttarakhand, India

[2] KIIT School of Pharmacy, KIIMS, KIIT Deemed to be University, Odisha, India

[3]Amrapali Institute of Pharmacy and Sciences, Haldwani, Uttarakhand, India

[4]Global Institute of Pharmaceutical Education and Research, Kashipur, Uttarakhand, India

sisir.iicb@gmail.com

Abstract

The colon is the attractive target for the delivery of nano formulation due to distinct advantages such as near-neutral pH, longer transit time, and reduced enzymatic activity. Colon-specific drug delivery not only increases the bioavailability of the drug at the target site but also reduces the side effects and multiple dosing. In recent studies, colon-targeted drug delivery systems are gaining importance to treat local pathologies of the colon such as anorectal problems, Crohn's disease, inflammatory bowel disease, and colonic cancer and also for the systemic delivery of protein and peptide drugs. This is because the peptide and protein drugs get destroyed or inactivated in the acidic environment of the stomach or by pancreatic enzymes in the small intestine. Various approaches such as prodrugs, pH-dependent, time-dependent and microflora-activated systems have been developed to achieve successful colonic delivery, the chemical, enzymatic and mucoadhesive barriers within the gastrointestinal (GI) tract for colon-specific drug delivery. Among the different approaches, an attempt has been made to depict vital uses of polymers, specifically biodegraded by colonic bacterial enzymes holds greater promise for the colon-targeted nano drug delivery systems.

Keywords

Gastrointestinal, Nano Drug Delivery, Colon Targeting, Probiotics, Chitosan

Contents

Materials Research Forum LLC
https://doi.org/10.21741/9781644902998-5

1. Introduction

Although oral delivery has become a widely accepted route of administration of therapeutic drugs, the gastrointestinal tract presents several formidable barriers to drug delivery. Various peptide and protein drugs are inactivated in the acidic environment of the stomach or by pancreatic enzymes in the small intestine. Hence colonic drug delivery has gained increased importance not just for the delivery of drugs for the treatment of local diseases associated with the colon but also for its potential for the delivery of proteins and therapeutic peptides [1]. The advantages of targeting drugs specifically to the colon include reduced incidence of systemic adverse effects, a reduced dose of the drug, and delivery of the drug in its intact form as close to the target site as possible [2]. To achieve successful colonic delivery, a drug needs to be protected from absorption in the environment of the upper gastrointestinal tract (GIT) and then be abruptly released into the proximal colon, which is considered the optimum site for colon-targeted delivery of drugs. Colon targeting is naturally of value for the topical treatment of diseases of the colon such as Crohn's disease, ulcerative colitis, colorectal cancer, and amoebiasis [3]. To combat these problems, we need to know the detailed anatomy and physiology of the colon.

2. Anatomy of the colon

2.1 Superficial anatomy

The colon is a part of the gastrointestinal tract. The colon is a tube-like organ in the abdominal cavity. The human gastrointestinal tract is divided mainly into the stomach, small intestine, and large intestine. The large intestine extending from the ileocaecal junction to the anus is divided into three parts namely the colon, rectum, and anal canal. The colon is approximately five feet (1.5 meters) in length, begins at the ileocecal valve, and ends at the rectosigmoid junction. The colon is made up of the caecum, the ascending colon, the hepatic flexure, the transverse colon, the splenic flexure, the descending colon, and the sigmoid colon [4, 5].

2.2 Blood supply

Arterial blood supply to the colon from cecum to splenic flexure is through the superior mesenteric artery which gives rise to the ileocolic, right colic, and middle colic arteries. The left and sigmoid colon is supplied by the inferior mesenteric artery which gives rise to the left colic and sigmoidal arteries. There can be several anatomic variations in the colic arteries including absent middle colic artery, absent right colic artery, common trunk for right and ileocolic artery, and the presence of an Arc of Riolan between the middle and left colic artery. The venous drainage is by the inferior mesenteric vein draining into the splenic vein, and the superior mesenteric vein joins the splenic vein to form the hepatic portal vein that then enters the liver. Lymphatic drainage from the entire colon is to the paraaortic lymph nodes that then drain into the cisterna chyli. The lymph from the remaining rectum and anus can either follow the same route or drain to the internal iliac and superficial inguinal nodes. The pectinate line only roughly marks this transition [4, 6].

2.3 Histology

The colonic wall consists of a simple columnar, lamina propria, muscularis mucosa, submucosa, muscularis propria formed by an inner circular and outer longitudinal layer of smooth muscle, and serosa. The wall of the colon is composed of four layers: the serosa, the muscularis externa, the submucosa, and the mucosa. The serosa is the exterior coat and consists of areolar tissue that is covered by a single layer of squamous mesothelial cells. The major muscular coat of the colon is muscularis externa. This is composed of an inner circular layer of fibers that surrounds the bowel and an outer longitudinal layer. The submucosa is the layer of connective tissue that lies immediately beneath the mucosa. Lining the lumen of the colon, the mucosa is divided into epithelium, lamina propria, and muscularis mucosae. Closely spaced crypts extend down into the surface of the mucosa. The muscularis mucosae consist of a layer of smooth muscle and separate the submucosa from the lamina propria. The lamina propria supports the epithelium and occupies space between the crypts and beneath the crypts. Within the lamina propria are located blood capillaries and lymphatic lacteals. The space also acts as a reservoir for macrophages, neutrophils, eosinophils, lymphocytes, and plasma cells, which locally produce immunoglobulin A antibodies. The epithelium consists of a line of single layers of cells, which lines the crypts and covers the surface of the mucosa. Three major cell types found in the epithelium are the columnar absorptive cells, goblet (mucous cells), and enteroendocrine cells. Adjacent columnar absorptive cells are attached near their apical margins by a junctional complex. Mucous production is the function of goblet cells and the proportion of goblet cells increases in the elderly [6, 7].

2.4 Nerve supply

The Sympathetic nerve supply to the colon extends downwards from the coeliac plexus and its extensions on the antero-lateral aspects of the aorta, on each side, are usually one, two, or three fine bundles of nerve fibres, to which the name intermesenteric nerves has been applied. These nerves are joined laterally by rami from the upper lumbar ganglia, or adjoining parts of the lumbar sympathetic cord, of their own side. There are usually two or three slender strands of nerve fibres passing across the aorta to connect the intermesenteric nerves of the two sides; these strands are often to be seen both above and below the point of origin of the inferior mesenteric artery. The inferior mesenteric plexus is formed by a number of sets of fibres arising from the medial sides of the right and left intermesenteric nerves, and these fibres form a dense network around the inferior mesenteric artery about half to three-quarters of an inch beyond its origin. From the dense plexus around the stem of the inferior mesenteric artery perivasular fibres can be traced along all the branches of this vessel. Below the origin of the inferior mesenteric plexus three descending bundles are usually to be found, which converge to form the hypogastric plexus. The two lateral trunks are direct continuations of the intermesenteric nerves after they have contributed fibres to the intermesenteric plexus.

The colon receives parasympathetic fibres from splanchnic nerves (sacral outflow). On examining closely, the lower part of the hypogastric and pelvic plexuses a small bundle of fibres can be seen on their ventral aspect on each side. The two bundles are easily separated from the main plexuses and can be traced upwards, where they converge to meet and fuse to the left side of the hypogastric plexus. The small trunk, formed by the fusion of the two bundles, can be followed upwards and to the left over the left common iliac artery to the inferior mesenteric plexus, which joins about one to one and a half inches distal to the origin of the artery from the aorta. On joining the inferior mesenteric plexus, the fibres mix with the sympathetic fibres, but with care, it can be seen that these parasympathetic fibres extend as a perivascular plexus mainly onto the left colic artery and its ascending and descending divisions [4, 5].

3. Physiology of colon

3.1 Absorption of water and electrolytes and propulsion of unabsorbed fecal waste

The major functions of the colon are the absorption of water and electrolytes and the propulsion of stored unabsorbed fecal waste for evacuation. Approximately one liter of fluid chyme enters the cecum each day with an average of only 100cc excreted in the feces. By the time the chyme has reached to colon, most nutrients and 90 % of the water has been absorbed by the body. At this point, some electrolytes like sodium, magnesium, and chloride are left as well as indigestible parts of ingested food (e.g., a large part of ingested amylose, protein which has been shielded from digestion heretofore, and dietary fiber, which is largely indigestible carbohydrate in either soluble or insoluble form). As the chyme moves through the large intestine, most of the remaining water is removed, while the chyme is mixed with mucus and bacteria (known as gut flora) and becomes feces. The ascending colon receives fecal material as a liquid. The muscles of the colon then move the watery waste material forward and slowly absorb all the excess water. The stools become semi-solid as they move along into the descending colon. The bacteria break down some of the fiber for their own nourishment and create acetate, propionate, and butyrate as waste products, which in turn are used by the cell lining of the colon for nourishment [2, 7].

3.2   Secretion of mucus

The internal thin liners of colon contain epithelial cells that secrete mucus. This mucus moisturizes and lubricates the colon lining. This lining protects the colon wall and nerve tissues.

3.3   Bacterial growth

Gut microbiome live and grow along the colon lining. Using fluid and food intake, bacteria actually manufacture the nutrients that sustain their environment and their food supply.

3.4   Manufacture of some vitamins and electrolytes

Bacteria change proteins into amino acids and break these amino acids down further into indole and skatole (which gives stools their odor), hydrogen sulfide, and fatty acids. Bacterial action also synthesizes some vitamins (K and some B), and electrolytes, and breaks down bilirubin into a pigment that gives stools their brown color.

3.5   Production of lubrication

Bacteria ferment soluble fiber into a lubricating gel that is incorporated into the stool mass as it is formed. This gel helps to make stools soft, flexible and ease of expulsion. Some of this gel also coats the exterior of the stools and is used by the colon to moisturize the colon lining. This lubrication helps in defecation of the colon.

3.6   Defense against infection

Healthy intestinal bacteria help to groom the colon and keep it clean so that infections do not develop. They also help to fight the growth of infectious bacteria.

4.   Significance of colon-targeted drug delivery

The targeted drug delivery to the colon would ensure direct treatment at the disease target site. It enables lower dosing of the drug. Although, fewer systemic side effects are observed by targeted drug delivery to the colon. It helps to delay drug absorption. Drugs that cause gastric irritation or ulceration can be directed to the colon using targeted delivery approaches. Formulations for colonic delivery are also suitable for the delivery of drugs that are polar and/or susceptible to chemical and enzymatic degradation in the upper gastrointestinal tract, highly affected by hepatic metabolism, in particular, therapeutic proteins and peptidase. Drug loss is prevented by extensive first-pass metabolism. Lower daily cost to the patient due to fewer dosage units required by the patient [5, 8].

5. Approaches to colonic drug delivery

5.1 Drug release based on variation of pH

In the stomach, pH ranges between 1 and 2 during fasting but increases up to 4 after eating. The pH is about 6.5 in the proximal small intestine and about 7.5 in the distal small intestine. The pH declines significantly from the ileum to the colon. It is about 6.4 in the caecum. However, pH values as low as 5.7 have been measured in the ascending colon in healthy volunteers. The pH in the transverse colon is 6.6, and in the descending colon 7.0. The use of pH-dependent polymers is based on these differences in pH levels [7].

Various pH-dependent coating polymers include cellulose acetate phthalate (CAP), polyvinyl acetate phthalate (PVAP), hydroxypropyl methylcellulose phthalate (HPMCP), and methacrylic acid copolymers, commonly known as Eudragit [7, 9]. Eudragit products are pH-dependent methacrylic acid polymers containing carboxyl groups. The number of esterified carboxyl groups affects the pH level at which dissolution takes place. Eudragit S is soluble above pH 7 and Eudragit L above pH 6. Eudragit S coatings protect well against drug liberation in the upper parts of the gastrointestinal tract and have been used in preparing colon-specific formulations. Eudragit S coatings have been used to target the anti-inflammatory drug 5-aminosalicylic acid (5-ASA) in single-unit formulations on the large intestine [10].

5.2 Drug release based on gastrointestinal transit time (Pulsatile drug delivery)

Pulsatile release systems are formulated to undergo a lag-time of the predetermined span of time of no release, followed by a rapid and complete release of loaded drugs(s). The approach is based on the principle of delaying the time of drug release until the system transits from mouth to colon. A lag time of 5 hours is usually considered sufficient since small intestine transit is about 3-4 hours, which is relatively constant and hardly affected by the nature of the formulation administered. This system offers many advantages over conventional oral drug delivery systems like patient compliance, reduced dosage, reduced dosage frequency, avoidance of side effects, avoidance of peak valley fluctuation, and nearly constant drug level at the target site [11, 12].

Enteric coated time-release press coated tablets are developed. It consists of three components, a drug layer containing core tablet, the press coated swellable hydrophobic polymer layer, and an enteric coating layer [13]. The tablet does not release the drug in the stomach due to the acid resistance of the outer enteric coating layer. After gastric emptying, the enteric coating layer rapidly dissolves, and the intestinal fluid begins to slowly erode the press-coated polymer layer. When the erosion front reaches the core tablet, rapid drug release occurs since the erosion process takes a long time as there is no drug release period (lag phase) after gastric emptying (Fig. 1).

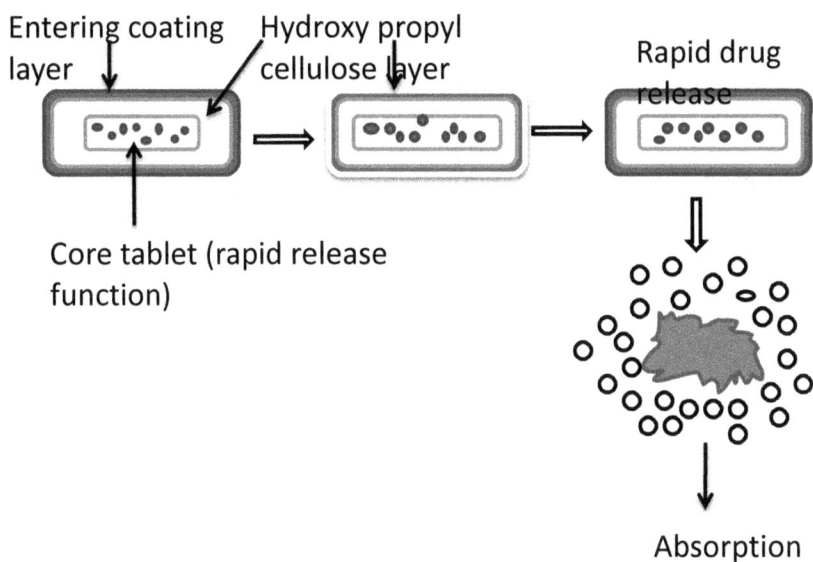

Figure 1: Time controlled drug release system [13]

5.3   Drug release based on the presence of colonic micro flora

These systems are based on the exploitation of the specific enzymatic activity of the microflora (enterobacteria) present in the colon [1]. The colonic bacteria are predominately anaerobic in nature and secrete enzymes that are capable of metabolizing substrates such as carbohydrates and proteins that escape the digestion in the upper GI tract [6-8]. Bacterial count in the colon is much higher around 1011-1012 CFU/ml with some 400 different species which are fundamentally aerobic, predominant species such as Bacteroides, Bifidobacterium and Eubacterum etc., whose major metabolic process occurring in the colon are hydrolysis and reduction.

The vast microflora fulfills its energy needs by fermenting various types of substrates that have been left undigested in the small intestine, e.g., di- and tri-saccharides, polysaccharides, etc. For this fermentation, the microflora produces a vast number of enzymes like glucuronidase, xylosidase, arabinoside, sulfatase, galactosidase, nitroreductase, azoreducatase, deaminase, sulfoxide reductase and urea dehydroxylase [17]. Because of the presence of the biodegradable enzymes only in the colon, the use of biodegradable polymers for colon-specific drug delivery seems to be a more site-specific approach as compared to other approaches. These polymers shield the drug from the environments of the stomach and small intestine and are able to deliver the drug to the colon. On reaching the colon, they undergo assimilation by micro-organisms, degradation by enzymes, or break down of the polymer backbone leading to a subsequent reduction in their molecular weight and thereby loss of mechanical strength. They are then unable to hold the drug entity any longer [14, 15, 16].

On this basis prodrug approach that is based on the formation of a bio-reversible covalent linkage between drug and carrier is utilized for colon targeted delivery. The type of linkage that is formed between drug and carrier would decide the triggering mechanism for the release of drug in the colon. The various covalently linked prodrugs reported for colon-specific drug delivery are summarized in the following section [17, 18] (Fig. 2).

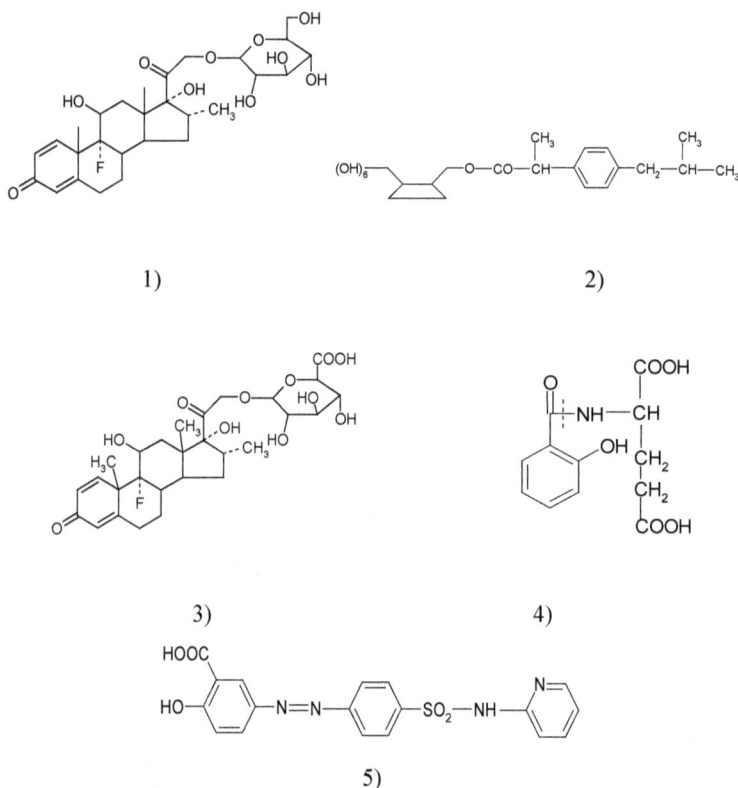

Figure 2: Conjugates for targeted colonic delivery [17]

1. Dexamethason-21-β-glucoside conjugate, 2. α Cyclodextrin ester conjugate of ibuprofen, 3. Dexamethasone- β-D-glucuronide conjugate, 4. Salicyl-glutamic acid conjugate, 5. Sulfasalazine (sulfa pyridine azo conjugate)

5.3.1 Glycoside conjugates

Steroid glycosides and the unique glycosidase activity of the colonic microflora form the basis of a new colon-targeted drug delivery system. Drug glycosides are hydrophilic and thus, poorly absorbed from the small intestine. Once such a glycoside reaches the colon, it can be cleaved by bacterial glycosidases, releasing the free drug to be absorbed by the colonic mucosa. The major glycosidases identified in human feces are b-D-galactosidase, b-D-glucosidase, a-L-arabinofuranosidase, b-D-xylopyranosidase. Friend and Chang have synthesized dexamethason-21-b-glucoside for delivery of these steroids to the colon [19].

5.3.2 Cyclodextrin conjugates

Cyclodextrins (CyDs) are cyclic oligosaccharides consisting of six to eight glucose units through a-1,4 glucosidic bonds and have been utilized for targeting drugs in the colon. An anti-inflammatory drug biphenylacetic acid (BPAA) as a model drug was selectively conjugated onto one of the primary hydroxyl groups of α, β, and γ-CyDs through an ester or amide linkage, and the *in vivo* drug release behavior of these prodrugs in rat gastrointestinal tract after oral administration was investigated. The CyD prodrugs were stable in rat stomach and small intestines and released BPPA specifically in the colon [20].

5.3.3 Dextran conjugates

Mcleod et al. synthesized dextran ester prodrugs of dexamethasone and methylprednisolone and proved the efficacy of the prodrugs for delivering drugs to the colon [21, 22]. In this study, methylprednisolone and dexamethasone were covalently attached to dextran by the use of a succinate linker.

5.3.4 Glucuronide conjugates

Bacteria of the lower GIT secrete b-glucuronidase and can deglucuronidate a variety of drugs in the intestine. Haeberlin *et al.* prepared a dexamethasone- b-D-glucuronide prodrug, which successfully delivered dexamethasone specifically to the colon [23]. Nolen et al. investigated the steady-state pharmacokinetics of corticosteroid delivery from glucuronide prodrugs in normal and colitic rats [24]. Two prodrugs, dexamethasone- b -D-glucuronide (DXglrd) and Budesonide- b -D-glucuronide (BUDglrd) were administered by intragastric infusion to conventional and colitic rats. In addition, dexamethasone and Budesonide were administered either intragastrically or subcutaneously to healthy and colitic rats and colon-specific delivery was assessed using the drug delivery index. In conventional rats, drug delivery indices for DXglrd ranged from about five to as high as 11 in the luminal contents relative to dexamethasone administered subcutaneously or intragastrically. Drug delivery index values were also elevated in the mucosa of both healthy and colitic rats following intragastric administration of DXglrd. Budesonide was delivered somewhat less effectively from BUDglrd to the rat large intestine than was dexamethasone from DXglrd.

5.3.5 Amino acid conjugates

Due to the hydrophilic nature of polar groups like -NH$_2$ and -COOH present in the proteins and their basic units of amino acids, their membrane permeability is reduced. Various prodrugs have been prepared by conjugation of drug molecules to these polar amino acids through amide linkage [25, 26].

5.3.6   Azo bond conjugates

Sulfasalazine is the prodrug in which sulfapyridine is linked to a salicylate by an azo bond. Chemically it is salicylazosulfapyridine (SASP). When taken orally, only a small proportion of the ingested dose is absorbed from the small intestine and the bulk of the sulfasalazine reaches the colon intact. There it splits at the azo bond by the colonic bacteria with the liberation of 5-ASA and sulfapyridine (SP) respectively [9].

By replacing the sulfapyridine molecule with others, a number of prodrugs of 5-ASA have been prepared e.g., 4-amino benzoyl-b-alanine in balsalazide and p-aminohippurate (4-amino benzoyl glycine) in ipsalazine [27]. The most interesting prodrug is olsalazine (OSZ), which is a dimer representing two molecules of 5-ASA that are linked via an azo bond. When olsalazine reaches the large intestine, it is cleaved releasing two molecules of 5-ASA for every mole of olsalazine administered [28].

5.4   Pressure-controlled drug-delivery systems

As a result of peristalsis, higher pressures are encountered in the colon than in the small intestine. Takaya has developed pressure-controlled colon delivery capsules prepared using ethyl cellulose, which is insoluble in water. In such systems, drug release occurs following disintegration of a water-insoluble polymer capsule as a result of pressure in the lumen of the colon. The thickness of the ethyl cellulose membrane is the most important factor for disintegration of the formulation. The system also appeared to depend on capsule size. When salivary secretion of caffeine after oral administration of pressure-controlled capsules was studied in human volunteers, a correlation was found between ethyl cellulose membrane thickness and the time of first appearance of caffeine in the saliva. Because of reabsorption of water from the colon, the viscosity of luminal content is higher in the colon than in the small intestine. It has therefore been concluded that drug dissolution in the colon could present a problem in relation to colon-specific oral drug delivery systems. In pressure-controlled ethyl cellulose single-unit capsules the drug is in a liquid. Lag times of three to five hours in relation to drug absorption were noted when pressure-controlled capsules were administered to human subjects. It was concluded that the capsules disintegrated in the colon because of increases in pressure [6, 16, 17].

5.5   Osmotically controlled system (ORDS-CT)

The OROS-CT can be used to target the drug locally to the colon for the treatment of disease or to achieve systemic absorption. The OROS-CT system can be a single osmotic unit or may incorporate as many as 5-6 push-pull units, each 4 mm in diameter, encapsulated within a hard gelatin capsule. Each bilayer push pull unit contains an osmotic push layer and a drug layer, both surrounded by a semi permeable membrane. An orifice is drilled through the membrane next to the drug layer [6, 17] (Fig. 3).

Materials Research Forum LLC

https://doi.org/10.21741/9781644902998-5

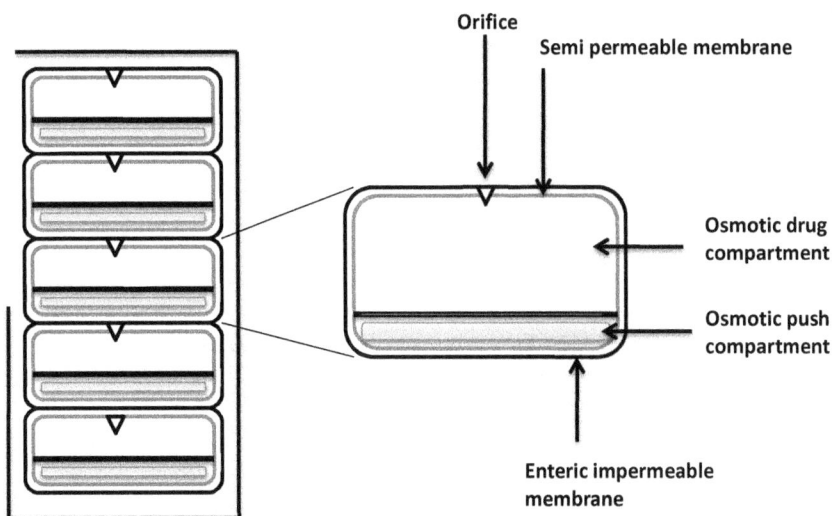

Figure 3: Osmotically controlled drug delivery [6]

Immediately after the OROS-CT is swallowed, the gelatin capsule containing the push-pull units dissolves. Because of its drug-impermeable enteric coating, each push-pull unit is prevented from absorbing water in the acidic aqueous environment of the stomach, and hence no drug is delivered. As the unit enters the small intestine, the coating dissolves in this higher pH environment (pH >7), water enters the unit, causing the osmotic push compartment to swell, and concomitantly creates a flowable gel in the drug compartment. Swelling of the osmotic push compartment forces drug gel out of the orifice at a rate precisely controlled by the rate of water transport through the semi permeable membrane. OROS-CT units can maintain a constant release rate for up to 24 hours in the colon or can deliver drug over a period as short as 4 h [29].

6. Polymers for colon targeted delivery

Natural polysaccharides like guar gum, pectin, chondroitin sulphate, dextran, chitosan, cyclodextrin, inulin and amylose have been used in control drug delivery systems, as they can remain intact in the upper GIT, but are degraded by microflora to release the drug (Fig. 4). Moreover, polysaccharide structural modifications or derivatives can enhance drug release behaviour, site-specificity, and stability. Polysaccharide mucoadhesive ness may be beneficial for drug absorption through extended contact between drug delivery carriers and the mucosal surface. Polysaccharide-based drug delivery systems provide additional advantages such as low toxicity, low cost, immunogenicity, biodegradability, and biocompatibility. As a result, the microbiota triggered polysaccharide-based system offers a viable approach for colon-targeted nano-drug

Materials Research Forum LLC
https://doi.org/10.21741/9781644902998-5

delivery systems. The rate of drug release is affected by the type and concentration of polysaccharides.

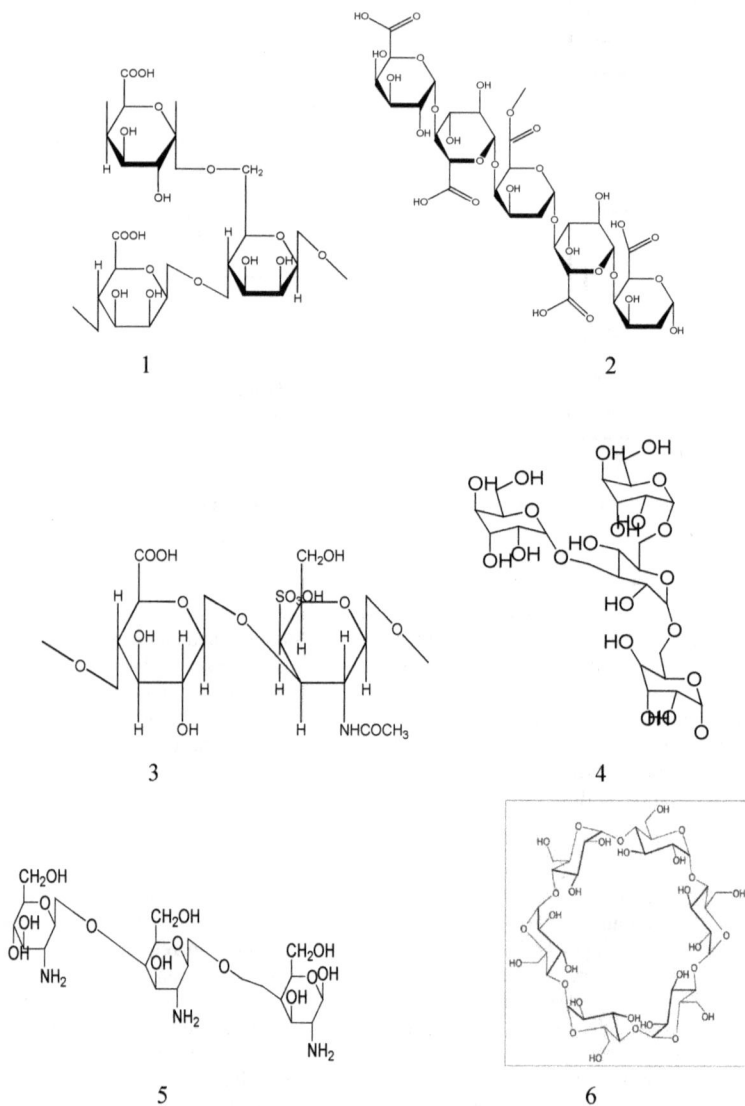

Materials Research Forum LLC
https://doi.org/10.21741/9781644902998-5

7

8

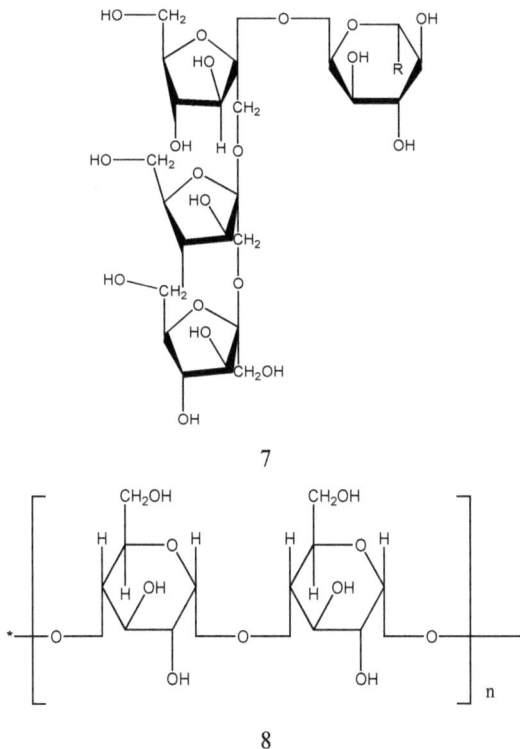

Figure 4: Polymers for colon targeting

1. Guar gum, 2. Pectin, 3. Chondroitin sulphate, 4. Dextran, 5. Chitosan, 6. Cyclodextrin, 7. Inulin, 8. Amylose. [30, 31, 32, 33, 34]

6.1   Guar gum

Guar gum is derived from the seeds of the *cyomopsis tetragonolobus* (Fam. Leguminosae). Chemically, guar gum is a polysaccharide composed of the sugar's galactose and mannose. The backbone is a linear chain of β 1, 4-linked mannose residues to which galactose residues are 1,6-linked at every second mannose, forming short side branches. Guar gum is used in colon-targeted drug delivery systems due to its drug release retarding property and susceptibility to microbial degradation in the large intestine [35].

Wong, *et al.* studied the dissolution of dexamethasone and budesonide from guar gum-based formulations using reciprocating cylinder dissolution apparatus (USP Dissolution Apparatus III) and observed that the drug release in simulated colonic fluid was markedly increased at galactomannanase concentrations >0.01 mg/ml [36].

Krishnaiah, *et al.* in their study, performed the pharmacokinetic evaluation of guar gum-based colon-targeted tablets of mebendazole against an immediate release tablet in six healthy human volunteers. Colon-targeted tablets showed delayed t $_{max}$ (9.4±1.7 h) and absorption time, and decreased C $_{max}$ (25.7±2.6 μg/ml) and absorption rate constant when compared to the immediate release tablets. The results of the study indicated that the guar gum-based colon-targeted tablets of mebendazole did not release the drug in the stomach and small intestine but delivered the drug to the colon resulting in a slow absorption of the drug and making the drug available for local action in the colon [37].

Core tablets containing 5-aminosalicylic acid (5-ASA) were prepared by wet granulation with starch paste and were compression coated with coating formulations containing different quantities of guar gum. The study confirmed that selective delivery of 5-ASA to the colon can be achieved using guar gum as a carrier in the form of compression coating over the drug core [38].

6.2 Pectin

Pectin is a linear, heterogeneous polysaccharide that is mainly composed of galacturonic acid and its methyl ester. These are predominantly linear polymers of mainly (1→4) linked D-galacturonic acid residue interrupted by 1,2-linked L-rhamnose residue with a few hundred to about one thousand building blocks per molecule. It is one of the major sources of dietary fiber and is extracted from fruit and vegetable cell walls. [33] Calcium pectinate, the insoluble salt of pectin was used by Rubeinstein, *et al.* for colon-targeted drug delivery of Indomethacin formulation [39].

Dupuis, *et al.* used zinc pectinate beads for colonic delivery of ketoprofen and reported similar performance when compared to calcium pectinate in hard capsules, but significant differences when the same pellets were compared encapsulated in enteric hard caspsules. This study revealed that Zinc pectinate beads could protect drugs entrapped sufficiently from the upper gastrointestinal conditions and drug release will be controlled by pectin degradation with colonic microflora. Zinc pectinate beads in enteric hard capsules are promising as a carrier for specific colonic delivery of drugs after oral administration [40].

6.3 Chondroitin Sulfate

Chondroitin sulphate is a mucopolysaccharide, which consists of D-glucuronic acid linked to N-acetyl-D-galactosamide [41]. It is degraded by the anaerobic bacteria of the large intestine mainly by *Bacteroids thetaiotaoimicron* and *B. ovatus.* The high-water solubility of chondroitin sulphate put hurdles in the formulation of colon targeted drug delivery systems and hence crosslinking has been reported in the literature to alleviate this problem. Chondroitin sulphate was cross-linked with 1, 12 diaminododecane using dicyclohexylcarbodiimide as a catalyst and formulated in a matrix with indomethacin as a drug marker. The indomethacin release kinetics from the various formulations was analyzed in PBS with and without rat caecal content at 37°C under a carbon dioxide atmosphere and it was concluded that the release of indomethacin was dependent upon the biodegradation action of the caecal content [42].

6.4   Dextran

It is a polysaccharide consisting of linear chains of α-D glucose molecules, 95 % of the chains consist of 1:6-α-linked linear glucose units while the side chains consist of 1:3-α-linked moieties [43]. They are obtained from microorganisms of the family *Lactobacillus (Leuconostoc mesenteroides)*. Dextrans are colloidal, hydrophilic, and water-soluble substances that are inert for the biological system and also do not affect cell viability. Dextranases are the enzyme that hydrolyzes glycosidic linkages of the dextrans. Anaerobic Gram-negative intestinal bacteria show dextranase activity of the colon, especially by the bacteroides of the colon. Glucocorticoid-dextran ester prodrugs have been prepared and proved efficacious in delivering drugs to the colon [44-46].

6.5   Cyclodextrin

Cyclodextrins (CyDs) are cyclic oligosaccharides consisting of six to eight glucose units through 1,4 glucosidic bonds and have been utilized to improve certain properties of drugs such as solubility, stability, and bioavailability of drugs. The interior of these molecules is relatively lipophilic and the exterior relatively hydrophilic, they tend to form inclusion complexes with various drug molecules. They are known to be barely capable of being hydrolyzed and only slightly absorbed in passage through the stomach and small intestine; however, they are fermented by colonic microflora into small saccharides and thus absorbed in the large intestine [47].

6.6   Inulin

It is a naturally occurring polysaccharide found in many plants such as garlic, onion, artichoke, and chicory. Chemically, it belongs to the glucofructans and consists of a mixture of oligomers and polymers containing 2-60 (or more) β-2-1 linked D-fructose molecules. Most of these fructose chains have a glucose unit as the initial moiety. It is not hydrolyzed by the endogenous secretions of the human digestive tract. [48] Inulin has been incorporated into Eudragit RS films for preparation of mixed films that resisted degradation in the upper GI tract but digested in the human fecal medium by the action of bifidobacteria and bacteroides [49].

6.7   Amylose

It is a poly (1-4-α -D-glucopyranose that consists of D-glucopyranose residues linked by α- (1-4) bonds. Those substances, present naturally in the diet, have the advantages of being safe, nontoxic, and easily available. These are resistant to pancreatic α-amylase, but are degraded by colonic bacterial enzyme. Mixed films of amylose and ethyl cellulose as coatings have shown a great potential as colon delivery carriers [50].

6.8   Natural Polymer conjugates

Natural polymers can be made colon-specific by specific reactions. Satheesh Madhav, *et al.* synthesized a novel colon targeting polymer conjugate by using sodium carboxymethyl cellulose (CMC) and tyrosine for colon targeting [51]. Sodium carboxy methyl cellulose was subjected to synthesizing its methyl ester hydrochloride derivative by using tyrosine, sodium nitrite, methanol, and conc. hydrochloric acid. It was diazotized and the azo derivative was coupled with sodium CMC solution under alkaline conditions. It was further subjected to evaluating its colon targeting property by an *in vitro* method using rat fecal matter. The *in vitro* release of sodium carboxymethyl cellulose azo polymer was evaluated by precipitating with gelatin and estimated by a nephelometer. The novel sodium carboxymethyl cellulose azo derivative showed promising colon specificity for a period of 90 minutes in a controlled manner.

Similarly, an azo polymer has been synthesized using agar and cysteine for colon targeting by Satheesh Madhav, *et al.* [52]. The *in-vitro* release was evaluated in rat fecal matter using Shimadzu 1800 UV spectrophotometer at 488 nm. The *in vitro* release results reveal that the novel agar-azo derivative has 100 % drug release in 90 min. Thus, the conclusion was drawn that the synthesized agar azo derivative can be used as an excipient for formulating colon-targeted drug delivery systems.

7. Role of natural probiotic formulation which main gut and colon environment

Probiotics are small microorganisms that deliver health benefits to their host. Our bodies need all sorts of microorganisms to keep things working and keep us healthy and probiotics play a role in this. Probiotics are the good bacteria that you actually need in your system. They support healthy body systems from mouth to gut and help to control harmful microorganisms like germs. At the right levels, Probiotics aid digestion and improve nutrient absorption. Probiotics occur naturally in fermented foods and cultured milk but it can also find manufactured probiotic supplements.

According to the Food and Agricultural Organization of the United Nations and World Health Organization, probiotics may be defined as 'the living microorganisms which when administered in sufficient amounts confer health benefits on the host. The mechanisms involved in probiotics are the manipulation of intestinal microbial communities, suppression of pathogens, immunomodulation, stimulation of epithelial cell proliferation and differentiation, and fortification of the intestinal barrier [53,54].

Probiotics may manipulate the intestinal microbial communities and suppress the growth of pathogens by inducing the host's production of β-defensin and IgA. Probiotics have the ability to strengthen the intestinal barrier by maintaining tight junctions and inducing mucin production. Probiotic-mediated immunomodulation may occur through mediation of cytokine secretion through signaling pathways such as NFκB and MAPKs which can also affect proliferation and differentiation of immune cells (such as T cells) or epithelial cells. Gut motility and nociception may be modulated by the regulation of pain receptor expression and secretion of neurotransmitters. A proliferation-inducing ligand viz. hsp, heat shock protein, IEC, intestinal epithelial cell, Ig, immunoglobulin, MAPK, mitogen-activated protein kinase, NFκB, nuclear factor-kappaB, pIgR, polymeric immunoglobulin receptor, STAT, signal transducer and activator of transcription, Treg, T regulatory cell [55].

8. Limitations and challenges in colon targeted drug delivery

a) One challenge in the development of colon-specific drug delivery systems is to establish an appropriate dissolution testing method to evaluate the designed system *in vitro*. This is due to the rationale after a colon specific drug delivery system is quite diverse.

b) As a site for drug delivery, the colon offers a near neutral pH, reduced digestive enzymatic activity, a long transit time, and increased responsiveness to absorption enhancers; however, the targeting of drugs to the colon is very complicated. Due to its location in the distal part of the alimentary canal, the colon is particularly difficult to access. In addition to that the wide range of pH values and different enzymes present throughout the gastrointestinal tract, through which the dosage form has to travel before reaching the target site, further complicating the reliability and delivery efficiency.

c) Successful delivery through this site also requires the drug to be in solution form before it arrives in the colon or alternatively, it should dissolve in the luminal fluids of the colon, but this can be a limiting factor for poorly soluble drugs as the fluid content in the colon is much lower and it is more viscous than in the upper part of the GI tract

d) In addition, the stability of the drug is also a concern and must be taken into consideration while designing the delivery system. The drug may potentially bind in a nonspecific way to dietary residues, intestinal secretions, mucus, or faecal matter.

e) The resident microflora can also affect colonic performance via metabolic degradation of the drug. Lower surface area and relative 'tightness' of the tight junctions in the colon can also restrict drug transport across the mucosa and into the systemic circulation.

In view of the above limitations, nano formulations have been trying to develop to target the colonic-specific treatment [5, 7, 8].

9.   Colon-targeted nano drug delivery

Targeted nanoparticles are those that contain ligands on their surface and are capable of specifically recognizing cells. Targeted nanoparticles could take advantage of differentially expressed molecules on the surface of tumor cells, effectively releasing cytotoxic drugs. Targeted nanoparticles exploit antigens differentially expressed on the surface of cancer cells, such as integrin and folic acid receptors [56]. Nano particles have generated substantial advances in pharmacology, decreasing the side effects of cytotoxic drugs and improving their efficacy, solubility, pharmacokinetics and biodistribution of anti-cancer drugs. several nano-formulations such as Quantum dots, Iron oxide, Carbon nanotubes, Poly Lactic-co-Glycolic Acid Nanoparticles, Liposomes, Dendrimers, Silica Nanoparticles, Nanoemulsion, Core-Shell Polymeric Nano-Formulations, Gold nanoparticles, etc. have been targeted to the colon [57-58].

9.1   Parameters influencing colon-targeted nano drug delivery

Many parameters influence the development of the colon targeted drug delivery system as mentioned below [59-62].

9.1.1   Anatomical/physiological factors

The human large intestine is approximately 1.5 m long and forms the ascending, transverse, and descending colon with a small distal part forming the rectum. The colon is 2 to 3 inches in diameter, and its lumen is lined with mucus. The physiology of the colon differs significantly from other segments of the gastrointestinal tract (GIT). Moreover, the physiology and the physical properties of the colonic contents also differ between the ascending, transverse, descending, and sigmoidal colon. In addition, there exists variability in the movement of food and dosage forms across the colon, which may present a challenge in the development of colonic drug delivery systems. Another physiological factor that affects colonic drug delivery and bioavailability is the variation in pH of the GIT. Significant intra- and inter-subject variability in the pH of the GIT has been observed during disease states, fasted/fed states, sexes, and ages in humans. Factors such as the viscosity and volume of colonic fluids, the presence of microbial enzymes, and the resulting colonic metabolism are other important factors influencing colon-targeted nano drug delivery.

9.1.2 Intestinal-colonic transit time

The intestinal-colonic transit time influences the performance of colon-targeted nano-drug delivery and the colonic bioavailability of drugs. The transit times are also influenced by colonic disease states such as ulcerative colitis (UC) and Crohn's disease (CD). Patients with UC are known to have shorter colonic times (~24 h) compared to healthy subjects (~52 h). The transit of dosage forms generally depends on the time of administration, the presence/absence of food, and the type of dosage form.

9.1.3 Colonic fluid volume

The average human food intake is approximately 1.5 kg/day and mainly consists of undigested proteins, carbohydrates, and fats. These food components may serve as substrates for the microbial enzymes in the colon. The colon has a high water-absorbing capacity and can absorb about 90 % of the water entering the colon. The colonic fluid volume is calculated to be in the range of 1 to 44 ml with an average volume of approximately 13 ml. Due to this low volume of colonic fluids, the dissolution of drugs from the dosage forms becomes challenging and may affect the local drug bioavailability.

9.1.4 Viscosity of Colonic Luminal Contents

Due to a higher water-absorbing capacity, the viscosity of the colonic luminal contents is higher than upper GIT contents and presents a challenge for the dissolution of targeted nano drug delivery systems. Moreover, the viscosity of the contents progressively increases as it transits from the ascending colon towards the descending colon, resulting in reduced drug dissolution and mucosal absorption. Viscosity also influences the penetration of the drug into the disease-causing bacteria in the colon.

9.1.5 Colonic pH

The pH varies significantly between different regions of the GIT. For example, the pH of gastrointestinal contents can be as low as 1 to 2 in the stomach and rise to 7.5 in the distal small intestine. The pH then declines from the end of the small intestine to the colon and gradually increases once again in the colon. The pH of the colon may be influenced by a carbohydrate-rich diet. This is due to the fermentation of polysaccharides by colonic bacteria and the subsequent formation of short-chain fatty acids.

Similarly, polysaccharide-based drugs may also alter colonic pH. Laxative drugs like lactulose are known to be fermented by colonic bacteria to produce lactic acid and reduce colonic pH. Gastrointestinal disease states such as UC have also been found to influence the colonic pH. The pH of the colon affects the pharmacokinetic and pharmacodynamic behavior of a targeted nano-drug delivery system by controlling the solubility of drugs in the colonic fluid.

9.1.6 Colonic enzymes and metabolism

The colon is known to consist of over 400 different species of aerobic and anaerobic microorganisms like *Escherichia coli* and *Clostridium* species, respectively. These bacteria contain several hydrolytic and reductive metabolizing enzymes such as azoreductase, α-arabinosidase, β-galactosidase, β-glucuronidase, β-xylosidase, deaminase, nitroreductase, urea and hydroxylase. The colonic enzymes catalyze a range of reactions, including the metabolism of xenobiotics (*e.g.*, drugs) and other biomolecules (*e.g.*, bile acid), deactivation of harmful metabolites as well as carbohydrate and protein fermentation. Polysaccharides such as chitosan, guar gum, pectin, *etc.*, are commonly employed as release rate-controlling components in colon-targeted dosage forms. These polysaccharides are known to be resistant to gastric and intestinal enzymes but are metabolized by anaerobic bacteria in the colon. Drugs are also known to be susceptible to biotransformation by colonic enzymes. The metabolism of drugs by the colonic enzymes may result in the formation of metabolites that are pharmacologically active or inactive.

9.1.7 Formulation factors

The formulation factors that influence colonic drug delivery and bioavailability include the physicochemical properties, dose, and the dosage form factors of drugs. Due to the lower amount (1 to 44 ml) of colonic fluid available for dissolution, the solubility and the dose of a drug become important factors for its colonic bioavailability. Although the highly potent drug budesonide (dose, 9 mg) has lower aqueous solubility, it is absorbed well in the colon and is used successfully in the treatment of UC. Mesalamine has a significantly higher solubility (3.64 mg/ml) compared to budesonide (0.24 mg/ml); however, it also has a significantly higher dose (4.8 g daily) which becomes a rate-limiting factor for its colonic absorption.

9.1.8 Mucus

The small intestine has a loose and unattached mucosal layer, while the stomach and colon have a two-layered mucus system composed of an outer loosely adherent layer and an inner firmly adherent layer. It is claimed that mucus is the primary barrier against oral drug bioavailability due to the difficulty of diffusing into the mucus layer. The inner mucus layer facilitates the absorption and improves the drug uptake efficiency, justifying the dual role of mucus in the absorption/desorption of orally administered drugs. Goblet cells are continuously secreted mucus; consequently, it is cleared in each part of the GIT (50 to 270 min). The fluid environment of the colon allows digestive enzymes to contact with food, and this aids in transit the intestinal contents along the GIT without damaging the epithelial lining and it facilitates the nutrients and drugs solubility and absorption.

9.2   Colon targeted nano formulations using natural polymers: Chitosan

Chitosan is a functional linear polymer obtained from the alkaline deacetylation of chitin. Chitosan is a linear polysaccharide, comprised of randomly repeated units of N-acetyl-D-glucosamine and D-glucosamine linked by β (1→4) glycosidic bonds. Chitosan is a nontoxic, biodegradable, biocompatible, and bioactive polymer. Chitosan is used as an excipient and drug carrier in drug delivery systems. Chitosan is used for colon-targeted drug delivery because it has a tendency to dissolve in the acidic pH of the stomach but gets swollen in the intestinal pH. Chitosan is the most valuable derivative of chitin, the second most abundant naturally occurring polymer following cellulose. Chitin can be found in the exoskeleton of insects and crustaceans, in addition to the cell walls of certain fungi. Chitosan, the partially deacetylated form of chitin, can be obtained by chitin deacetylase enzyme or chemical alkaline treatment using concentrating NaOH.

Interaction between chitosan and colon mucosa mucus sticks to most particles, as a tenacious barrier prevents their penetration to the epithelial surface. The mucus layer is continuously secreted and degraded or digested. Mucins consist of highly glycosylated proteins (>106 Da) that are the main components of mucus covering epithelia. They are involved with several functions such as forming a viscous matrix that controls the adhesion of the mucosal layer to the cells as well as drug targeting and pathogens penetration [63-65].

9.2.1 Chitosan derivatives in colon targeted drug delivery system

Exploration of biochemical mechanisms of chitosan polymer has received immense attention in developing nanocarriers as drug delivery systems for anticancer agents. In terms of colorectal cancer targeting, chitosan can form complexes with anionic drugs, reduce the premature release of entrapped drugs at the upper GIT and release the encapsulant at the colon part of the body due to its biodegradability via glycosidic linkage lysis by means of the specified enzyme of the colonic micro flora.

9.2.1.1 Trimethyl chitosan

Trimethyl chitosan (TMC) is synthesized through the reaction of chitosan with methyl iodide in a strong alkaline environment at elevated temperatures. The iodide ion is then substituted by chloride by means of an ion exchange process. N-Methylated chitosan with the hydrophobic groups $N(CH_3)_2$ and the hydrophilic groups $N^+(CH_3)_3$ is amphiphilic, thus it can be self-assembled into vesicles and also is considered as a suitable derivative for nanoparticles preparation. TMC has been proven to be useful in enhancing the paracellular transport of targeted drug delivery systems through increasing the retention time of encapsulant on the mucosal surface as well as binding the protonated amino groups on the C-2 position with negatively-charged mucus and by opening the tight junctions of epithelial cells. In addition, the degree of quaternization increases the permeation-enhancing effect. Several studies have applied TMC due to its solubility in a wide range of pH (1–9), thus the advantageous properties of the polymer are maintained regardless of the varying pH values along GIT [65-67].

9.2.1.2 Carboxymethyl chitosan

Carboxymethyl chitosan is developed by introducing a carboxymethyl group in the structure of chitosan as a hydrophilic modification. It is synthesized by carboxymethylation of the amine and hydroxyl moieties of chitosan. This modification increases chitosan solubility in neutral and basic solutions without negatively affecting other important characteristics. Carboxymethylation reaction takes place especially either at C-6 hydroxyl groups or at the NH_2 moiety that produces water-soluble N,O–carboxymethyl chitosan compounds and comprises either a primary amino group ($- NH_2$) or as secondary amine ($-NH-CH_2COOH$). Carboxymethyl chitosan has been synthesized with different degrees of carboxymethyl substitution and molecular weights to prepare nanoparticles by means of ionotropic gelation with calcium ions for targeted anticancer drug delivery.

9.2.1.3 N-Succinyl-chitosan

N-succinyl-chitosan, which is an acyl chitosan derivative obtained by the introduction of succinyl groups into the chitosan N-terminal of the glucosamine units. Polyion complexes can be generated between the NH^{3+} and -COOH- groups in the succinyl chitosan. The degree of succinylation can

Materials Research Forum LLC
https://doi.org/10.21741/9781644902998-5

be easily modulated by adjusting reaction conditions through succinic anhydride. N-succinyl-chitosan exhibits a good ability to react with several kinds of agents due to the reactive functional $-NH_2$ and $-COOH$ groups. This derivative has shown good water solubility in a wide pH range with biocompatibility *in vitro* and *in vivo*, low toxicity, long-term retention in the systemic circulation, and high accumulation in tumor tissue.

9.2.1.4 Hyaluronic acid-conjugated chitosan

Hyaluronic acid, a linear glycosaminoglycan consists of alternating β-1,4 linked d-glucuronic acid (GlcUA) and β-1,3 linked N-acetyl-d-glucosamine (GlcN Ac) units, having potential for tumor-targeting moiety through virtue of its high affinity for the cell surface receptors CD44 and RHAMM which, in a cancer stage-specific manner, over expressed in colorectal carcinoma. These lead to enhancing the antitumor efficiency of antitumor agents and increasing drug accumulation in tumor cells by facilitating targeted drug delivery via CD44 and RHAMM. The deposition of hyaluronic acid preferentially on the chitosan-based nanoparticles' surface can improve cells targeting overexpressed receptors, enhance *in vivo* stability, and prolong the circulation times of therapeutic agents.

9.2.1.5 PEG-conjugated chitosan

Polyethylene glycol (PEG) has gained wide recognition as an ideal graft-forming polymer because of its solubility in aqueous and organic solvents as well as having weak immunogenicity, low toxicity, good biodegradability, and biocompatibility. PEGylation is a suitable method for introducing new physicochemical properties to chitosan such as the solubility at a wide range of pH values. It is also a critical way to develop colloidal system NPs as biomedical carriers and can improve the colloid stability of the particle system. PEGylation occurs by the substitution of amino groups of chitosan through the n-succinimidyl ester (NHS) derivative (known as NHS-amine coupling) of PEG. PEG chains prevent the adhesion of opsonins existing in the blood by generating a barrier layer surrounding the particles that can be "invisible" to phagocytic cells. Due to these benefits, chitosan-g-PEG was applied as a potential copolymer for colon cancer drug delivery.

9.2.1.6 Folic acid-conjugated chitosan

Folate receptors are considered as one of the most expressed among receptors on many cancer cell types, besides, their expression levels are limited in normal cells. The cell lines HT-29 and HCT-116 have over levels of folate receptors that can lead to easier interaction with folate-targeted nanoparticles; more release and more uptake of folic acid conjugates at colorectal cancer cells. Folic acid is stable, inexpensive, and has a high affinity for folate receptors. Folic acid can conjugate covalently with chitosan molecules via folate's γ-carboxyl moiety. In particular, the conjugates enter the cells through endocytosis via folate receptors and are then transported to the cell cytoplasm via vesicular trafficking. Later, the unligated folate receptor returns to the cell surface once more to transport more folic acid conjugates. Folic acid-conjugated colloidal systems NPs were used in several studies to improve drug uptake and drug accumulation into colorectal cancer cells.

9.3 Case studies: Applications of nano formulations targeting colon

Sharat, *et al.* enhanced the selectivity and cytotoxicity of poorly water-soluble curcumin by fabricating polymeric nanoparticles (NPs). The localized delivery of CU to the colon was achieved by employing a combination of Eudragit S100 (ES100), as a pH-sensitive polymer and polylactic-co-glycolic acid (PLGA), as a biodegradable polymer. The curcumin loaded dual-functional NPs were prepared by nanoprecipitation method and optimized using Box-Behnken experimental design. The *in vitro* cytotoxicity study of Curcumin Nanoparticles (CU-NPs) in CT26 murine colon carcinoma cells showed higher cytotoxicity in comparison to free drugs. The IC50 values of free curcumin and NPs containing curcumin was found to be 1.43±0.08 µg/ml and 0.25± 0.12 µg/ml, respectively y (p <0.05). They suggested that the dual functional polymeric NPs exhibited a remarkably promising carrier system for effective delivery of such poorly water-soluble dietary phytochemicals in colon cancer [68].

Gao, *et al.* prepared DFNPs using the oil-in-water emulsion method. PLGA was selected as sustained-release polymer, and ES100 and EL30D-55 as pH responsive polymers. The morphology and size distribution of NPs were measured by SEM and DLS technique. The DFNPs were evaluated for the colon targeting, DiR was encapsulated as a fluorescent probe into NPs. Fluorescent distribution of NPs was investigated. The therapeutic potential and *in vivo* transportation of NPs in gastrointestinal tract were evaluated in a colitis model. SEM images and zeta data indicated the successful preparation of DFNPs. This formulation exhibited high loading capacity. Drug release results suggested DFNPs released less than 20 % at the first 6 h in simulated gastric fluid (pH1.2) and simulated small intestine fluid (pH 6.8). A high amount of 84.7 % sustained release from NPs in simulated colonic fluid (pH 7.4) was beyond 24 h. DiR-loaded NPs demonstrated a much higher colon accumulation, suggesting effective targeting due to functionalization with pH and time-dependent polymers. DFNPs could significantly ameliorate the colonic damage by reducing DAI, macroscopic score, histological damage and cell apoptosis. Our results also proved that the potent anti-inflammatory effect of DFNPs is contributed by decrease of NADPH, gene expression of COX-2 and MMP-9 and the production of TNF-α, IL-17, IL-23 and PGE2. DFNPs were proven to exert protective effects through inhibiting the inflammatory response, which could be developed as a potential colon-targeted system [69].

Alaa, *et al.* formulated the solid lipid nanoparticle (SLNs) for the release of 5-fluorouracil (5-FU) inside the colonic medium for local treatment of colon cancer. SLNs were prepared by double emulsion-solvent evaporation technique (w/o/w) using triglyceride esters, Dynasan™ 114 or 118 along with soyalecithin as the lipid parts. Different formulation parameters; including type of Dynasan, soyalicithin: Dynasan ratio, drug: total lipid ratio, and polyvinyl alcohol (PVA) concentration were studied with respect to particle size and drug entrapment efficiency. Results showed that formula 8 (F8) with the composition of 20 % 5-FU, 27 % Dynasan™ 114, and 53 % soyalithicin and F14 (20 % 5-FU, 27% Dynasan™ 118, and 53 % soyalithicin), which were stabilized by 0.5 % PVA, as well as F10 with similar composition as F8 but stabilized by 2% PVA were considered the optimum formulae as they combined small particle sizes and relatively high encapsulation efficiencies. F8 had a particle size of 402.5 nm ± 34.5 with a polydispersity value of 0.005 and an encapsulation efficiency of 51 %, F10 had a 617.3 nm ± 54.3 particle size with 0.005 polydispersity value and 49.1 % encapsulation efficiency, whereas formula F14 showed a particle size of 343 nm ± 29 with 0.005 polydispersity and an encapsulation efficiency of 59.09 %. DSC and FTIR results suggested the existence of lipids in the solid crystalline state. An incomplete biphasic prolonged release profile of the drug from the three formulae was observed in phosphate buffer pH 6.8 as well as a simulated colonic medium containing rat caecal contents. A burst release with magnitudes of 26, 32, and 28.8 % cumulative drug released were noticed in the first hour of samples incubated in phosphate buffer pH 6.8 for both F8, F10, and F14, respectively, followed by a slow-release profile reaching 50, 46.3, and 52 % after 48 h [70].

Bakry, *et al.* designed a novel nanoparticulate drug-delivery vehicle for oral use capable of colon-specific release. A modified double-emulsion solvent evaporation method was used in the preparation of pH-responsive Eudargit RS100 polymeric nanoparticles, loaded with 5-FU/LV combination (5-FU/LV-loaded Eudargit S100 NPs). Our optimized drug-loaded NP showed a pH-responsive drug release and exhibited significantly more cytotoxic actions in cancer-cell lines than free drugs findings open the way for conducting clinical trials for colon malignancies treated with nanoparticles [71].

Bangale, *et al.* developed stable nanoparticulate formulation for sustained release of Prednisolone. Chitosan nanoparticles were prepared by ionic gelation method using tripolyphosphate as a cross-linking agent. Different nanoparticulate formulations were prepared by using 32 factorial designs in which varying the concentration of chitosan (0.1 to 0.3 %), and concentration of tripolyphosphate (0.02 to 0.03 %) as two factors. The effect of these factors on the particle size, % entrapment efficiency and *in vitro* drug release was evaluated to develop an optimized formulation. Particle size, % entrapment efficiency and *in vitro* release of optimized formulation were found to be 168.1 nm, 78.53, and 70.80 % respectively. ANOVA study applied with $p < 0.01$ suggests that the model is significant and contour, and the surface response & overlay plot was contracted to optimize the formulation. Optimized formulation (C-10) showed sustained drug release at the end of 11th hour compared to other formulations. Based on the release kinetic model, the drug release data fit well to the Higuchi model ($r^2 = 0.9935$) indicating the diffusion limited drug release from nanoparticles. Drug release mechanism according to the Korsmeyer-Peppas model was found to have anomalous transport (n = 0.5847). Scanning electron microscopy (SEM) revealed that the nanoparticles were spherical in shape and there was no crystallization of the drug and other excipients. Drug excipients compatibility confirmed by FTIR study [72].

Tozaki, *et al.* developed colon-specific insulin delivery with chitosan capsules. *In vitro* drug release experiments from chitosan capsules containing 5(6)-carboxyfluorescein (CF) were carried out by the rotating basket method with slight modifications. The intestinal absorption of insulin was evaluated by measuring the plasma insulin levels and its hypoglycaemic effects after oral administration of the chitosan capsules containing insulin and additives. Little release of CF from the capsules was observed in an artificial gastric juice (pH 1), or in an artificial intestinal juice (pH 7). However, the release of CF was markedly increased in the presence of rat caecal contents [73].

Kaur, et al. formulated and developed colon-targeted mucopenetrating metronidazole nanoparticles. Metronidazole-loaded chitosan nanoparticles with a pH-sensitive polymer, hydroxypropyl methylcellulose phthalate (HPMCP), were prepared by ionic gelation technique and then coated with Eudragit S100 by the solvent evaporation method. The nanoparticles were optimized using the one variable at a time (OVAT) approach. Further, the nanoparticles were evaluated by scanning electron microscopy (SEM) and zeta sizer, as well as for in-vitro release. Muco-adhesion was evaluated by modified bioadhesion detachment force measurement balance and muco-penetration of fluorescein isothiocyanate (FITC) labeled optimized nanoparticles were determined by microscopic technique. Morphological assessment results revealed smooth, spherical particles with homogeneous distribution and a polydispersity index (PDI) of 0.213. The optimized formulation showed particle size of 202 ± 27 nm, zeta potential of 26.9 ± 2.4 mV as well as an entrapment efficiency of 79 ± 5.4 %. There was a significant difference in drug release between coated (8.46 ± 2.49 %) and uncoated (28.96 ± 4.04 %) nanoparticles at the 5th h in simulated gastric conditions. Muco-adhesion data revealed that uncoated nanoparticles (14.98 × 103 dyne/cm^2) showed higher mucoadhesion detachment force compared to coated (12.34 × 103 dyne/cm2) nanoparticles. Muco-penetration results confirm the retention (for up to 12 h) of the developed formulation at the target site for enhanced therapeutic exposure of the entrapped drug. Eudragit S100 coating of chitosan-HPMCP nanoparticles promotes efficient drug targeting and thus provides a strategy for treating mucosal infections [74].

Sanjay, *et al.* formulated and evaluated the nanocapsules (NC), an oral system designed to achieve site-specific and instant drug release in the colon for effective treatment of IBD. The nanoprecipitation method was used to prepare polymeric NC of PD with pH-responsive polymer Eudragit S100. The effect of several formulation variables such as surfactant, oil, and polymer on the PD-NC properties (average size, drug release rate, and drug entrapment) was investigated. An *in vitro* drug release study was done by changing the pH method and an *in vivo* study on rats was done to ascertain the efficiency of PD-NC to release drugs specifically in the colon. The optimized formulations led to the preparation of PD-NC with a mean size of 567.87 nm and a high encapsulation efficiency of 90.21 %. *In vitro* studies reveal that NC releases the drug after a 4.5 h lag time corresponding to time to reach the colonic region, and *in vivo* studies show that NC releases the drug after a 3 h lag time in rats corresponding to arrival in the colon. The above NC formulation of PD is the targeted drug to the colon and may provide an effective way of treatment of colonic disease [75].

Conclusion

The colonic region of the GIT has become an increasingly important site for drug delivery and absorption. Colon-targeted drug delivery offers considerable therapeutic benefits to patients in terms of both local and systemic treatment. Drug targeting to the diseased colon is advantageous in reducing the systemic side effects, lowering the dose of the drug supplying the drug only when it is required, and maintaining the drug as possible for colon-targeted drug delivery. There is a need to develop a novel approach, which can result in the delivery of drugs in a safe, effective, and less expensive manner with minimum fluctuation in terms of the release of drugs at the target.

References

[1]   M. Ashford, J. Fell, Targeting drugs to the colon: Delivery systems for oral administration, J. Drug Target. 2 (1994) 241–257. https://doi.org/10.3109/10611869408996806.

[2]   R. Kinget, W. Kalala, L. Vervoort, G. Van Den Mooter, Colonic drug targeting, J. Drug Target. 6 (1998) 129–149. https://doi.org/10.3109/10611869808997888.

[3]   S.K. Bajpai, M. Bajpai, R. Dengre, Chemically treated hard gelatin capsules for colon-targeted drug delivery: A novel approach, J. Appl. Polym. Sci. 89 (2003) 2277–2282. https://doi.org/10.1002/app.12478.

[4]   M.H. Irving, B. Catchpole, ABC of colorectal diseases. Anatomy and physiology of the colon, rectum, and anus., Bmj. 304 (1992) 1106–1108. https://doi.org/10.1136/bmj.304.6834.1106.

[5]   Effect of bran on colonic targeting of a novel drug delivery system, Gastroenterology. 108 (1995) A612. https://doi.org/10.1016/0016-5085(95)26762-x.

[6]   R. S, V.K. D, S.R. K, Self-Nanoemulsifying Drug Delivery System, Int. J. Pharm. Sci. Rev. Res. 79 (2023). https://doi.org/10.47583/ijpsrr.2023.v79i02.026.

[7]   M. Sharadha, D. V. Gowda, N. Vishal Gupta, A.R. Akhila, An overview on topical drug delivery system – updated review, Int. J. Res. Pharm. Sci. 11 (2020) 368–385. https://doi.org/10.26452/ijrps.v11i1.1831.

[8]   A.H. El-Kamel, A.A.M. Abdel-Aziz, A.J. Fatani, H.I. El-Subbagh, Oral colon targeted delivery systems for treatment of inflammatory bowel diseases: Synthesis, in vitro and in vivo assessment, Int. J. Pharm. 358 (2008) 248–255. https://doi.org/10.1016/j.ijpharm.2008.04.021.

[9]   K. Shahane, M. Kshirsagar, S. Tambe, D. Jain, S. Rout, M.K.M. Ferreira, S. Mali, P. Amin, P.P. Srivastav, J. Cruz, R.R. Lima, An Updated Review on the Multifaceted Therapeutic Potential of Calendula officinalis L., Pharmaceuticals. 16 (2023) 611. https://doi.org/10.3390/ph16040611.

[10]   K.W. Schroeder, W.J. Tremaine, D.M. Ilstrup, Coated Oral 5-Aminosalicylic Acid Therapy for Mildly to Moderately Active Ulcerative Colitis, N. Engl. J. Med. 317 (1987) 1625–1629. https://doi.org/10.1056/nejm198712243172603.

[11]   S.N. Politis, D.M. Rekkas, Recent Advances in Pulsatile Oral Drug Delivery Systems, Recent Pat. Drug Deliv. Formul. 7 (2013) 87–98. https://doi.org/10.2174/1872211311307020001.

[12]   T. Bussemer, I. Otto, R. Bodmeier, Pulsatile drug-delivery systems, Crit. Rev. Ther. Drug Carrier Syst. 18 (2001) 433–458. https://doi.org/10.1615/critrevtherdrugcarriersyst.v18.i5.10.

[13]   E. Fukui, N. Miyamura, K. Uemura, M. Kobayashi, Preparation of enteric coated timed-release press-coated tablets and evaluation of their function by in vitro and in vivo tests for colon targeting, Int. J. Pharm. 204 (2000) 7–15. https://doi.org/10.1016/S0378-5173(00)00454-3.

[14]   L. Yang, J.S. Chu, J.A. Fix, Colon-specific drug delivery: New approaches and in vitro/in vivo evaluation, Int. J. Pharm. 235 (2002) 1–15. https://doi.org/10.1016/S0378-5173(02)00004-2.

[15]   M. Ramanathan, Formulation and Evaluation of Colon Targeted Matrix Tablets of Ibuprofen, Asian J. Pharm. Res. Dev. 6 (2018) 9–19. https://doi.org/10.22270/ajprd.v6i2.366.

[16]   S. Jose, K. Dhanya, T. Cinu, J. Litty, A. Chacko, Colon targeted drug delivery: Different approaches, J. Young Pharm. 1 (2009) 13. https://doi.org/10.4103/0975-1483.51869.

[17]   M.K. Chourasia, S.K. Jain, Polysaccharides for Colon Targeted Drug Delivery, Drug Deliv. J. Deliv. Target. Ther. Agents. 11 (2004) 129–148. https://doi.org/10.1080/10717540490280778.

[18] O. Carrette, C. Favier, C. Mizon, C. Neut, A. Cortot, J.F. Colombel, J. Mizon, Bacterial enzymes used for colon-specific drug delivery are decreased in active Crohn's disease, Dig. Dis. Sci. 40 (1995) 2641–2646. https://doi.org/10.1007/BF02220454.

[19] D.R. Friend, G.W. Chang, A Colon-specific Drug-Delivery System Based on Drug Glycosides and the Glycosidases of Colonic Bacterial, J. Med. Chem. 27 (1984) 261–266. https://doi.org/10.1021/jm00369a005.

[20] K. Minami, F. Hirayama, K. Uekama, Colon-specific drug delivery based on a cyclodextrin prodrug: Release behavior of biphenylylacetic acid from its cyclodextrin conjugates in rat intestinal tracts after oral administration, J. Pharm. Sci. 87 (1998) 715–720. https://doi.org/10.1021/js9704339.

[21] A.D. McLeod, D.R. Friend, T.N. Tozer, Synthesis and chemical stability of glucocorticoid-dextran esters: potential prodrugs for colon-specific delivery, Int. J. Pharm. 92 (1993) 105–114. https://doi.org/10.1016/0378-5173(93)90269-L.

[22] A.D. McLeod, L. Tolentino, T.N. Tozer, Glucocorticoid-dextran conjugates as potential prodrugs for colonspecific delivery: Steady-state pharmacokinetics in the rat, Biopharm. Drug Dispos. 15 (1994) 151–161. https://doi.org/10.1002/bdd.2510150207.

[23] B. Haeberlin, W. Rubas, H.W. Nolen, D.R. Friend, In Vitro Evaluation of Dexamethasone-β-D-Glucuronide for Colon-Specific Drug Delivery, Pharm. Res. An Off. J. Am. Assoc. Pharm. Sci. 10 (1993) 1553–1562. https://doi.org/10.1023/A:1018956232628.

[24] H.W. Nolen, R.N. Fedorak, D.R. Friend, Steady-state pharmacokinetics of corticosteroid delivery from glucuronide prodrugs in normal and colitic rats, Biopharm. Drug Dispos. 18 (1997) 681–695. https://doi.org/10.1002/(SICI)1099-081X(199711)18:8<681::AID-BDD56>3.0.CO;2-A.

[25] J. Nakamura, K. Asai, K. Nishida, H. Sasaki, A Novel Prodrug of Salicylic Acid, Salicylic Acid-Glutamic Acid Conjugate Utilizing Hydrolysis in Rabbit Intestinal Microorganisms, Chem. Pharm. Bull. 40 (1992) 2164–2168. https://doi.org/10.1248/cpb.40.2164.

[26] J. Nakamura, M. Kido, K. Nishida, H. Sasaki, Hydrolysis of salicylic acid-tyrosine and salicylic acid-methionine prodrugs in the rabbit, Int. J. Pharm. 87 (1992) 59–66. https://doi.org/10.1016/0378-5173(92)90227-S.

[27] R.P. Chan, D.J. Pope, A.P. Gilbert, P.J. Sacra, J.H. Baron, J.E. Lennard-Jones, Studies of two novel sulfasalazine analogs, ipsalazide and balsalazide, Dig. Dis. Sci. 28 (1983) 609–615. https://doi.org/10.1007/BF01299921.

[28] C.P. Willoughby, J.K. Aronson, H. Agback, N.O. Bodin, S.C. Truelove, Distribution and metabolism in healthy volunteers of disodium azodisalicylate, a potential therapeutic agent for ulcerative colitis, Gut. 23 (1982) 1081–1087. https://doi.org/10.1136/gut.23.12.1081.

[29] D.R. Swanson, B.L. Barclay, P.S.L. Wong, F. Theeuwes, Nifedipine gastrointestinal therapeutic system, Am. J. Med. 83 (1987) 3–9. https://doi.org/10.1016/0002-9343(87)90629-2.

[30] E. Schacht, A. Gevaert, E.R. Kenawy, K. Molly, W. Verstraete, P. Adriaensens, R. Carleer, J. Gelan, Polymers for colon specific drug delivery, J. Control. Release. 39 (1996) 327–338. https://doi.org/10.1016/0168-3659(95)00184-0.

[31] V.R. Sinha, R. Kumria, Polysaccharides in colon-specific drug delivery, Int. J. Pharm. 224 (2001) 19–38. https://doi.org/10.1016/S0378-5173(01)00720-7.

[32] V.R. Sinha, R. Kumria, Binders for colon specific drug delivery: An in vitro evaluation, Int. J. Pharm. 249 (2002) 23–31. https://doi.org/10.1016/S0378-5173(02)00398-8.

[33] O.A. Cavalcanti, G. Van Den Mooter, I. Caramico-Soares, R. Kinget, Polysaccharides as excipients for colon-specific coatings. Permeability and swelling properties of casted films, Drug Dev. Ind. Pharm. 28 (2002) 157–164. https://doi.org/10.1081/DDC-120002449.

[34] H. Rajpurohit, P. Sharma, S. Sharma, A. Bhandari, Polymers for colon targeted drug delivery, Indian J. Pharm. Sci. 72 (2010) 689–696. https://doi.org/10.4103/0250-474X.84576.

[35] Y.V.R. Prasad, Y.S.R. Krishnaiah, S. Satyanarayana, In vitro evaluation of guar gum as a carder for colon-specific drug delivery, J. Control. Release. 51 (1998) 281–287. https://doi.org/10.1016/S0168-3659(97)00181-8.

[36] D. Wong, S. Larrabee, K. Clifford, J. Tremblay, D.R. Friend, USP dissolution apparatus III (reciprocating cylinder) for screening of guar-based colonic delivery formulations, J. Control. Release. 47 (1997) 173–179. https://doi.org/10.1016/S0168-3659(97)01633-7.

[37] Y.S.R. Krishnaiah, P. Veer Raju, B. Dinesh Kumar, V. Satyanarayana, R.S. Karthikeyan, P. Bhaskar, Pharmacokinetic evaluation of guar gum-based colon-targeted drug delivery systems of mebendazole in healthy volunteers, J. Control. Release. 88 (2003) 95–103. https://doi.org/10.1016/S0168-3659(02)00483-2.

[38] Y.S.R. Krishnaiah, S. Satyanarayana, Y. V. Rama Prasad, Studies of guar gum compression-coated 5-aminosalicylic acid tablets for colon-specific drug delivery, Drug Dev. Ind. Pharm. 25 (1999) 651–657. https://doi.org/10.1081/DDC-100102221.

[39] A. Rubinstein, R. Radai, M. Ezra, S. Pathak, J.S. Rokem, In Vitro Evaluation of Calcium Pectinate: A Potential Colon-Specific Drug Delivery Carrier, Pharm. Res. An Off. J. Am. Assoc. Pharm. Sci. 10 (1993) 258–263. https://doi.org/10.1023/A:1018995029167.

[40] G. Dupuis, O. Chambin, C. Génelot, D. Champion, Y. Pourcelot, Colonic drug delivery: Influence of cross-linking agent on pectin beads properties and role of the shell capsule type, Drug Dev. Ind. Pharm. 32 (2006) 847–855. https://doi.org/10.1080/03639040500536718.

[41] A. Rubinstein, D. Nakar, A. Sintov, Chondroitin sulfate: A potential biodegradable carrier for colon-specific drug delivery, Int. J. Pharm. 84 (1992) 141–150. https://doi.org/10.1016/0378-5173(92)90054-6.

[42] A. Rubinstein, D. Nakar, A. Sintov, Colonic drug delivery: Enhanced release of indomethacin from cross-linked chondroitin matrix in rat cecal content, Pharm. Res. 9 (1992) 276–278. https://dol.org/10.1023/a:1018910128452.

[43] E. Harboe, C. Larsen, M. Johansen, H.P. Olesen, Macromolecular prodrugs. XIV. Absorption characteristics of naproxen after oral administration of a dextran T-70-naproxen ester prodrug in pigs, Int. J. Pharm. 53 (1989) 157–165. https://doi.org/10.1016/0378-5173(89)90239-1.

[44] A.D. McLeod, R.N. Fedorak, D.R. Friend, T.N. Tozer, N. Cui, A glucocorticoid prodrug facilitates normal mucosal function in rat colitis without adrenal suppression, Gastroenterology. 106 (1994) 405–413. https://doi.org/10.1016/0016-5085(94)90599-1.

[45] L. Gautam, S.K. Prajapati, P. Shrivastava, S.P. Vyas, Bioinspired and biomimetic conjugated drug delivery system(s): A biohybrid concept combining cell(s) and drug delivery carrier(s), Smart Polym. Nano-Constructs Drug Deliv. (2023) 465–483. https://doi.org/10.1016/b978-0-323-91248-8.00009-x.

[46] A.D. McLeod, D.R. Friend, T.N. Tozer, Glucocorticoid–dextran conjugates as potential prodrugs for colon-specific delivery: Hydrolysis in rat gastrointestinal tract contents, J. Pharm. Sci. 83 (1994) 1284–1288. https://doi.org/10.1002/jps.2600830919.

[47] V.J. Stella, R.A. Rajewski, Cyclodextrins: Their future in drug formulation and delivery, Pharm. Res. 14 (1997) 556–567. https://doi.org/10.1023/A:1012136608249.

[48] J. Thibault, J. Fockedey, P. Dysseler, B. Quemener, P. Coussement, D. Hoffem, Determination of Inulin and Oligofructose in Food Products (Modified AOAC Dietary Fiber Method), Complex Carbohydrates in Foods. (1999). https://doi.org/10.1201/9780203909577.ch17.

[49] L. Vervoort, R. Kinget, In vitro degradation by colonic bacteria of inulinHP incorporated in Eudragit RS films, Int. J. Pharm. 129 (1996) 185–190. https://doi.org/10.1016/0378-5173(95)04322-5.

[50] J.H. Cummings, S. Milojevic, M. Harding, W.A. Coward, G.R. Gibson, R.L. Botham, S.G. Ring, E.P. Wraight, M.A. Stockham, M.C. Allwood, J.M. Newton, In vivo studies of amylose-and ethylcellulose-coated [13C]glucose microspheres as a model for drug delivery to the colon, J. Control. Release. 40 (1996) 123–131. https://doi.org/10.1016/0168-3659(95)00186-7.

[51] M. Ojha, N.S. Madhav, A. Singh, Synthesis and evaluation of sodium carboxymethyl cellulose azo polymer for colon specificity, Int. Curr. Pharm. J. 1 (2012) 209–212. https://doi.org/10.3329/icpj.v1i8.11252.

[52] M. Ojha, N.S. Madhav, A. Singh, Synthesis and evaluation of sodium carboxymethyl cellulose azo polymer for colon specificity, Int. Curr. Pharm. J. 1 (2012) 209–212. https://doi.org/10.3329/icpj.v1i8.11252.

[53] S. Muzammil, J. Neves Cruz, R. Mumtaz, I. Rasul, S. Hayat, M.A. Khan, A.M. Khan, M.U. Ijaz, R.R. Lima, M. Zubair, Effects of Drying Temperature and Solvents on In Vitro Diabetic Wound Healing Potential of Moringa oleifera Leaf Extracts, Molecules. 28 (2023) 710. https://doi.org/10.3390/molecules28020710.

[54]   P. Markowiak, K. Ślizewska, The role of probiotics, prebiotics and synbiotics in animal nutrition, Gut Pathog. 10 (2018). https://doi.org/10.1186/s13099-018-0250-0.

[55]   S. Nandi, S. Ahmed, A. Saxena, A.K. Saxena, Exploring the Pathoprofiles of SARS-COV-2 Infected Human Gut–Lungs Microbiome Crosstalks, Probiotics, Prebiotics, Synbiotics, and Postbiotics. (2023) 217–235. https://doi.org/10.1007/978-981-99-1463-0_12.

[56]   S. Hua, E. Marks, J.J. Schneider, S. Keely, Advances in oral nano-delivery systems for colon targeted drug delivery in inflammatory bowel disease: Selective targeting to diseased versus healthy tissue, Nanomedicine Nanotechnology, Biol. Med. 11 (2015) 1117–1132. https://doi.org/10.1016/j.nano.2015.02.018.

[57]   D. Ganguly, A. Choudhury, S. Majumdar, Nanotechnology Approaches for Colon Targeted Drug Delivery System: A Review, J. Young Pharm. 15 (2023) 233–238. https://doi.org/10.5530/jyp.2023.15.32.

[58]   F.S. Alves, J.N. Cruz, I.N. de Farias Ramos, D.L. do Nascimento Brandão, R.N. Queiroz, G.V. da Silva, G.V. da Silva, M.F. Dolabela, M.L. da Costa, A.S. Khayat, J. de Arimatéia Rodrigues do Rego, D. do Socorro Barros Brasil, Evaluation of Antimicrobial Activity and Cytotoxicity Effects of Extracts of Piper nigrum L. and Piperine, Separations. 10 (2023) 21. https://doi.org/10.3390/separations10010021.

[59]   A. Imanparast, L. Hamzehzadeh, A. Pasdar, New Approaches to Colorectal Cancer Treatment; An Overview, Cancer Press. 2 (2016) 36. https://doi.org/10.15562/tcp.25.

[60]   M.H. Sarfraz, M. Zubair, B. Aslam, A. Ashraf, M.H. Siddique, S. Hayat, J.N. Cruz, S. Muzammil, M. Khurshid, M.F. Sarfraz, A. Hashem, T.M. Dawoud, G.D. Avila-Quezada, E.F. Abd_Allah, Comparative analysis of phyto-fabricated chitosan, copper oxide, and chitosan-based CuO nanoparticles: antibacterial potential against Acinetobacter baumannii isolates and anticancer activity against HepG2 cell lines, Front. Microbiol. 14 (2023) 1188743. https://doi.org/10.3389/fmicb.2023.1188743.

[61]   S. Kaur, R.K. Narang, G. Aggarwal, Formulation and development of colon-targeted mucopenetrating metronidazole nanoparticles, Trop. J. Pharm. Res. 16 (2017) 967–973. https://doi.org/10.4314/tjpr.v16i5.1.

[62]   M. Kumar, U. Kumar, A.K. Singh, Therapeutic Nanoparticles: Recent Developments and Their Targeted Delivery Applications, Nano Biomed. Eng. 14 (2022) 38–52. https://doi.org/10.5101/nbe.v14i1.p38-52.

[63]   H. Choukaife, S. Seyam, B. Alallam, A.A. Doolaanea, M. Alfatama, Current Advances in Chitosan Nanoparticles Based Oral Drug Delivery for Colorectal Cancer Treatment, Int. J. Nanomedicine. 17 (2022) 3933–3966. https://doi.org/10.2147/IJN.S375229.

[64]   I.M. Ibrahim, Advances in Polysaccharide-Based Oral Colon-Targeted Delivery Systems: The Journey So Far and the Road Ahead, Cureus. (2023). https://doi.org/10.7759/cureus.33636.

[65]   R.B.M. de Almeida, D.B. Barbosa, M.R. do Bomfim, J.A.O. Amparo, B.S. Andrade, S.L. Costa, J.M. Campos, J.N. Cruz, C.B.R. Santos, F.H.A. Leite, M.B. Botura, Identification of a

Novel Dual Inhibitor of Acetylcholinesterase and Butyrylcholinesterase: In Vitro and In Silico Studies, Pharmaceuticals. 16 (2023) 95. https://doi.org/10.3390/ph16010095.

[66]   B.A. Cisterna, N. Kamaly, W. Il Choi, A. Tavakkoli, O.C. Farokhzad, C. Vilos, Targeted nanoparticles for colorectal cancer, Nanomedicine. 11 (2016) 2443–2456. https://doi.org/10.2217/nnm-2016-0194.

[67]   K.E. Wong, S.C. Ngai, K.G. Chan, L.H. Lee, B.H. Goh, L.H. Chuah, Curcumin nanoformulations for colorectal cancer: A review, Front. Pharmacol. 10 (2019). https://doi.org/10.3389/fphar.2019.00152.

[68]   A. Saraf, N. Dubey, N. Dubey, M. Sharma, Curcumin loaded eudragit s100/plga nanoparticles in treatment of colon cancer: Formulation, optimization, and in-vitro cytotoxicity study, Indian J. Pharm. Educ. Res. 55 (2021) S428–S440. https://doi.org/10.5530/ijper.55.2s.114.

[69]   C. Gao, S. Yu, X. Zhang, Y. Dang, D.D. Han, X. Liu, J. Han, M. Hui, Dual functional eudragit® s100/l30d-55 and plga colon-targeted nanoparticles of iridoid glycoside for improved treatment of induced ulcerative colitis, Int. J. Nanomedicine. 16 (2021) 1405–1422. https://doi.org/10.2147/IJN.S291090.

[70]   A.E.B. Yassin, M.D. Khalid Anwer, H.A. Mowafy, I.M. El-Bagory, M.A. Bayomi, I.A. Alsarra, Optimization of 5-fluorouracil solid-lipid nanoparticles: A preliminary study to treat colon cancer, Int. J. Med. Sci. 7 (2010) 398–408. https://doi.org/10.7150/ijms.7.398.

[71]   B. Ibrahim, O.Y. Mady, M.M. Tambuwala, Y.A. Haggag, PH-sensitive nanoparticles containing 5-fluorouracil and leucovorin as an improved anti-cancer option for colon cancer, Nanomedicine. 17 (2022) 367–381. https://doi.org/10.2217/nnm-2021-0423.

[72]   G. V. Shinde, G.S. Bangale, D.G. Umalkar, K.S. Rajesh, ChemInform Abstract: Supercritical Fluids: A Potential Approach for Novel Drug Formulation, ChemInform. 42 (2011) no-no. https://doi.org/10.1002/chin.201126263.

[73]   S. Narade, Y. Pore, Optimization of ex vivo permeability characteristics of berberine in presence of quercetin using 3 2 full factorial design, J. Appl. Pharm. Sci. 9 (2019) 73–82. https://doi.org/10.7324/JAPS.2019.90111.

[74]   H. Tozaki, J. Komoike, C. Tada, T. Maruyama, A. Terabe, T. Suzuki, A. Yamamoto, S. Muranishi, Chitosan Capsule for colon-specific drug delivery: Improvement of insulin absorption from the rat colon, J. Pharm. Sci. 86 (1997) 1016–1021. https://doi.org/10.1021/js970018g.

[75]   S. Kaur, R.K. Narang, G. Aggarwal, Formulation and development of colon-targeted mucopenetrating metronidazole nanoparticles, Trop. J. Pharm. Res. 16 (2017) 967–973. https://doi.org/10.4314/tjpr.v16i5.1.

Nanoparticle Toxicity and Compatibility Materials Research Forum LLC
Materials Research Foundations 161 (2024) 153-181 https://doi.org/10.21741/9781644902998-6

Chapter 6

Bioinspired Nano-Engineering for Plasmon-Enhanced Biosensing Applications

Seemesh Bhaskar[1]*, and Sai Sathish Ramamurthy[2]*

[1]Department of Electrical and Computer Engineering, Holonyak Micro and Nanotechnology Laboratory, Carl R. Woese Institute for Genomic Biology, University of Illinois at Urbana−Champaign, Urbana, IL, 61801, USA

[2] STAR Laboratory, Central Research Instruments Facility (CRIF), Department of Chemistry, Sri Sathya Sai Institute of Higher Learning, Prasanthi Nilayam, Puttaparthi, Anantapur, Andhra Pradesh, India – 515134

seemeshb@illinois.edu, rsaisathish@sssihl.edu.in

Abstract

Recently, interdisciplinary applications are being benefited from functional nanomaterials synthesized using pragmatic bioinspired approaches. Although such bioinspired nano-engineering is gaining the interest of the scientific community, their relevance in the development of plasmon-enhanced biosensing frameworks is seldom discussed. In this chapter, the utility of biopolymers and proteins as excellent candidates for nano-engineering is presented. Case studies where the resulting nanomaterials via frugal approaches are used for surface plasmon-coupled emission (SPCE) based biosensing technologies are highlighted. The advantages of biopolymers such as lycoat, kollidon® and gelucire® as well as proteins such as sesame, castor, sericin and silkworm protein for nanofabrication is presented along with their potential towards development of SPCE based biosensing platforms.

Keywords

Nanomaterials, Bioinspired Nanosynthesis, Plasmonics, Surface Plasmon-Coupled Emission, Biosensing

Contents

1. Introduction

Nanoscience and nanotechnology remain a fascinating research field where the materials are explored (synthesis and applications) in nano regimes (1-100 nm). At such a tiny scale, the materials exhibit dramatically unexpected characteristics which are harnessed by the scientific community for different technological applications. Nano-engineered materials have enabled the miniaturization of electronics, leading to faster and more energy-efficient devices. Consequently, the development of smartphones, computers, and other modern technologies are influenced by the increasing number of scientists working in the field of nanotechnology [1–3]. The materials when broken down to nano dimensions generate mechanical, electrical, thermal as well as physicochemical properties significantly different from the bulk counterparts. Because of this, the nano-engineered materials are investigated all the way from aerospace engineering, drug delivery, electronics, smart textiles, energy storage, self-cleaning surfaces, to development of efficient probes for detection of disease biomarkers such as micro-RNA, DNA, amino acids, exosomal RNA and other molecules and ions [4–8].

The effectiveness of the conventionally studied biosensors has been greatly improved with nano-engineered materials. Such explorations have revolutionized the biosensing portfolio where the high surface to volume ratio of the nano-engineered materials is extensively studied in different biosensing platforms including electrochemical, plasmonics, photonics, biochemical to name a few [9–15]. Among them the field of plasmonics deals with the interaction of light (electromagnetic radiation) with the free electron density of the plasmonic materials (metals). By and large, the term 'plasmon' is often referred to as the quantum of plasma oscillation of electrons, which in other words is also the collective charge density oscillation of electrons in a sold material of interest [16–19]. When the plasmons are localized in space around a nanoparticle, it is referred to as localized surface plasmon resonance (LSPR). In a metal-dielectric interface, the excited plasmons tend to propagate along the interface, and the hence developed plasmons are referred to as surface plasmon polaritons (here the polariton is the coupled oscillation of light with the plasmons of the metal surface) [20–24]. It is worth noting that the plasmonic effects are predominantly observed in the nanoscale objects as compared to their bulk counterparts, thereby leading to novel opto-electronic properties. This is intuitive as the size of metallic nanoparticles are in the size range comparable to that of the wavelength of light. Such explorations of interaction of photons (light radiation) with the plasmonic materials has been utilized in different fields of research including sensing, imaging, microscopy, energy harvesting to name a few [18,19,21,25,26].

Fundamentally, plasmonic nanostructures demonstrate enhanced absorption, scattering, interference, emission of electromagnetic field (EM) radiation. Moreover, it also results in the entrapment and concentration of EM fields in the nanojunctions and nano-vortices. As a result of this these nanostructures are referred to as nanoantennas which can effectively couple the EM field intensity to the surface plasmons of the underlying substructure [27–29]. An interesting aspect of engineering nanomaterials is the feasibility to tailor the size and shape of the nanomaterials, which further enables tuning the absorption and scattering characteristics. The molecular interactions with the plasmonic structures have been explored using relevant chemistry principles and spectroscopy methods, under the broad domain of surface plasmon resonance (SPR). This methodology is commonly used in the development of biosensors with effective molecular binding kinetics (where the changes in the refractive index at the metal-dielectric interface are studied). The subwavelength structural properties of plasmonic NPs enables overcoming the diffraction limit and augmenting the resolutions in different imaging methodologies [18,27,30]. Distinct types of plasmonic materials with unique functionalization have been developed for different applications including superlens imaging, biosensing, drug delivery, photovoltaics, photothermal therapy, metamaterials, telecommunications to name a few. Although a number of technologies and applications of plasmonic nanomaterials are being reported, the utility of bioinspired routes for obtaining the plasmonically active nanomaterials is seldom discussed [31–34]. Although the plasmonic nanomaterials used in the treatment and therapy are often functionalized with biocompatible biochemical molecules, the nanomaterials used in biosensing platforms are often made via toxic chemicals that are not biocompatible [12,13,35]. Moreover, the methodologies used are not economical as elevated temperature and specific pressure conditions are necessary to synthesize plasmonic NPs. In this background, this chapter focuses to discuss the latest developments in the domain of plasmonics where bioinspired materials are used for development of biosensing platforms. As it is not practical to discuss all the several types of such plasmonic platforms and nano-engineering approaches, we have focused the discussion on the latest developments in the innovative field of surface plasmon-coupled emission (SPCE) [28,36–39].

It is instructive to briefly discuss the aspects of metal-enhanced fluorescence (MEF) to present an inaugural foundation for the understanding of SPCE. The phenomenon that occurs when the light emitting species (typically the fluorophores) are in the proximal vicinity of the plasmonic nanomaterials, on account of altered absorption and scattering characteristics of the combined plasmophore (plasmon+fluorophore) is called the MEF effect [40–42]. Typically, noble metals like platinum, silver and gold are used in developing the MEF platforms with novel substrate engineering tools. The emitter molecules couple with the LSPR of the plasmonic nanoparticles and generate enhanced excitation and emission coefficients, under appropriate light irradiation conditions. A large enhancement in the EM field intensity is observed in the nanogap regions and in the sharp nano-corners or vortices, well known as hotspots [43–46]. The MEF effect has been utilized for development of bioassays, cellular imaging, nanoscale probing, molecular detection, and environmental monitoring tools based on luminescence. Despite the considerable success of the MEF platforms, the major limitations include the need for precise positioning of the fluorophores from the metallic surface and optimization of the desired enhancement effects. The inadequacy in such approaches can result in the experimental artifacts leading to misleading results. While the MEF is all about the exploration of light-matter interactions using nanomaterials, the metallic thin films are instead used in the case of SPCE, presenting several advantages over the MEF biosensing platforms. Although there are different SPCE platforms developed for different applications, in this chapter a few case studies are specifically chosen and discussed to highlight the importance of bioinspired nano-engineering for plasmon-enhanced biosensing applications.

2.   Bioinspired nano-engineering: Relevance

To present a comprehensive understanding of the relevance of studying the subject 'bioinspired nano-engineering for plasmon-enhanced biosensing applications,' we present a Scopus analysis as part of Figure 1. The number of publications reported from 2000 to 2022, spanning 22 years, is presented as obtained from the Scopus analysis. It is worth noting that the number of publications increased linearly for searches pertaining to 'biosensing' and 'plasmon+biosensing.' An exponential increase in the number of publications reported has been observed for searches related to 'nano-engineering' and 'bioinspired + nano-engineering', with the steepness of the curve being prominent from 2016-2018 onwards. This is of significance because such a trend reemphasizes the fact that in the past 10 years there has been a significant interest in the scientific community to explore nano-engineering aspects, especially in a bioinspired perspective [13,33,35,47,48].

Further, one may also note that while the number of publications for the broad term 'biosensing' is 604141, the same for plasmon-based biosensing technology is 108816 (Figure 1a and 1b). This reduction in the number of publications is reasonable because of the several other sensing methodologies available for achieving the same purpose of biosensing (such as electrochemical, piezoelectric, thermal, microbial, and enzymatic biosensors to name a few). Although the plasmon-based biosensors are a subset of the broad field of biosensors, it is notable that 18 % of all the existing biosensors are categorized as plasmon-based biosensors, hence validating the relevance of documenting the research outcomes in this domain. Moreover, in recent years bioinspired nano-engineering is gathering more attention on account of several advantages and great health safety. However, it is important to mention that only 5 % of the different types of nano-engineering pursued are being categorized as bioinspired nano-engineering (Figure 1c and 1d). Therefore, although there is an augmented growth in the awareness of utilizing the green chemistry principles for go-green based bioinspired nano-engineering methodologies, there is still room for improvement, to bring down the toxicity induced by the nano-engineering protocols [34,44].

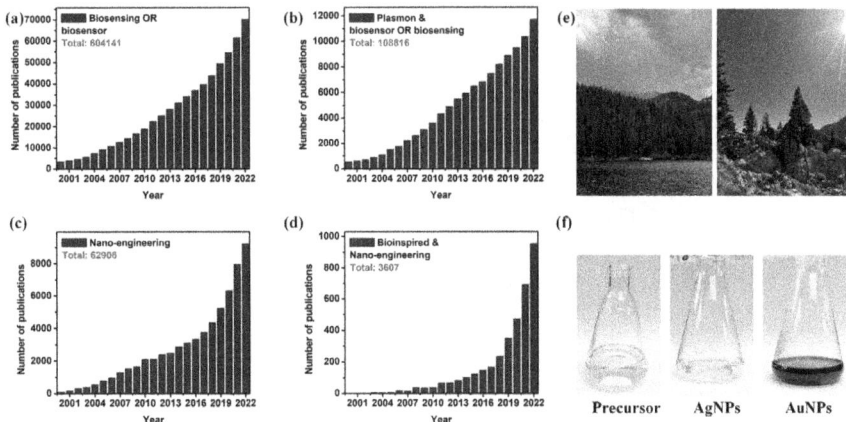

Figure 1. Scopus analysis of the number of publications reported in different domains of scientific research with key words: (a) Biosensing, (b) plasmon & biosensing, (c) nano-engineering, (d) bioinspired & nano-engineering. (e) Representative pictures of the natural resources that may be used in the synthesis and production of nanomaterials [picture credit Kodchakorn Khemtonglang, Nanosensors group, University of Illinois Urbana-Champaign]. (f) Visible colour change observed upon the formation of silver (Ag) and gold (Au) nanoparticles in a typical nanosynthesis methodology.

Figure 5e presents a representative image of the natural resources available for synthesis of the nanomaterials. Recently, there is a rising demand for nanomaterials using plant-based and microorganism-based sources due to their non-hazardous nature towards the environment. Such eco-friendly biomaterials are expected to catalyze the emergence of nano enabled as well as bioinspired smart materials for application in different industries [49–53]. The sustainable routes for nanosynthesis assist in guaranteeing the balanced approach towards providing both the technological use in terms of plasmonic biosensing as well as in maintaining the quality of environment (soil, air, and water quality) with minimal usage of toxic chemicals [40,48,54,55]. A few representative images of the plasmonic nanoparticle solution are shown in Figure 5f. It is worth noting that although the interest in developing the bioinspired routes for nanosynthesis is on the raise, the research progress and commercialization of this area is very marginal on account of the difficulty in understanding the complicated and dynamic attributes of nanomaterial-biomaterial interactions. In this context, the careful evaluation of the synthesized bioinspired nanomaterials in terms of physicochemical properties is of utmost importance for the evolution of green nanotechnology with economic and ecological impact.

3.   Surface plasmon-coupled emission (SPCE): Plasmonics-based biosensing technology

The nano-engineering that is typically carried out for plasmonics based biosensing applications is presented in Figure 2. The left panel shows the different types of plasmonic interfaces that are being developed using the diverse types of nanomaterials. When the electron free density is localized in space around a single nanoparticle, then it is said to comprise the localized surface plasmon resonance (LSPR), under resonant conditions with appropriate incident light coupling. The incident light coupling with the propagating plasmons of metal thin film is achieved using appropriate phase matching conditions and the resulting phenomenon is often referred to as surface plasmon resonance (SPR). This method is used for different SPR based imaging techniques (SPRi) [19–21,24,56].

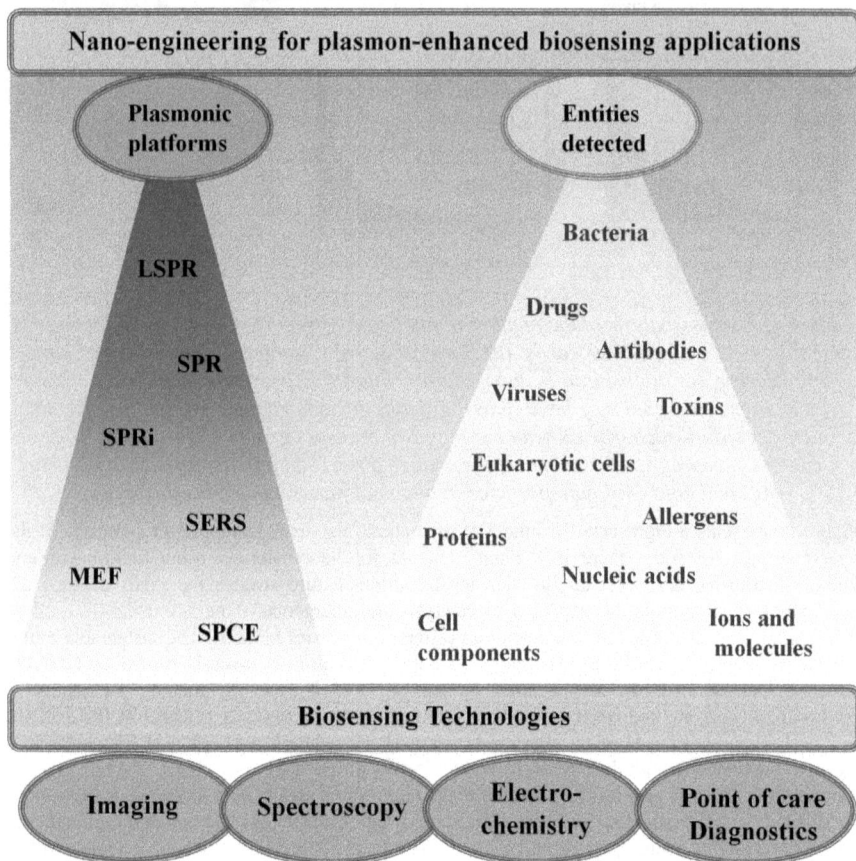

Figure 2. The facets of nano-engineering for plasmon-enhanced biosensing applications. The different plasmonic platforms as well the entities detected are presented. The biosensing technologies that are developed by interfacing the detectable species over the plasmonic platforms are shown in the bottom.

The exploration of plasmonic nanomaterials in the vicinity of fluorescent molecules results in enhanced intensity, which is referred to as metal-enhanced fluorescence (MEF). Further, yet another important field of plasmonic sensing is the surface enhanced Raman scattering (SERS) technique where the Raman active molecules highlight significantly high intensity stretching bands while placed in the proximity of plasmonic nanoparticles. All these techniques have been used in the development of several types of biosensing technologies for sensing bacteria, viruses, drugs, proteins and several other molecules and ions (called as analytes of interest) [13,47,57–59].

Although it is believed that the initial studies of realizing directional emission from a metal-dielectric interface with the dielectric embedded with the fluorophore were reported in 1979, the clear understanding of the phenomenon was not established until Lakowicz and co-workers scrutinized the SPCE technique in the year 2003 [29,41,42]. The important and unique feature of SPCE is the highly directional and polarized emission, accompanied with low background noise, high spectral resolution. The optical setup of a typical SPCE experimentation is presented in Figure 3. By and large, the SPCE experiments are carried out in two configurations, namely, kretschmann (KR) configuration and reverse kretschmann (RK) configuration for satisfying the wave-vector matching conditions [36,60,61]. In order to maintain simplicity, in this chapter we are discussing a few case studies highlighting the utility of SPCE only in RK configuration for biosensing applications.

As shown in Figure 3, in the RK configuration, the laser excitation is performed from the sample side and the emission is collected from the curved surface of the prism. On the other hand, the sample is excited from the prism side in the KR configuration and the emission is collected from the same side [9,32,62,63]. The figures show a typical case in which the rhodamine moiety is used as a fluorophore and the emission is collected from the prism side via a long wave pass filter and a polarizer. The directional emission is highly polarized carrying the characteristic of the SPR phenomenon. On account of this the overall collection efficiency can be enhanced by a factor of 10 as compared to that of the free space emission [64,65]. In addition to this, it has been found that photobleaching and the photodestruction can be largely reduced, especially while operating with biological samples.

The experimental results discussed in this chapter are carried out using two detectors. are acceptable. One is using the conventional Ocean optics detector and the other is the low-cost ubiquitously available smartphone-based detection system. The mobile phone based biosensing is carried out in two separate ways: luminosity and chromaticity. In the former, the luminous intensity of the emission is extracted using the color grab app and correlated with the SPCE enhancements. In the latter, the color of the emission is extracted using the RGB (Red Green Blue) color space and correlated with the spectral shift observed in the SPCE experiments. The angle of SPCE is related to several factors including the wavelength of the luminophores used, thickness of the metal as well as the dielectric, the refractive index of the surrounding medium at the metal-dielectric interface [22,23,66,67]. It is observed that such properties of SPCE aid in distinguishing different analytes of interest in an ensemble, thereby facilitating multiplexing. In this chapter, we shall discuss the bioinspired nano-engineering performed over the SPCE substrate using two dissimilar sources of biomaterials, namely proteins and biopolymers. In the recent past, the focus of nanoscientists has been driven towards utilizing biocompatible and biodegradable materials for nanosynthesis. In this perspective, utilizing such biomaterials for nano-engineering and further for biosensing applications.

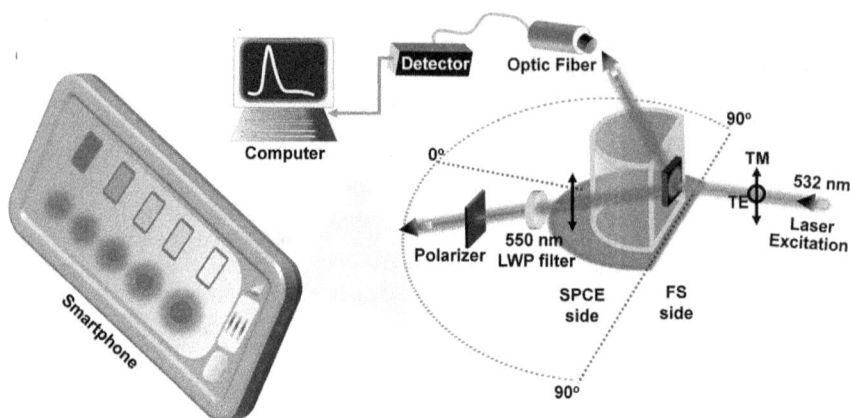

Figure 3. Conceptual schematic of surface plasmon-coupled emission technology. The laser source, polarizer, BK7 prism, filter, metal thin film, optic fibre for collection of emission along with the detector comprises the general optical elements of the setup. The detection is carried out using the Ocean optics spectrophotometer as well as the mobile phone-based detection system. Reprinted with permission from Reference [45] Copyright 2023, American Chemical Society.

4.   Bioinspired nano-engineering using polymers in SPCE

By and large, different bioinspired methods are used in the nanosynthesis. For instance, UV light, sonication, microwave, IR light, heating, and several other simple methodologies that do not demand the use of additional resources (except the biopolymers) have been explored for the synthesis of nanomaterials. In these situations, the performance of the final device has been evaluated and compared with that comprising the nanomaterials synthesized using the toxic chemical agents. In this section a few case studies where the biopolymers are used for the nanosynthesis via go-green approach is discussed. The application thus realized by interfacing the obtained nanomaterials over the metallic thin film in the SPCE platform is presented highlighting their utility in biosensing applications.

4.1   Sensing Hg^{2+} ion using Ag nanocubes obtained via lycoat biopolymer

In SPCE, there is a demand for utilizing nanomaterials with sharp corners and vortices [68,69]. Nanocubes have been explored for realizing high fluorescence enhancements, enabling excellent biosensing platforms [70,71]. However, there is a tradeoff between the requirement of sharp nanogeometries for photoplasmonics applications and the need to utilize bioinspired methods for nanosynthesis. Although different biopolymers are used in different applications on account of their biocompatible nature, rarely are they studied in nanoscience for biosensing applications. In this section, a case study is presented where the biopolymer, lycoat is used for the synthesis of plasmonic AgNPs with cubic geometry [68]. Such Ag nanocubes assist in the realization of large EM field intensity while interfaced with metallic thin films. Such generation of hotspots is facilitated by the coupling between the LSPR of the metallic nanocubes and the propagating plasmons of the metal thin film. The lycoat polymer is a high energy absorber and hence the abundantly available UV light has been used in the nanosythesis [68,72]. It has been observed that a simple mixture of metallic ions and the biopolymer and exposure to the UV light (high energy radiation) results in the generation of plasmonic nanoparticles. Depending on the time of UV exposure and the concentration of the bipolymer and the metal ion used, the size and the shape of the resultant nanomaterial can be tuned [68].

Figure 4a shows the schematic of the sensor development. The mercury ions possess the ability to react with the AgNPs as shown in Figure 4a, resulting in the generation of metal ions (due to the redox chemical reaction between the zero valent Ag (part of AgNPs) and divalent Hg^{2+} ions) [68]. Consequently, the results observed from the changes in the absorbance profile of the AgNPs with the addition of Hg^{2+} ions is presented in Figure 4b. The Ag nanocubes obtained with the lycoat based polymer were mixed with the Hg^{2+} ions at various concentrations and studied over the SPCE platform and the results are presented in Figure 4d. The change in the color of the solution is shown in Figure 4c. The attomolar sensitivity has been achieved using the SPCE sensing platform on account of the performance of Ag nanocubes over the SPCE substrate [68]. The additional details about the biosensor's selectivity towards the other bivalent cations and the elaborated future scope of this research work are discussed in the associated publication [68]. The biosensing was also carried out using smartphone-based detector system. The shade cards corresponding to the different luminosity values are shown in Figure 4e. It it observed that there is an excellent correlation between the luminosity values and the SPCE enhancements, thereby presenting a cost-effective biosensing tool [68].

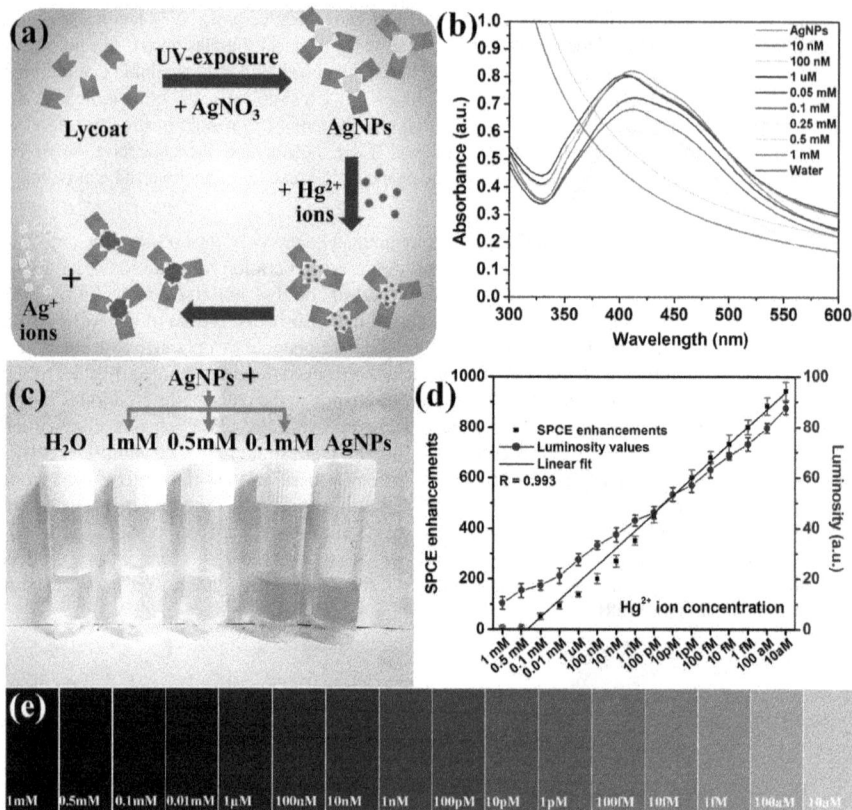

Figure 4. (a) Conceptual illustration of the steps involved in the disintegration of AgNPs to Ag$^+$ ions upon the addition of Hg^{2+} ions. (b) UV−vis absorbance spectra with different concentration of Hg^{2+} ions addition to the solution containing AgNPs. (c) Visual change in the color of the solution of AgNPs observed with addition of different concentrations of Hg^{2+} ions. (d) Correlation between the SPCE enhancements (gathered via spectrophotometer) and the luminosity data (gathered via smartphone). (e) Shade cards corresponding to different luminosity values shown in Figure 8d. Reprinted with permission from Reference [68], Copyright 2021, American Chemical Society.

4.2 Sensing tyrosine molecule using AgAu nanohybrids obtained via kollidon®

In this section, a case study of utilizing a bimetallic nanohybrid for sensing application using SPCE framework is discussed. In the domain of SPCE explorations, the use of plasmonic AuNPs results in predominant quenching of the emission due to the Ohmic losses. To overcome this quenching, different methodologies have been employed, where the sharp edged nanostars as well as metal-dielectric nanohybrids have been studied. However, the SPCE enhancements have not been exceedingly higher in these situations. In this context, the use of bimetallic nanohybrids where the positive aspects of both the metals of the bimetallic nanohybrid is utilized has been explored extensively. Especially, it has been observed that AgAu nanohybrids with sharp nano-curvatures assist in the realization of plasmon passages where the loss in the Au and the low chemical stability of Ag (due to surface oxidation) has been overcome [16,32,34,54]. To synthesize AgAu nanohybrids different methods have been developed, albeit at the expense of using complicated procedures, toxic chemicals, and expensive equipment. In this regard, the Kollidon® which is a cross polymer formed by polymerization of monomer N-vinyl pyrrolidone has been utilized as a biocompatible material for AgAu nanohybrid synthesis. Although Kollidon® has found relevance in pharmaceuticals on account of the bioavailability of the drugs, they are seldom studied for nanosynthesis and related applications [54].

To obtain shape directed nanomaterials, typically several parameters need to be adjusted including the concentration of the starting materials, reducing and capping agents, temperature, specific solvents, pH, surfactant concentration to name a few. Incorporation of these factors dramatically increases the overall cost involved in the nanosynthesis. Considering these observations, a direct method where abundantly available resources are utilized for nanosynthesis is necessary and needed of the hour [54]. In this context, a direct mixture of Kollidon® and metal ions such as Ag^+, and Au^{3+} in an aqueous solution result in the formation of plasmonic AgAu nanohybrids when subjected to UV exposure. This is accomplished because of the high energy absorption capability of the Kollidon® polymer. That is upon light exposure, the light induced, or photo-induced electron transfer occurs in the solution from the Kollidon® to the free moving metallic ions in the solution, thereby reducing them to bimetallic nanorod shaped nanohybrids [54]. These AgAu nanohybrids when interfaced with the metallic thin film / the SPCE substrate assist in the generation of photo-plasmonic hotspots. Interestingly, the rod-shaped morphology sustains the lightning rod effect which is like the effect where the antenna assists in collecting the lighting conditions and grounding the same [54]. These two aspects of lightning rod effect enhanced plasmonic coupling and the bimetallic nanohybrid resulted in over 1000-fold enhancement in the fluorescence signal intensity. This high signal intensity was further used for detecting tyrosine molecules due to its biological relevance, and the results are presented in Figure 5. A consistent shift in the emission maxima was observed for increasing concentrations of the tyrosine molecule as seen in Figure 5a. The calibration plot with a wide sensing range for tyrosine detection using SPCE framework is shown in Figure 5b. Since the SPCE signal showed a steady shift in the emission maxima, the most appropriate configuration for mobile phone-based sensing is dependent on the RGB value-based chromaticity plot as shown in Figure 5c. As the sensing can be performed using the smartphone-based detector as well as the conventional Ocean optics detector, the developed platform is highly conducive for development of biosensors for resource limited settings [54].

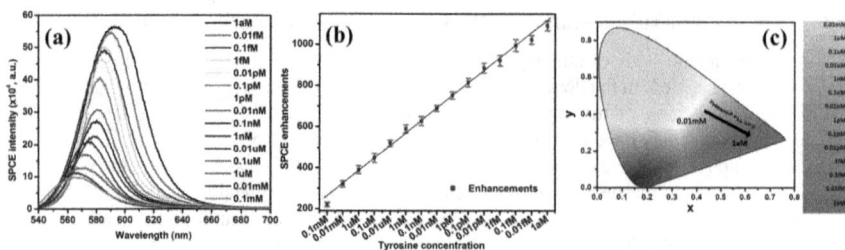

Figure 5. Biosensing performed using Kollidon® based AgAu nanohybrids. (a) SPCE signal intensity for different concentrations of tyrosine from 0.1 mM to 1 aM. (b) Plasmon-coupled emission enhancements for different concentrations of tyrosine from 0.1 mM to 1 aM. (c) Chromaticity diagram presenting a shift in color of the emission observed with different tyrosine concentrations. Reprinted with permission from Reference [54], Copyright 2022, Elsevier.

4.3 Sensing cysteine using AgAu nanohybrids obtained via gelucire®

Another important polymer that has been used in the synthesis of AgAu nanohybrid is the Gelucire® which is a member of the saturated polyglycolized glycerides (consisting of mono, di and tri glycerides and mono, di-fatty acids esters of polyethylene glycols) [40,48]. These polymers consist of a surfactant and a fatty acid part. According to French language, 'gélule' meaning capsule and 'cire' meaning wax are the key components of the name Gelucire®. It contains water-soluble surfactants, water-dispersible surfactants as well as the matrix former. Although this biopolymer has been utilized for several applications including topical formulations for drugs, in the sustained drug release, thermoplastic applications, and solubility enhancer in pharmaceuticals, their utility in nanosynthesis and related applications is seldom explored [40,48]. A similar approach described in the previous section was used for the synthesis of AgAu nanohybrids, where the mixture of the biopolymer and the metal ions were subjected to UV irradiation for defined intervals of time. Depending on the time of UV exposure, the size, and the shape of the resultant AgAu nanohybrids were tailored.

The formation of the bimetallic nanohybrid has been confirmed using UV-Vis absorption, DLS (Dynamic Light Scattering) size distribution as well as the microscopic characterization tools such as SEM (Scanning Electron Microscopy). All the SPCE experiments were carried out in spacer, cavity, and extended cavity nanointerfaces, the details of which are presented in the related publication. These nanointerfaces differ in the manner the nanomaterials are engineered over the SPCE substrate [40]. Such analysis assists in the comprehensive understanding of the photo-plasmonic coupling aspect at the metal-dielectric nanointerface. It was observed that since the AuAg is made of the AuNPs, the quenching effects were dominant in the cavity nanointerface. The extended cavity nanointerface resulted in the highest SPCE enhancements over 1000-fold compared to that observed in the conventional fluorescence spectrophotometer [40].

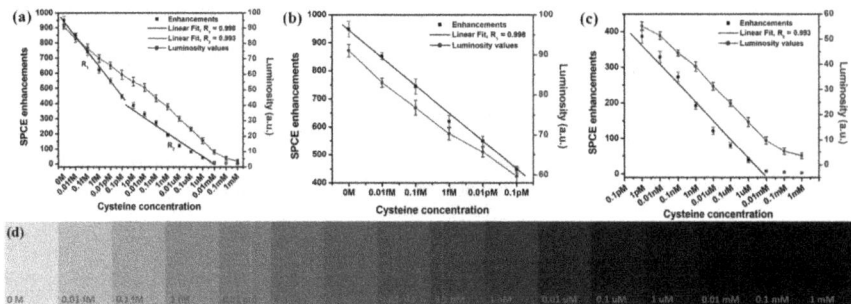

Figure 6. Biosensing performed using Gelucire® based AgAu nanohybrids. Overlap of SPCE enhancements (obtained via conventional detector) and the luminosity (obtained via smartphone detector) given in the left and right y-axis respectively for (a) complete range of cysteine concentration, (b) in the concentration range of 0.01 fM to 1 mM and (c) in the concentration range of 1pM to 1 mM. (d) The gray-scale shade cards corresponding to the luminosity values. Reprinted with permission from Reference [40], Copyright 2022, Elsevier.

The high fluorescence enhancements were used for the biosensing platform's development, and the results are shown in Figure 6. The cysteine molecule comes across as a biologically relevant amino acid with a powerful role in maintaining normal functionalities of human health. The close interaction between the ARS (Alizarin Red-S)-Cu^{2+} ion ensemble and the cysteine in presence of BR (Britton-Robinson) was utilized for sensor development [40]. The addition of cysteine to the ARS-Cu^{2+} ion mixture resulted in the quenching of the ARS fluorescence, which was used for the indirect detection of the cysteine molecules. As seen in Figure 6, the increase in the concentration of the cysteine decreased the SPCE enhancements presenting two linear ranges for the detection of cysteine. Unlike the research work described in the previous section, here the luminosity value was extracted from the photographs of the emission collected using the mobile phone. The extracted luminosity values were plotted against the SPCE enhancements, and the results are presented in Figure 6, with an excellent correlation between both. Although humans are endowed with the ability to distinguish between several colors through naked eyes, the ability to distinguish between different intensities of the same color is seldom successful, leading to biased outcomes. In this context, luminosity values extracted using the ColorGrab App serves as a standard to overcome such trivial errors [40].

5. Bioinspired nano-engineering using proteins in SPCE

In this section, a few case studies where the proteins were used for synthesis of nanomaterials are described. Firstly, the generation of plasmonic monometallic nanomaterials and further the use of bimetallic nanomaterials for generation of photo-plasmonic hotspots in the SPCE platform is presented. The performance of the sensor developed for detecting the SPCE reporter molecule rhodamine is presented to start with. The detection of one molecule and one ion is further elaborated to comprehensively present the utility of bioinspired nano-engineering for biosensing applications.

5.1   Sesame protein-based Ag nanocubes for femtomolar RhB sensing

Diverse types of proteins extracted from plants, animals, insects, and other sources are being explored for the synthesis of plasmonic nanomaterials. The sesame (Sesamum indicum Linn) is one of the condiments that is directly consumed as food ingredient in condiments and snacks and has been explored in scientific research for applications including production of edible films, medication, and nano-beads immobilization. Here, we present the case study where the nanomaterials synthesized using sesame protein have shown the potential towards development of a biosensing technique [55].

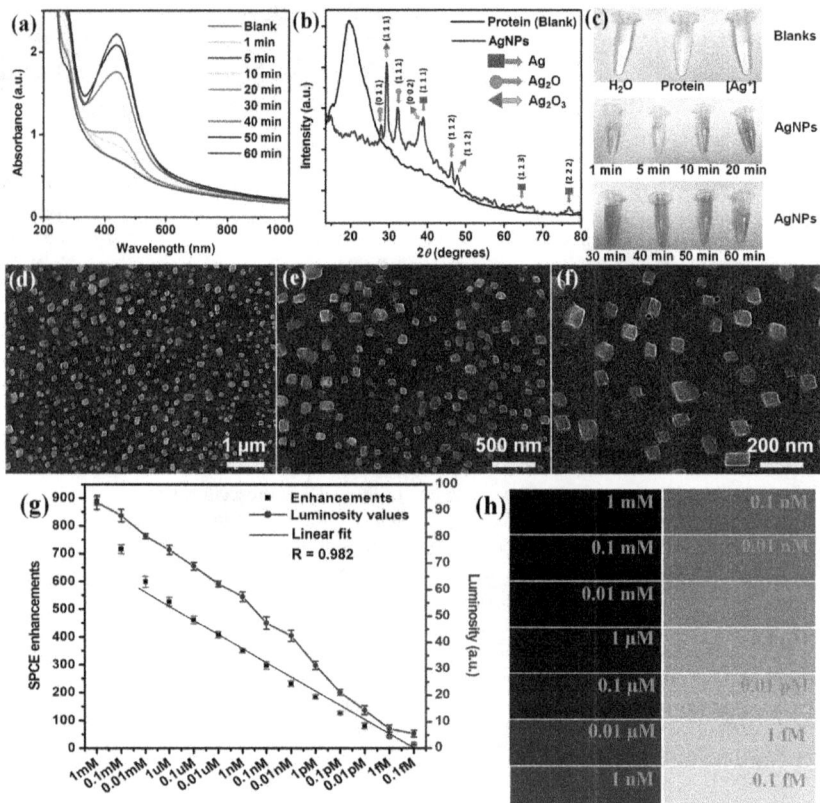

Figure 7. (a) UV–vis-NIR absorbance of AgNPs. (b) XRD pattern of sesame protein and AgNPs. (c) Visual change in the color of the solution upon formation of AgNPs. (d-f) Multiple SEM images with different resolutions of AgNPs obtained at 30 min UV-exposure. (g) Overlap of SPCE enhancements and luminosity values shown for RhB detection along with (h) respective shade cards on the right side. Reprinted with permission from Reference [55], Copyright 2022, Elsevier.

Materials Research Forum LLC
https://doi.org/10.21741/9781644902998-6

As described in the previous sections, the proteins also can reduce the solution of metal ions into their respective zero-valent nanoparticles, upon exposure to UV light via photo-induced electron transfer. The sesame protein was extracted from the de-oiled sesame seed cake. The complete process of extraction of the protein is presented in the associated publication [55]. The UV-vis absorbance of the obtained nanoparticles along with the XRD patterns are shown in Figure 7a and 7b, respectively. The strong bands corresponding to that of the LSPR of AgNPs are seen in the absorbance spectra. An increase in the absorbance band's intensity is correlated with the increase in the number of generated nanoparticles. While the blank sesame protein showed an amorphous pattern in the XRD, the corresponding nanoparticles showed characteristic bands specific to that of the AgNPs [55] Further, the change in the color of the solution observed via naked eyes is shown in Figure 7c. These preliminary results semi-quantitatively confirm the formation of plasmonic AgNPs. Additional characterizations such as SEM, AFM (Atomic Force Microscopy), TEM (Transmission Electron Microscopy), are often performed to confirm the size and shape of the obtained nanomaterial [55]. The representative SEM images are presented in Figures 7d to 7f. It can be observed that the sesame protein-based Ag nanoparticles exhibited a cubic nano-morphology, which inherits the potential to generate extreme plasmonic hotspots in the nanogap generated between the nanocubes and metallic thin film in SPCE platform [55]. Upon interfacing with SPCE thin films, >850-fold enhancements in the fluorescence signal intensity were observed. The SPCE reporter molecule, rhodamine, was detected by decreasing the concentration of the same in the extended cavity nanointerface. It has been observed that femtomolar limit of detection can be achieved [55]. The performance of the biosensing tool is also validated using a smartphone-based detection method as shown in Figures 7g-h.

5.2 Castor protein-based Ag nanorods for attomolar RhB sensing

Another protein underused in the synthesis of nanomaterials is the castor protein. It is well-known for its nutritional quality and used after appropriate processing in the assorted products. The plasmonic AgNPs were synthesized using castor protein as the only reducing and capping agent in the presence of UV-exposure, using a frugal, rapid, and economical method [16]. Unlike the sesame protein, the castor protein generated rod shaped plasmonic AgNPs which have enormous potential to sustain lightning rod effect-based hotspots in SPCE platform. Since modest SPCE enhancements were realized using castor protein-based Ag nanorods, a hybrid nano-engineering approach was incorporated to further boost the SPCE enhancements. To this end, the SPCE substrate was pre-coated with a monolayer of graphene oxide. Recently, diverse types of low-dimensional materials including graphene analogues, MoS_2, WS_2, BN are being explored for plasmonic sensing application [73–75]. Consequently, the photo-plasmonic coupling between the LSPR of the Ag nanorods and the SPPs (Surface Plasmon Polaritons) of metal thin film is coupled via the π-plasmons of the graphene oxide sandwich [76–78]. The SPCE enhancements increased from 700-fold to >1000-fold on account of this hybrid nano-engineering [16]. The graphene oxide used in the experiments was characterized using the TEM instrument, and the data is shown in Figure 8a-c.

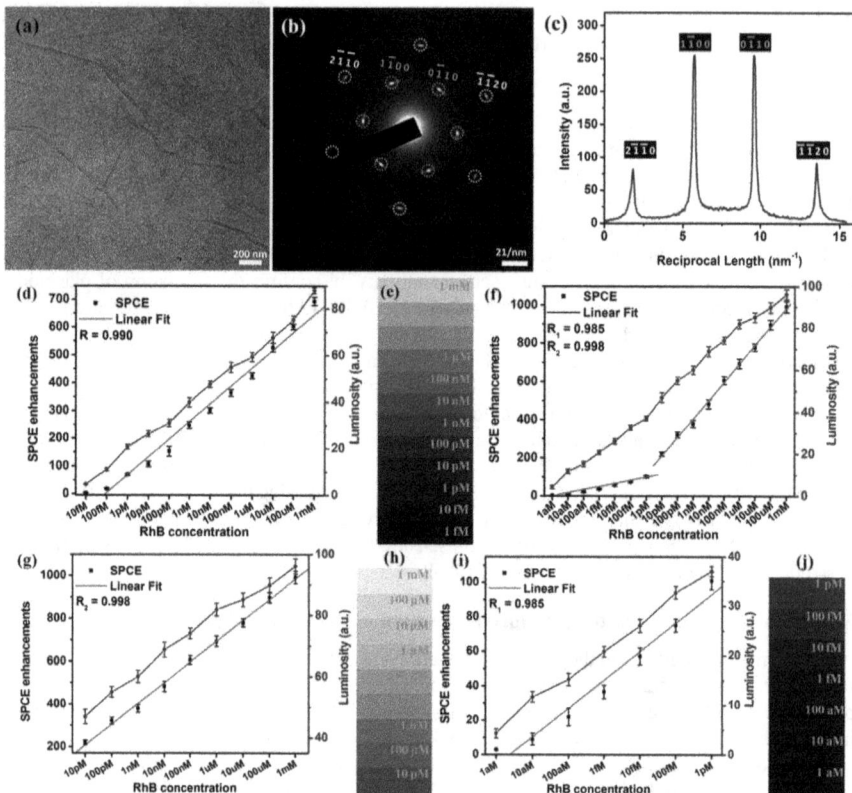

Figure 8. (a) TEM images of GO sheet coated with AgNRs(b) SAED pattern at the center region of (a) showing bright spots labelled with Miller-Bravais indices presenting bright spots characteristics of silver (c) Intensity profile across the line-profile as presented in (b) which includes inner and outer diffractions dots. (d) SPCE enhancements (left y-axis) and luminosity (right y-axis) for 30 min UV-exposed samples in the ext. cavity nanointerface for AgNPs in 10 fM to 1 mM RhB concentration. The corresponding shade cards are shown in (e). SPCE enhancements (left y-axis) and luminosity (right y-axis) for 30 min UV-exposed samples in the ext. cavity nanointerface for AgNPs and GO in (f). Pre-functionalized silver slide in 10aM to 1 mM RhB concentration, (b) pre-functionalized silver slide with GO in 10pM to 1 mM RhB concentration, (c) pre-functionalized silver slide with GO in 1aM to 1pM RhB concentration. (d) The corresponding shade cards obtained with mobile phone based analytical detection. Reprinted with permission from Reference [16], Copyright 2023, Elsevier.

Nanoparticle Toxicity and Compatibility   Materials Research Forum LLC
Materials Research Foundations 161 (2024) 153-181   https://doi.org/10.21741/9781644902998-6

The large electron transparency observed in the TEM image at lower resolution indicates the predominant presence of monolayer of graphene oxide in the sample. The selected area electron diffraction patterns (Figure 8b) and the associated line profile in Figure 8c shows that the inner diffraction rings are brighter than the outer ones [16]. This confirms the existence of monolayer of graphene oxide in the sample as compared to the bi-, tri- and multilayer formations. Figure 8d and 8e present the performance of the sensor for detecting the RhB molecule without the use of graphene oxide monolayer. It is seen that the linear sensing ranges from 1 mM to 100 fM concentration was observed. Further, the higher SPCE enhancements obtained using the graphene oxide coated SPCE substrate displayed larger sensing range with better limit of detection (10 aM) [16]. In this context, it is worth noting that the incorporation of graphene oxide film assisted in increasing the performance of the biosensing platform. The pre-functionalization of graphene oxide over the SPCE substrate not only assist in generation of better hotspots, but also aids in better coupling with the fluorescent moieties such as RhB molecules, as differently oriented molecules (high π–π stacking interactions with the RhB molecule) can also couple with the propagating plasmons of metal thin film [16,22,73].

5.3 Sericin protein based AgAu cubes for attomolar mefanamic acid

In the previous sections, the sharp edges Ag nanoparticles were used for generating high fluorescence enhancements which was further used in sensing the SPCE reporter molecule [44]. It was observed that one approach to enhance the fluorescence signal intensity is by engineering graphene oxide over the SPCE substrates. This and the following sections focus on other techniques that can be utilized for augmenting the fluorescence signal. Biocompatible, non-hazardous sericin protein (obtained from *Bombyx mori*) was used for synthesis of bimetallic AgAu nanohybrids (Figure 9a). Modulating the UV-exposure time resulted in different shapes and sizes of nanohybrids. It was observed that 30 minutes UV exposure resulted in sharp edged AgAu nanocubes and the obtained SEM images are shown in Figure 9b-c. Interfacing such nanocubes over the SPCE substrate assisted in high fluorescence enhancements on account of multiple factors: (i) strong inter-plasmonic coupling between the Ag and Au species comprising the AgAu nanohybrids, (ii) powerful inter-plasmonic coupling between the AgAu nanohybrids and the Ag thin film (SPCE substrate), (iii) intense plasmonic coupling of AgAu nanocubes with the emitter dipoles. The AgAu nanohybrids with a cubic geometry are an ideal candidate for generation of plasmonic hotspots as it directly aids the metal-enhanced fluorescence (MEF) phenomenon [44].

Mefenamic acid is a fenamate class of non-steroidal anti-inflammatory, analgesic, antipyretic drugs that are well-known for their pharmacological functionality [44]. Its influence on cyclooxygenase (COX) pathways is reported to result in disparity in the sex hormone production and take steps as an endocrine disrupter. Several disease conditions are associated with the irregulated levels of mefenamic acid in the biological systems in addition to resulting in harmful effects over the endocrine system. Unfortunately, due to the large-scale production occurring in conjugation with the inappropriate treatment of the chemical in the sewage systems, the chemical is often disclosed in the surface and groundwater systems. On account of its detrimental effects on the environment, the mefenamic acid is being detected in several systems to get reliable understanding of the concentration levels [44]. The direct interaction of mefenamic acid with the rhodamine B molecule results in quenching of the fluorescence of the dye molecules. This phenomenon is used for developing the sensor system for detecting mefenamic acid, and the results are presented in Figure 9d-g, where the detection is done using the conventional and mobile phone-

Materials Research Forum LLC
https://doi.org/10.21741/9781644902998-6

based detection systems. The World Health Organization (WHO) has defined the point-of-care (POC) diagnostics as "ASSURED"— affordable, sensitive, specific, user-friendly, rapid, equipment-free, and delivered to the end users. In this context, there is a growing demand for developing research in such a manner that it would focus on the use of simplified biosensor platforms [16,33,54]. One such system that is user-friendly, incorporated with easy readout systems, and of global reach are smartphones which are often used by medical health professionals in hospital settings. In this context, the smartphone based SPCE platforms provide opportunities to miniaturize the optical detection system, thereby rendering it as a facile, simple, and rapid biosensing tool for layperson's use coupled with advanced computing technologies to analyze the complex data for provide feedback the patients (for immediate hospitalization or self-quarantine) [44,79].

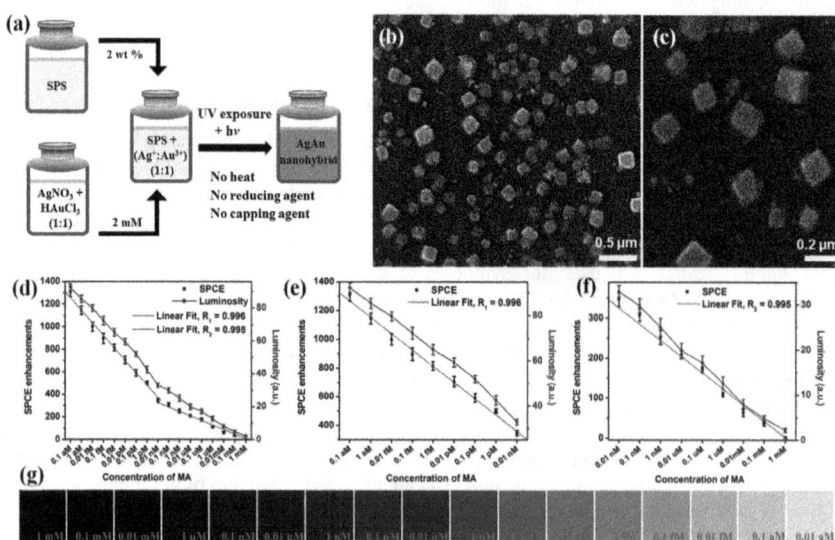

Figure 9. (a) Schematic representation of AgAu nanohybrid synthesis using SER protein solution (SPS) through a single-step process. (b,c) SEM images of the 30 min UV-exposed samples at lower and higher magnifications, respectively. (d) Overlap of the SPCE enhancements (left y-axis) obtained using the Ocean optics detector and the luminosity values (right y-axis) extracted using the mobile phone-based detector. (e) Concentration range of mefenamic acid (MA) taken for sensing in the range from 0.1 aM to 0.01 nM. (f) Concentration range of MA taken for sensing in the range from 0.01 nM to 1 mM. (g) The gray-scale shade cards corresponding to the luminosity values. Reprinted with permission from Reference [44], Copyright 2022, American Chemical Society.

5.4 Silkworm protein based AgAu nanohybrids for attomolar zinc ion sensing

In this section, we discuss the case study where the silkworm protein extracted from the silk pupae is used for the nanosynthesis. Silkworms that are discarded as waste is silk production industry are known as spent worms and contain 30–40% proteins, oil and carbohydrates [34]. The Ag, Au and AgAu nanohybrids were synthesized using the UV-exposure method and the resulting nanomaterials were interfaced with the SPCE platform to evaluate the photo-plasmonic coupling efficiency. It was observed that the AgAu nanohybrids with a semi-cubic shape yielded the highest SPCE enhancement [34]. The high fluorescence enhancements were used for the detection of zinc ion by exploiting the simple chemistry between the ARS-Zn^{2+} and the results are presented in Figure 10. The conceptual schematic of the sensor development is presented in Figure 10a. The blank sample with ARS and without Zn^{2+} yielded 54-fold SPCE enhancements with the addition of AgAu nanohybrids. Further, ARS and Zn^{2+} presented only 32-fold fluorescence enhancements. The ensemble of all three resulted in 1000-fold fluorescence enhancements, which was further utilized for detection of Zn^{2+} ions as shown in Figures 10b-d. The decrease in the SPCE intensity profile as well as the decrease in the overall fluorescence enhancements with a decrease in the Zn^{2+} ions concentration is shown in Figure 10b and 10c [34]. The response of the sensor against other potential ions were also evaluated as shown in Figure 10e by keeping the concentration of all the ions at a fixed concentration. Further, concentration of the other potentially interfering ions was taken at 100 times more concentration than that of Zn^{2+} ions to verify the interference (Figure 10f). The results shown in Figure 10e and 10f indicate the high selectivity of the developed sensor towards Zn^{2+} ions. Further, the results pertaining to the spiking studies also validate the high reliability and reproducibility of the developed biosensor [34].

Materials Research Forum LLC
https://doi.org/10.21741/9781644902998-6

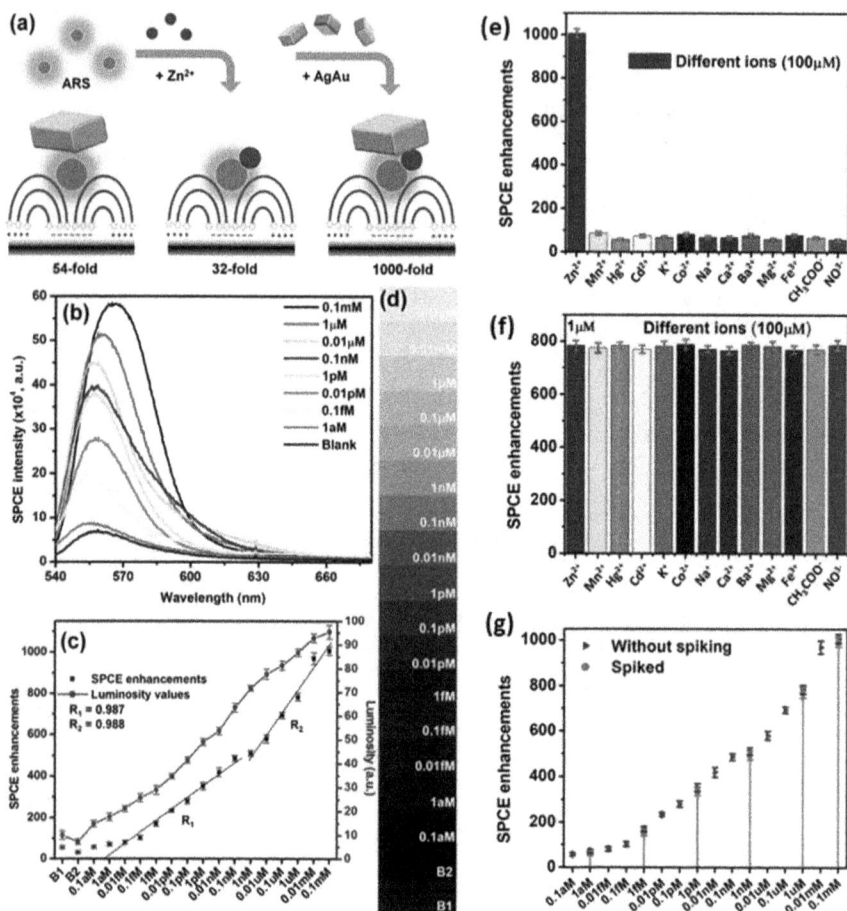

Figure 9. (a) Conceptual schematic representation of various combinations: ARS with AgAu nanohybrids (with 54-fold enhancement), ARS–Zn²⁺ ion complex (with 32-fold enhancement), and ARS–Zn²⁺ ion complex with AgAu nanohybrids in the ext. cavity nanointerface (with 231-fold enhancement). ARS is shown as red fluorescent molecules and AgAu NPs as cuboidal nanostructures in line with SEM images. (b) SPCE spectra for alternate concentration of Zn²⁺ ions added to the ensemble. (c) Overlap of SPCE enhancements (left y-axis) obtained from an Ocean Optics detector and the luminosity values (right y-axis) extracted from the mobile phone-based detector. (d) Gray-scale shade cards corresponding to the luminosity values. (a) SPCE enhancements for different

ions at 100 μM concentration. These ions were taken in the ext. cavity interface as part of the dye layer and the presented values are the average of three independent measurements (triplicates). (b) SPCE enhancements obtained for all the interfering ions under study (100 μM) taken in 100 times the concentration of Zn^{2+} (1 μM) for understanding the interference from other ions. Even in 100 times the concentration of interference ions, the emission enhancements from Zn^{2+} remained unaffected. (c) Overlap of emission enhancements obtained without spiking (as presented in Figure 5c) and from spiking experiments (red color dots shown with red lines connecting them to the corresponding Zn^{2+} ion concentration in the x-axis for ease of understanding). For spiking studies, 100 μM, 1 μM, 1 nM, 1 pM, 1 fM, and 1aM concentrations were chosen. Reprinted with permission from Reference [34]*, Copyright 2021, American Chemical Society.*

At this juncture it is worth noting that industrialization has presented many scientific breakthroughs for advancing the human lifestyle [51,52,80–83]. However, the widespread elimination of industrial waste in copious amounts in the form of byproducts has been rising especially in the sectors connected to the sugar, paper, aquaculture, brewer, and silk industries. The byproducts generated in these industries are typically discarded as waste without making an attempt to add value to the waste, inspite of them being rich in proteins and polysaccharides [84,85]. In this regard, awareness needs to be enriched where the wastes and byproducts can be used for inter-disciplinary applications, some of which are discussed in this chapter via case studies.

Conclusion

In this chapter we presented a few case studies emphasizing the importance of exploring biopolymers, proteins and related biomaterials for the generation of plasmonic nanomaterials. Although biopolymers and proteins are widely used in several interdisciplinary research fields, it is surprising to see that they are seldom employed for nanosynthesis of plasmonic materials via go-green approaches. In the recent past, several technologies such as MEF, SERS and SPCE are being explored for detection of several biomarkers and analytes of environmental and biological significance. In this perspective, here we present the pertinence of investigating bioinspired nanomaterials for biosensing applications, by considering SPCE platform as proof-of-concept. The rapid dissemination of smartphones and economic digital transformation, mobile phone-based detection technologies are gaining large. The smartphone enabled SPCE platform for detecting different analytes of interest is highlighted in this regard. Hence, the chapter is expected to support innovations in this direction where translational and convergent research via bioinspired routes would be utilized for development of biosensing platforms.

Consent for publicaton

Authors provide consent for the publication

Conflict of interest

Authors have no conflict of interest

Author details

ORCID IDs of corresponding authors*:

Seemesh Bhaskar: http://orcid.org/0000-0003-2714-3776

Sai Sathish Ramamurthy: http://orcid.org/0000-0001-9957-9259

Acknowledgement

Authors acknowledge support from Tata Education and Development Trust [TEDT/MUM/HEA/SSSIHL/2017- 2018/0069-RM-db], Prasanthi Trust, Inc., U.S. (22-06-2018), DST-Technology Development Program (IDP/MED/ 19/2016), Life Sciences Research Board (LSRB), and DST-Inspire Fellowship (IF180392), Govt. of India. S.B. is supported by a postdoctoral fellowship from the Carl R. Woese I nstitute for Genomic Biology. SB thanks Prof. Brian T. Cunningham, Department of Electrical and Computer Engineering, USA, Illinois Cancer Center, USA, Woese Institute for Genomic Biology, USA, Holonyak Micro and Nanotechnology Laboratory, USA, University of Illinois at Urbana-Champaign, USA for research support and guidance. We especially acknowledge SSSIHL-CRIF for extending the usage of the required instrumentation facility. Guidance from Bhagawan Sri Sathya Sai Baba is gratefully acknowledged.

References

[1] S. Malik, K. Muhammad, Y. Waheed, Nanotechnology: A Revolution in Modern Industry, Molecules. 28 (2023). https://doi.org/10.3390/molecules28020661

[2] S. Bayda, M. Adeel, T. Tuccinardi, M. Cordani, F. Rizzolio, The history of nanoscience and nanotechnology: From chemical-physical applications to nanomedicine, Molecules. 25 (2020). https://doi.org/10.3390/molecules25010112

[3] P. Thakur, A. Thakur, Introduction to Nanotechnology, in: Synthesis and Applications of Nanoparticles, 2022. https://doi.org/10.1007/978-981-16-6819-7_1

[4] M.T. Barako, V. Gambin, J. Tice, Integrated nanomaterials for extreme thermal management: A perspective for aerospace applications, Nanotechnology. 29 (2018). https://doi.org/10.1088/1361-6528/aaabe1

[5] H. Xin, B. Namgung, L.P. Lee, Nanoplasmonic optical antennas for life sciences and medicine, Nat Rev Mater. (2018). https://doi.org/10.1038/s41578-018-0033-8

[6] W.J. Stark, P.R. Stoessel, W. Wohlleben, A. Hafner, Industrial applications of nanoparticles, Chem Soc Rev. 44 (2015) 5793–5805. https://doi.org/10.1039/c4cs00362d

[7] I. Khan, K. Saeed, I. Khan, Nanoparticles: Properties, applications and toxicities, Arabian Journal of Chemistry. (2017). https://doi.org/10.1016/j.arabjc.2017.05.011

[8] F.J. Heiligtag, M. Niederberger, The fascinating world of nanoparticle research, Materials Today. 16 (2013) 262–271. https://doi.org/10.1016/j.mattod.2013.07.004

Materials Research Forum LLC
https://doi.org/10.21741/9781644902998-6

[9]   A. Rai, S. Bhaskar, K.M. Ganesh, S.S. Ramamurthy, Hottest Hotspots from the Coldest Cold: Welcome to Nano 4.0, ACS Appl Nano Mater. 5 (2022). https://doi.org/10.1021/acsanm.2c02556

[10]   M. Moronshing, S. Bhaskar, S. Mondal, S.S. Ramamurthy, C. Subramaniam, Surface-enhanced Raman scattering platform operating over wide pH range with minimal chemical enhancement effects: Test case of tyrosine, Journal of Raman Spectroscopy. 50 (2019). https://doi.org/10.1002/jrs.5587

[11]   P.J. Arathi, B. Seemesh, R.K.R. G., S.K. P., V. Ramanathan, Disulphide linkage: To get cleaved or not? Bulk and nano copper based SERS of cystine, Spectrochim Acta A Mol Biomol Spectrosc. 196 (2018). https://doi.org/10.1016/j.saa.2018.02.010

[12]   S.A. Mazari, E. Ali, R. Abro, F.S.A. Khan, I. Ahmed, M. Ahmed, S. Nizamuddin, T.H. Siddiqui, N. Hossain, N.M. Mubarak, A. Shah, Nanomaterials: Applications, waste-handling, environmental toxicities, and future challenges - A review, J Environ Chem Eng. 9 (2021). https://doi.org/10.1016/j.jece.2021.105028

[13]   M. Aflori, Smart nanomaterials for biomedical applications—a review, Nanomaterials. 11 (2021). https://doi.org/10.3390/nano11020396

[14]   M. Freeda, T.D. Subash, Comparision of Photoluminescence studies of Lanthanum, Terbium doped Calcium Aluminate nanophosphors (CaAl2O4: La, CaAl2O4: Tb) by sol-gel method., in: Mater Today Proc, 2017. https://doi.org/10.1016/j.matpr.2017.02.134

[15]   K. Shahane, M. Kshirsagar, S. Tambe, D. Jain, S. Rout, M.K.M. Ferreira, S. Mali, P. Amin, P.P. Srivastav, J. Cruz, R.R. Lima, An Updated Review on the Multifaceted Therapeutic Potential of Calendula officinalis L., Pharmaceuticals. 16 (2023) 611. https://doi.org/10.3390/ph16040611

[16]   K.M. Ganesh, A. Rai, S. Bhaskar, N. Reddy, S.S. Ramamurthy, Plasmon-enhanced fluorescence from synergistic engineering of graphene oxide and sharp-edged silver nanorods mediated with castor protein for cellphone-based attomolar sensing, J Lumin. 260 (2023). https://doi.org/10.1016/j.jlumin.2023.119835

[17]   S. Bhaskar, Biosensing Technologies: A Focus Review on Recent Advancements in Surface Plasmon Coupled Emission, Micromachines (Basel). 14 (2023). https://doi.org/10.3390/mi14030574

[18]   S.A. Maier, Plasmonics: Fundamentals and applications, 2007. https://doi.org/10.1007/0-387-37825-1

[19]   E. Ozbay, Plasmonics: Merging photonics and electronics at nanoscale dimensions, Science (1979). 311 (2006). https://doi.org/10.1126/science.1114849

[20]   S.A. Maier, H.A. Atwater, Plasmonics: Localization and guiding of electromagnetic energy in metal/dielectric structures, J Appl Phys. 98 (2005). https://doi.org/10.1063/1.1951057

[21]   D.K. Gramotnev, S.I. Bozhevolnyi, Plasmonics beyond the diffraction limit, Nat Photonics. 4 (2010). https://doi.org/10.1038/nphoton.2009.282

[22]   S. Bhaskar, S.M. Lis S, S. Kanvah, S. Bhaktha B. N., S.S. Ramamurthy, Single-Molecule Cholesterol Sensing by Integrating Silver Nanowire Propagating Plasmons and Graphene Oxide π-Plasmons on a Photonic Crystal-Coupled Emission Platform, ACS Applied Optical Materials. 1 (2023). https://doi.org/10.1021/acsaom.2c00026

[23]   S. Bhaskar, M. Moronshing, V. Srinivasan, P.K. Badiya, C. Subramaniam, S.S. Ramamurthy, Silver Soret Nanoparticles for Femtomolar Sensing of Glutathione in a Surface Plasmon-Coupled Emission Platform, ACS Appl Nano Mater. 3 (2020). https://doi.org/10.1021/acsanm.0c00470

[24]   J.A. Schuller, E.S. Barnard, W. Cai, Y.C. Jun, J.S. White, M.L. Brongersma, Plasmonics for extreme light concentration and manipulation, Nat Mater. 9 (2010). https://doi.org/10.1038/nmat2630

[25]   A. Rai, S. Bhaskar, K.M. Ganesh, S.S. Ramamurthy, Engineering of coherent plasmon resonances from silver soret colloids, graphene oxide and Nd2O3 nanohybrid architectures studied in mobile phone-based surface plasmon-coupled emission platform, Mater Lett. 304 (2021). https://doi.org/10.1016/j.matlet.2021.130632

[26]   G. V. Naik, V.M. Shalaev, A. Boltasseva, Alternative plasmonic materials: Beyond gold and silver, Advanced Materials. 25 (2013). https://doi.org/10.1002/adma.201205076

[27]   J.R. Lakowicz, Plasmonics in biology and plasmon-controlled fluorescence, Plasmonics. 1 (2006). https://doi.org/10.1007/s11468-005-9002-3

[28]   J.R. Lakowicz, Radiative decay engineering 5: Metal-enhanced fluorescence and plasmon emission, Anal Biochem. 337 (2005). https://doi.org/10.1016/j.ab.2004.11.026

[29]   J.R. Lakowicz, Principles of fluorescence spectroscopy, 2006. https://doi.org/10.1007/978-0-387-46312-4

[30]   J.B. Pendry, A.J. Holden, W.J. Stewart, I. Youngs, Extremely low frequency plasmons in metallic mesostructures, Phys Rev Lett. 76 (1996). https://doi.org/10.1103/PhysRevLett.76.4773

[31]   S. Bhaskar, S.S. Ramamurthy, Performance Enhancement of Light Emitting Radiating Dipoles (LERDs) Using Surface Plasmon-Coupled and Photonic Crystal- Coupled Emission Platforms, in: Organic and Inorganic Light Emitting Diodes: Reliability Issues and Performance Enhancement, 2023. https://doi.org/10.1201/9781003340577-8

[32]   A. Rai, S. Bhaskar, S.S. Ramamurthy, Plasmon-Coupled Directional Emission from Soluplus-Mediated AgAu Nanoparticles for Attomolar Sensing Using a Smartphone, ACS Appl Nano Mater. 4 (2021). https://doi.org/10.1021/acsanm.1c00841

[33]   S. Bhaskar, V. Srinivasan, S.S. Ramamurthy, Nd2O3-Ag Nanostructures for Plasmonic Biosensing, Antimicrobial, and Anticancer Applications, ACS Appl Nano Mater. (2022). https://doi.org/10.1021/acsanm.2c04643

[34]   A. Rai, S. Bhaskar, N. Reddy, S.S. Ramamurthy, Cellphone-Aided Attomolar Zinc Ion Detection Using Silkworm Protein-Based Nanointerface Engineering in a Plasmon-Coupled

Dequenched Emission Platform, ACS Sustain Chem Eng. 9 (2021).
https://doi.org/10.1021/acssuschemeng.1c05437

[35]   Y. Zhang, K. Poon, G.S.P. Masonsong, Y. Ramaswamy, G. Singh, Sustainable
Nanomaterials for Biomedical Applications, Pharmaceutics. 15 (2023).
https://doi.org/10.3390/pharmaceutics15030922

[36]   J.R. Lakowicz, Radiative decay engineering 3. Surface plasmon-coupled directional
emission, Anal Biochem. 324 (2004). https://doi.org/10.1016/j.ab.2003.09.039

[37]   S. Bhaskar, S.S. Ramamurthy, High refractive index dielectric TiO2and graphene oxide
as salient spacers for > 300-fold enhancements, in: Proceedings of 2021 IEEE International
Conference on Nanoelectronics, Nanophotonics, Nanomaterials, Nanobioscience and
Nanotechnology, 5NANO 2021, 2021. https://doi.org/10.1109/5NANO51638.2021.9491131

[38]   S. Bhaskar, S.S. Ramamurthy, Mobile Phone-Based Picomolar Detection of Tannic Acid
on Nd2O3 Nanorod-Metal Thin-Film Interfaces, ACS Appl Nano Mater. 2 (2019).
https://doi.org/10.1021/acsanm.9b00987

[39]   S. Bhaskar, D. Thacharakkal, S.S. Ramamurthy, C. Subramaniam, Metal-Dielectric
Interfacial Engineering with Mesoporous Nano-Carbon Florets for 1000-Fold Fluorescence
Enhancements: Smartphone-Enabled Visual Detection of Perindopril Erbumine at a Single-
molecular Level, ACS Sustain Chem Eng. 11 (2023).
https://doi.org/10.1021/acssuschemeng.2c04064

[40]   A. Rai, S. Bhaskar, K.M. Ganesh, S.S. Ramamurthy, Gelucire®-mediated heterometallic
AgAu nanohybrid engineering for femtomolar cysteine detection using smartphone-based
plasmonics technology, Mater Chem Phys. 279 (2022).
https://doi.org/10.1016/j.matchemphys.2022.125747

[41]   C.D. Geddes, J.R. Lakowicz, Metal-Enhanced Fluorescence, in: J Fluoresc, 2002.
https://doi.org/10.1023/A:1016875709579

[42]   J.R. Lakowicz, K. Ray, M. Chowdhury, H. Szmacinski, Y. Fu, J. Zhang, K. Nowaczyk,
Plasmon-controlled fluorescence: A new paradigm in fluorescence spectroscopy, Analyst. 133
(2008). https://doi.org/10.1039/b802918k

[43]   S. Bhaskar, P. Das, V. Srinivasan, S. Bhaktha B. N., S.S. Ramamurthy, Bloch Surface
Waves and Internal Optical Modes-Driven Photonic Crystal-Coupled Emission Platform for
Femtomolar Detection of Aluminum Ions, Journal of Physical Chemistry C. 124 (2020).
https://doi.org/10.1021/acs.jpcc.9b11092

[44]   S. Bhaskar, A. Rai, K.M. Ganesh, R. Reddy, N. Reddy, S.S. Ramamurthy, Sericin-Based
Bio-Inspired Nano-Engineering of Heterometallic AgAu Nanocubes for Attomolar Mefenamic
Acid Sensing in the Mobile Phone-Based Surface Plasmon-Coupled Interface, Langmuir. 38
(2022). https://doi.org/10.1021/acs.langmuir.2c01894

[45]   V.S.K. Cheerala, K.M. Ganesh, S. Bhaskar, S.S. Ramamurthy, S.C. Neelakantan,
Smartphone-Based Attomolar Cyanide Ion Sensing Using Au-Graphene Oxide Cryosoret

Materials Research Forum LLC
https://doi.org/10.21741/9781644902998-6

Nanoassembly and Benzoxazolium-Based Fluorophore in a Surface Plasmon-Coupled Enhanced Fluorescence Interface, Langmuir. (2023). https://doi.org/10.1021/acs.langmuir.3c00801

[46]   S. Muzammil, J. Neves Cruz, R. Mumtaz, I. Rasul, S. Hayat, M.A. Khan, A.M. Khan, M.U. Ijaz, R.R. Lima, M. Zubair, Effects of Drying Temperature and Solvents on In Vitro Diabetic Wound Healing Potential of Moringa oleifera Leaf Extracts, Molecules. 28 (2023) 710. https://doi.org/10.3390/molecules28020710

[47]   J.N. Anker, W.P. Hall, O. Lyandres, N.C. Shah, J. Zhao, R.P. Van Duyne, Biosensing with plasmonic nanosensors, Nat Mater. 7 (2008). https://doi.org/10.1038/nmat2162

[48]   A. Rai, S. Bhaskar, G. Kalathur Mohan, S.S. Ramamurthy, Biocompatible Gellucire ® Inspired Bimetallic Nanohybrids for Augmented Fluorescence Emission Based on Graphene Oxide Interfacial Plasmonic Architectures , ECS Trans. 107 (2022). https://doi.org/10.1149/10701.4527ecst

[49]   M. Nasrollahzadeh, M. Sajjadi, S.M. Sajadi, Z. Issaabadi, Green Nanotechnology, in: Interface Science and Technology, 2019. https://doi.org/10.1016/B978-0-12-813586-0.00005-5

[50]   A. Gałuszka, Z. Migaszewski, J. Namieśnik, The 12 principles of green analytical chemistry and the SIGNIFICANCE mnemonic of green analytical practices, TrAC - Trends in Analytical Chemistry. 50 (2013). https://doi.org/10.1016/j.trac.2013.04.010

[51]   R.A. Soni, M.A. Rizwan, S. Singh, Opportunities and potential of green chemistry in nanotechnology, Nanotechnology for Environmental Engineering. 7 (2022). https://doi.org/10.1007/s41204-022-00233-5

[52]   H.P.S. Abdul Khalil, A.H. Bhat, A.F. Ireana Yusra, Green composites from sustainable cellulose nanofibrils: A review, Carbohydr Polym. 87 (2012). https://doi.org/10.1016/j.carbpol.2011.08.078

[53]   F. Khan, M. Shariq, M. Asif, M.A. Siddiqui, P. Malan, F. Ahmad, Green Nanotechnology: Plant-Mediated Nanoparticle Synthesis and Application, Nanomaterials. 12 (2022). https://doi.org/10.3390/nano12040673

[54]   A. Rai, S. Bhaskar, K.M. Ganesh, S.S. Ramamurthy, Cellphone-based attomolar tyrosine sensing based on Kollidon-mediated bimetallic nanorod in plasmon-coupled directional and polarized emission architecture, Mater Chem Phys. 285 (2022). https://doi.org/10.1016/j.matchemphys.2022.126129

[55]   A. Rai, S. Bhaskar, P. Battampara, N. Reddy, S. Sathish Ramamurthy, Integrated Photo-Plasmonic coupling of bioinspired Sharp-Edged silver Nano-particles with Nano-films in extended cavity functional interface for Cellphone-aided femtomolar sensing, Mater Lett. 316 (2022). https://doi.org/10.1016/j.matlet.2022.132025

[56]   S. Bhaskar, R. Patra, N.C.S.S. Kowshik, K.M. Ganesh, V. Srinivasan, P. Chandran S, S.S. Ramamurthy, Nanostructure effect on quenching and dequenching of quantum emitters on surface plasmon-coupled interface: A comparative analysis using gold nanospheres and nanostars, Physica E Low Dimens Syst Nanostruct. 124 (2020). https://doi.org/10.1016/j.physe.2020.114276

[57]   S. Kumar, Z. Wang, W. Zhang, X. Liu, M. Li, G. Li, B. Zhang, R. Singh, Optically Active Nanomaterials and Its Biosensing Applications—A Review, Biosensors (Basel). 13 (2023). https://doi.org/10.3390/bios13010085

[58]   A.E. Nel, L. Mädler, D. Velegol, T. Xia, E.M.V. Hoek, P. Somasundaran, F. Klaessig, V. Castranova, M. Thompson, Understanding biophysicochemical interactions at the nano-bio interface, Nat Mater. 8 (2009). https://doi.org/10.1038/nmat2442

[59]   K. Illath, S. Kar, P. Gupta, A. Shinde, S. Wankhar, F.G. Tseng, K.T. Lim, M. Nagai, T.S. Santra, Microfluidic nanomaterials: From synthesis to biomedical applications, Biomaterials. 280 (2022). https://doi.org/10.1016/j.biomaterials.2021.121247

[60]   A.R. Sadrolhosseini, S. Shafie, Y.W. Fen, Nanoplasmonic sensor based on surface plasmon-coupled emission: Review, Applied Sciences (Switzerland). 9 (2019). https://doi.org/10.3390/app9071497

[61]   Surface Plasmon Enhanced, Coupled and Controlled Fluorescence, 2017. https://doi.org/10.1002/9781119325161

[62]   S. Bhaskar, A. Rai, G. Kalathur Mohan, S.S. Ramamurthy, Mobile Phone Camera-Based Detection of Surface Plasmon-Coupled Fluorescence from Streptavidin Magnetic Nanoparticles and Graphene Oxide Hybrid Nanointerface, ECS Trans. 107 (2022). https://doi.org/10.1149/10701.3223ecst

[63]   S. Bhaskar, N.C.S.S. Kowshik, S.P. Chandran, S.S. Ramamurthy, Femtomolar Detection of Spermidine Using Au Decorated SiO2 Nanohybrid on Plasmon-Coupled Extended Cavity Nanointerface: A Smartphone-Based Fluorescence Dequenching Approach, Langmuir. 36 (2020). https://doi.org/10.1021/acs.langmuir.9b03869

[64]   S. Bhaskar, A.K. Singh, P. Das, P. Jana, S. Kanvah, S. Bhaktha B N, S.S. Ramamurthy, Superior Resonant Nanocavities Engineering on the Photonic Crystal-Coupled Emission Platform for the Detection of Femtomolar Iodide and Zeptomolar Cortisol, ACS Appl Mater Interfaces. 12 (2020). https://doi.org/10.1021/acsami.0c07515

[65]   Y. Xiong, S. Shepherd, J. Tibbs, A. Bacon, W. Liu, L.D. Akin, T. Ayupova, S. Bhaskar, B.T. Cunningham, Photonic Crystal Enhanced Fluorescence: A Review on Design Strategies and Applications, Micromachines (Basel). 14 (2023). https://doi.org/10.3390/mi14030668

[66]   S. Bhaskar, P. Das, V. Srinivasan, S.B.N. Bhaktha, S.S. Ramamurthy, Plasmonic-Silver Sorets and Dielectric-Nd2O3 nanorods for Ultrasensitive Photonic Crystal-Coupled Emission, Mater Res Bull. 145 (2022). https://doi.org/10.1016/j.materresbull.2021.111558

[67]   S. Rathnakumar, S. Bhaskar, P.K. Badiya, V. Sivaramakrishnan, V. Srinivasan, S.S. Ramamurthy, Electrospun PVA nanofibers doped with titania nanoparticles in plasmon-coupled fluorescence studies: An eco-friendly and cost-effective transition from 2D nano thin films to 1D nanofibers, MRS Commun. 13 (2023). https://doi.org/10.1557/s43579-023-00342-5

[68]   S. Rathnakumar, S. Bhaskar, A. Rai, D.V.V. Saikumar, N.S.V. Kambhampati, V. Sivaramakrishnan, S.S. Ramamurthy, Plasmon-Coupled Silver Nanoparticles for Mobile Phone-

Based Attomolar Sensing of Mercury Ions, ACS Appl Nano Mater. 4 (2021).
https://doi.org/10.1021/acsanm.1c01347

[69]　R.B.M. de Almeida, D.B. Barbosa, M.R. do Bomfim, J.A.O. Amparo, B.S. Andrade, S.L. Costa, J.M. Campos, J.N. Cruz, C.B.R. Santos, F.H.A. Leite, M.B. Botura, Identification of a Novel Dual Inhibitor of Acetylcholinesterase and Butyrylcholinesterase: In Vitro and In Silico Studies, Pharmaceuticals. 16 (2023) 95. https://doi.org/10.3390/ph16010095

[70]　S. Bhaskar, S.S. Ramamurthy, Synergistic coupling of titanium carbonitride nanocubes and graphene oxide for 800-fold fluorescence enhancements on smartphone based surface plasmon-coupled emission platform, Mater Lett. 298 (2021).
https://doi.org/10.1016/j.matlet.2021.130008

[71]　S. Bhaskar, P. Das, M. Moronshing, A. Rai, C. Subramaniam, S.B.N. Bhaktha, S.S. Ramamurthy, Photoplasmonic assembly of dielectric-metal, Nd2O3-Gold soret nanointerfaces for dequenching the luminophore emission, Nanophotonics. 10 (2021).
https://doi.org/10.1515/nanoph-2021-0124

[72]　M.H. Sarfraz, M. Zubair, B. Aslam, A. Ashraf, M.H. Siddique, S. Hayat, J.N. Cruz, S. Muzammil, M. Khurshid, M.F. Sarfraz, A. Hashem, T.M. Dawoud, G.D. Avila-Quezada, E.F. Abd_Allah, Comparative analysis of phyto-fabricated chitosan, copper oxide, and chitosan-based CuO nanoparticles: antibacterial potential against Acinetobacter baumannii isolates and anticancer activity against HepG2 cell lines, Front Microbiol. 14 (2023) 1188743.
https://doi.org/10.3389/fmicb.2023.1188743

[73]　S. Bhaskar, N.S. Visweswar Kambhampati, K.M. Ganesh, M.S. Pa, V. Srinivasan, S.S. Ramamurthy, Metal-Free, Graphene Oxide-Based Tunable Soliton and Plasmon Engineering for Biosensing Applications, ACS Appl Mater Interfaces. 13 (2021).
https://doi.org/10.1021/acsami.1c01024

[74]　R. Dutta, T.D. Subash, N. Paitya, DC performance analysis of III–V/Si heterostructure double gate triple material PiN tunneling graphene nanoribbon FET circuits with quantum mechanical effects, J Comput Electron. 20 (2021). https://doi.org/10.1007/s10825-020-01649-5

[75]　R. Dutta, T.D. Subash, N. Paitya, InAs/Si Hetero-Junction Channel to Enhance the Performance of DG-TFET with Graphene Nanoribbon: an Analytical Model, Silicon. 13 (2021).
https://doi.org/10.1007/s12633-020-00546-7

[76]　A.N. Grigorenko, M. Polini, K.S. Novoselov, Graphene plasmonics, Nat Photonics. 6 (2012). https://doi.org/10.1038/nphoton.2012.262

[77]　M. Jablan, H. Buljan, M. Soljačić, Plasmonics in graphene at infrared frequencies, Phys Rev B Condens Matter Mater Phys. 80 (2009). https://doi.org/10.1103/PhysRevB.80.245435

[78]　L. Ju, B. Geng, J. Horng, C. Girit, M. Martin, Z. Hao, H.A. Bechtel, X. Liang, A. Zettl, Y.R. Shen, F. Wang, Graphene plasmonics for tunable terahertz metamaterials, Nat Nanotechnol. 6 (2011). https://doi.org/10.1038/nnano.2011.146

[79]　F.S. Alves, J.N. Cruz, I.N. de Farias Ramos, D.L. do Nascimento Brandão, R.N. Queiroz, G.V. da Silva, G.V. da Silva, M.F. Dolabela, M.L. da Costa, A.S. Khayat, J. de Arimatéia

Rodrigues do Rego, D. do Socorro Barros Brasil, Evaluation of Antimicrobial Activity and Cytotoxicity Effects of Extracts of Piper nigrum L. and Piperine, Separations. 10 (2023) 21. https://doi.org/10.3390/separations10010021

[80]   V. Polshettiwar, R.S. Varma, Green chemistry by nano-catalysis, Green Chemistry. 12 (2010). https://doi.org/10.1039/b921171c

[81]   V.C. Thipe, A.R. Karikachery, P. Çakılkaya, U. Farooq, H.H. Genedy, N. Kaeokhamloed, D.H. Phan, R. Rezwan, G. Tezcan, E. Roger, K. V. Katti, Green nanotechnology—An innovative pathway towards biocompatible and medically relevant gold nanoparticles, J Drug Deliv Sci Technol. 70 (2022). https://doi.org/10.1016/j.jddst.2022.103256

[82]   S.S. Salem, A mini review on green nanotechnology and its development in biological effects, Arch Microbiol. 205 (2023). https://doi.org/10.1007/s00203-023-03467-2

[83]   S. Karthick Raja Namasivayam, S. Srinivasan, K. Samrat, B. Priyalakshmi, R. Dinesh Kumar, A. Bharani, R. Ganesh Kumar, M. Kavisri, M. Moovendhan, Sustainable approach to manage the vulnerable rodents using eco-friendly green rodenticides formulation through nanotechnology principles – A review, Process Safety and Environmental Protection. 171 (2023). https://doi.org/10.1016/j.psep.2023.01.050

[84]   S. Bognár, P. Putnik, D.Š. Merkulov, Sustainable Green Nanotechnologies for Innovative Purifications of Water: Synthesis of the Nanoparticles from Renewable Sources, Nanomaterials. 12 (2022). https://doi.org/10.3390/nano12020263

[85]   S. Trombino, R. Sole, M.L. Di Gioia, D. Procopio, F. Curcio, R. Cassano, Green Chemistry Principles for Nano- and Micro-Sized Hydrogel Synthesis, Molecules. 28 (2023). https://doi.org/10.3390/molecules28052107

Nanoparticle Toxicity and Compatibility   Materials Research Forum LLC
Materials Research Foundations 161 (2024) 182-224   https://doi.org/10.21741/9781644902998-7

Chapter 7

Strategies for Enhancing Biocompatibility of Nanoparticles

Abhinay Thakur[1], Ashish Kumar[2]*

[1]Department of Chemistry, School of Chemical Engineering and Physical Sciences, Lovely Professional University, Phagwara, Punjab, 144411, India

[2]Department of Chemistry, Nalanda College of Engineering, Bihar Engineering University, Department of Science, Technology and Technical Education, Government of Bihar, Nalanda, Bihar, 803108, India

drashishchemlpu@gmail.com

Abstract

This book chapter explores the challenges of nanoparticles' biocompatibility in cardiovascular applications and proposes strategies to enhance their safety profile. It discusses interactions between nanoparticles and biological systems, highlighting factors like size, surface charge, and functionalization. The chapter examines cardiovascular toxicity, including adverse effects on heart function and vascular integrity. Cutting-edge strategies, such as surface modification with polyethylene glycol (PEG) and targeted nanoparticles, are explored to improve biocompatibility and reduce immunogenicity. Stimuli-responsive systems for controlled drug release are also discussed to enhance specificity and minimize off-target effects. These innovations hold promise for advancing nanoparticle applications in medicine.

Keywords

Nanoparticles, Biocompatibility, Drug Delivery, Surface Modification, Targeted Nanoparticles

Contents

1. Introduction

1.1 Nanoparticles in medicine

Nanoparticles have emerged as transformative agents in medicine due to their unique physicochemical properties and versatile applications. In drug delivery systems, nanoparticles offer opportunities to enhance bioavailability, specificity, and controlled release of therapeutic agents, leading to improved treatment outcomes and reduced side effects [1–5]. Moreover, in cancer therapy, nanoparticles enable selective targeting of tumor cells, either through active targeting with specific ligands or passive targeting utilizing the enhanced permeability and retention (EPR) effect in solid tumors as shown in Figure 1 [6]. The ability to deliver multiple therapeutic agents simultaneously via nanoparticles allows for combination therapy, which enhances treatment efficacy and reduces the risk of drug resistance. Additionally, in molecular imaging and diagnostics, nanoparticles act as contrast agents, providing enhanced sensitivity and specificity in various imaging modalities. Functionalized nanoparticles can also serve as molecular probes, enabling the detection of disease-specific markers for early diagnosis and personalized treatment approaches. The integration of therapeutic and diagnostic functionalities in theranostic nanoparticles facilitates real-time monitoring of treatment efficacy, optimizing therapeutic regimens for improved patient outcomes [7–9].

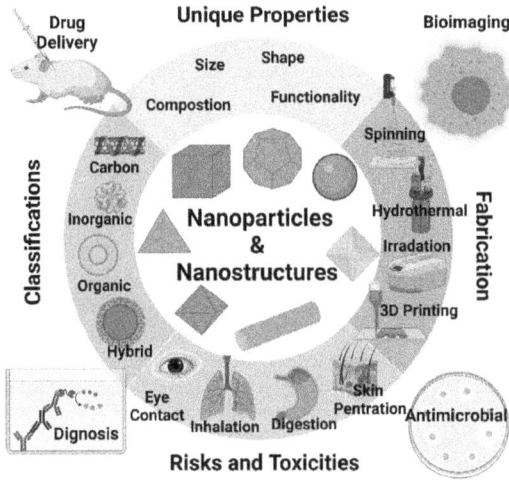

Figure 1: Use of nanoparticles and nanostructures in medicine. Adapted from Ref. [6] under CCBY 4.0.

Despite the vast opportunities nanoparticles offer in medicine, several challenges must be addressed to ensure their safe and effective clinical translation. Biocompatibility is a critical concern, as nanoparticles can trigger immune responses and potential clearance from the bloodstream, limiting their therapeutic effectiveness. Understanding the factors influencing nanoparticle clearance and conducting comprehensive preclinical studies are crucial for optimizing their pharmacokinetics and biodistribution [10,11]. Moreover, nanoparticles' small size and high reactivity raise concerns about potential toxicity, such as oxidative stress and cellular uptake leading to cytotoxicity. Understanding the factors contributing to organ-specific toxicity is vital for risk assessment and safe clinical use. Strategies to avoid nonspecific organ sequestration can enhance nanoparticle distribution to the target site. Immunogenicity is another challenge, as nanoparticles may induce antibody formation, potentially limiting their therapeutic efficacy upon repeated administration [12]. A comprehensive evaluation of nanoparticle long-term toxicity is essential to ensure patient safety during extended treatments. Addressing these challenges will facilitate the successful translation of nanoparticle-based medical technologies from bench to bedside, maximizing their potential in revolutionizing medicine and improving patient outcomes. Properly harnessed, nanoparticles hold the promise of revolutionizing medicine by providing innovative solutions for precise and effective drug delivery, non-invasive diagnostics, and personalized theranostic approaches, thereby ushering in a new era of medical advancement.

1.2 Importance of biocompatibility in nanoparticle applications

Nanoparticles have emerged as promising tools in various biomedical applications, ranging from targeted drug delivery to imaging and diagnostics. However, to successfully integrate nanoparticles into medical settings, ensuring their biocompatibility is of utmost importance. Biocompatibility refers to the ability of nanoparticles to interact safely with living organisms without eliciting adverse reactions or harm. The factors that determine nanoparticle biocompatibility, such as particle size, surface charge, and surface functionalization, play critical roles in their interactions with biological systems [13–16]. Firstly, particle size is a crucial parameter that affects nanoparticle biocompatibility. Small nanoparticles, typically in the range of tens to hundreds of nanometers, have extended circulation times in the bloodstream due to their ability to evade the immune system and prevent rapid clearance. This "stealth" behavior is particularly advantageous for drug delivery applications, where sustained circulation is essential for targeted and effective drug delivery. Smaller nanoparticles also possess enhanced cellular uptake and can penetrate cellular barriers more efficiently, improving their ability to reach target cells or tissues. However, extremely small nanoparticles may face challenges related to renal clearance, as they can be filtered and excreted through the kidneys. Therefore, finding the right balance between nanoparticle size and circulation time is crucial for optimizing their biocompatibility in medical applications [17–19]. Secondly, the surface charge of nanoparticles significantly influences their interactions with biological molecules and cells. Nanoparticles with positive or negative surface charges can interact with oppositely charged biomolecules, such as proteins or cell membranes. This interaction can affect cellular uptake, intracellular trafficking, and immune response. Neutral or slightly negatively charged nanoparticles tend to exhibit improved biocompatibility, as they are less likely to cause adverse interactions with cellular components. Additionally, the surface charge can influence the adsorption of proteins onto nanoparticle surfaces, forming a "protein corona" that can impact their biodistribution and biocompatibility. Modulating the surface charge of nanoparticles through surface functionalization or coatings can offer a strategy to enhance their biocompatibility in various biological environments [20–23].

Thirdly, surface functionalization involves modifying nanoparticle surfaces with specific molecules or ligands to impart desired functionalities. Functionalization can enhance nanoparticle biocompatibility by promoting stealth properties, reducing immune recognition, and enhancing target specificity. For instance, polyethylene glycol (PEG) is commonly used to functionalize nanoparticle surfaces, creating a hydrophilic and stealth-like surface that reduces protein adsorption and immune recognition. PEGylation has been widely adopted in drug delivery applications to improve nanoparticle circulation time and reduce immunogenicity. Moreover, functionalization with targeting ligands, such as antibodies or peptides, can enable nanoparticles to recognize specific cell receptors or disease biomarkers, leading to enhanced specificity and reduced off-target effects. However, poor biocompatibility of nanoparticles can have significant implications for medical applications [24–26]. Firstly, non-biocompatible nanoparticles can trigger immune reactions and induce inflammation in the body. The immune system may recognize nanoparticles as foreign invaders, leading to the production of antibodies against them. This immune response can result in the rapid clearance of nanoparticles from the bloodstream, limiting their therapeutic effectiveness. Additionally, immune reactions and inflammation at the site of nanoparticle administration can cause tissue damage and adverse effects on overall health. Secondly, poor biocompatibility can impact nanoparticle clearance and biodistribution in the body. Non-biocompatible nanoparticles may be rapidly cleared from the body through various routes, such as the liver, spleen, or kidneys. This rapid clearance can limit their circulation time, reducing their window of therapeutic effectiveness. Furthermore, non-biocompatible nanoparticles may accumulate in organs or tissues, leading to nonspecific toxicity and adverse effects. Thirdly, nanoparticles that interact unfavorably with cellular components can induce cellular toxicity. For instance, nanoparticles with high surface charge may disrupt cell membranes, leading to cellular damage and compromised cell function. Moreover, nanoparticles that are not biocompatible may cause oxidative stress and trigger intracellular signaling pathways that result in cell death or dysfunction [27–30].

1.3 Cardiovascular toxicity: A critical concern

Cardiovascular toxicity is a critical concern when considering the use of nanoparticles in diagnostic and therapeutic applications. While nanoparticles offer promising solutions for cardiovascular diseases, their potential adverse effects on the cardiovascular system must be carefully examined. Nanoparticles have found applications in cardiovascular diagnostics, serving as contrast agents for advanced imaging modalities that enable early disease detection and precise localization of pathologies [31–34]. In therapeutics, nanoparticles hold the potential for targeted drug delivery to diseased cardiovascular tissues, minimizing systemic exposure and reducing side effects. However, several factors contribute to cardiovascular toxicity associated with nanoparticles. One significant mechanism of cardiovascular toxicity involves the generation of reactive oxygen species (ROS) by certain nanoparticles. ROS are highly reactive molecules that induce oxidative stress, causing cellular damage and inflammation. Cardiovascular cells, such as endothelial cells and cardiomyocytes, are particularly susceptible to oxidative stress, leading to endothelial dysfunction and the progression of cardiovascular diseases. Additionally, nanoparticles can trigger an inflammatory response in the cardiovascular system, exacerbating cardiovascular inflammation and contributing to conditions like atherosclerosis. The physicochemical properties of nanoparticles influence their cardiovascular toxicity. Smaller nanoparticles with a larger surface area per unit mass exhibit increased interactions with biological components, leading to heightened ROS generation and inflammatory responses. Moreover, the surface coatings and functionalization of nanoparticles play a critical role in their biocompatibility. Coatings like polyethylene glycol (PEG) can improve nanoparticle stability and reduce interactions with blood components, potentially mitigating cardiovascular toxicity [35–37]. However, certain surface coatings may trigger immune responses or facilitate nanoparticle uptake by cardiovascular cells, increasing toxicity risks.

The dose and concentration of nanoparticles also impact their cardiovascular toxicity. High doses can overwhelm the body's antioxidant defense system, leading to ROS-induced damage. Similarly, high concentrations of nanoparticles in specific cardiovascular tissues may cause localized toxicity and inflammation. Biodistribution and targeting of nanoparticles can also influence their cardiovascular toxicity. Accumulation of nanoparticles in specific cardiovascular tissues may increase the risk of toxicity, and nanoparticles targeted to diseased tissues may be taken up by activated immune cells or contribute to inflammatory responses. To mitigate cardiovascular toxicity, researchers are exploring various strategies. Surface modification and coatings can improve nanoparticle biocompatibility [38]. PEGylation, for instance, can reduce protein adsorption and opsonization, minimizing immune recognition and macrophage uptake. Controlled release systems can limit nanoparticle exposure to cardiovascular tissues, enhancing drug efficacy while reducing toxicity. Precise targeting of nanoparticles to specific cardiovascular tissues can minimize off-target effects and reduce systemic exposure. Biodegradable nanoparticles that degrade over time can reduce their persistence in the cardiovascular system, lowering the risk of long-term toxicity. Comprehensive in vitro and in vivo testing is essential to evaluate nanoparticle biocompatibility and cardiovascular toxicity. Studying nanoparticle-cell interactions, ROS generation, and inflammatory responses can provide valuable insights into potential cardiovascular risks. By carefully addressing cardiovascular toxicity concerns and implementing safety measures, researchers can harness the full potential of nanoparticles to revolutionize cardiovascular medicine and improve patient outcomes. Rigorous testing and advancements in nanoparticle engineering will pave the way for safe and effective clinical translation, making nanoparticles a valuable tool in combating cardiovascular diseases.

2.   Fundamentals of nanoparticle biocompatibility

Nanoparticles have emerged as promising tools in various fields, including medicine, electronics, and environmental remediation, due to their unique physicochemical properties and versatile applications. In biomedical research, nanoparticles are being explored for drug delivery, molecular imaging, diagnostics, and theranostics. To realize the full potential of nanoparticles in these applications, understanding their biocompatibility is of utmost importance. Biocompatibility refers to the ability of nanoparticles to interact safely with living organisms without eliciting adverse reactions or harm. In this comprehensive discussion, we delve into the fundamentals of nanoparticle biocompatibility, exploring the interactions between nanoparticles and biological systems, the impact of particle size, the role of surface charge and functionalization, and the immune responses to nanoparticles.

2.1   Interactions between nanoparticles and biological systems

Interactions between nanoparticles and biological systems are complex and dynamic processes that determine the biocompatibility and safety of nanoparticle-based therapies. The formation of a protein corona on the nanoparticle surface when exposed to biological fluids plays a significant role in these interactions. The composition of the protein corona depends on the nanoparticle's physicochemical properties and the surrounding biological environment. This protein corona can modify the nanoparticle's surface properties, altering its charge, hydrophobicity, and cellular uptake. Moreover, the protein corona can influence the nanoparticle's behavior in vivo, affecting its biodistribution, cellular interactions, and clearance from the body. Opsonization and phagocytosis are crucial processes that occur when nanoparticles encounter the immune system in the bloodstream. The immune system may recognize nanoparticles as foreign entities and mark them for recognition and removal by phagocytic cells, such as macrophages. This opsonization and subsequent phagocytosis can clear nanoparticles from the bloodstream, limiting their circulation time and potentially reducing their effectiveness in drug delivery and imaging applications. To improve nanoparticle biocompatibility, strategies such as surface modification and PEGylation are employed to minimize opsonization and phagocytosis, thereby enhancing nanoparticle circulation in the bloodstream.

Cellular uptake is another important aspect of nanoparticle-biological interactions. Nanoparticles can enter cells through various mechanisms, including endocytosis, phagocytosis, and passive diffusion. The process of cellular uptake is influenced by the nanoparticle's size, shape, surface charge, and surface functionalization [39–41]. Once inside cells, nanoparticles can interact with different intracellular components, including organelles, cytoskeletal elements, and biomolecules. These interactions can trigger cellular responses, such as changes in gene expression, oxidative stress, and inflammation. Understanding nanoparticle-cell interactions is critical for evaluating their potential cytotoxicity and designing safe and effective nanoparticle-based therapies. To enhance biocompatibility, researchers focus on optimizing nanoparticle properties to minimize adverse effects and maximize therapeutic efficacy. Surface modification techniques, such as functionalizing nanoparticles with biocompatible materials like PEG, can improve their stealth properties, reduce immunogenicity, and enhance biocompatibility. Additionally, designing biomimetic nanoparticles that mimic the natural structures and functions of biological entities can reduce immune recognition and improve compatibility with biological systems. Moreover, the choice of materials for nanoparticle synthesis is crucial for biocompatibility. Biodegradable and biocompatible materials, such as polymers, lipids, and proteins, are preferred over non-biodegradable materials like metals. Biodegradable nanoparticles can be engineered to degrade into non-toxic byproducts, minimizing the risk of long-term accumulation and potential toxicity. To ensure the safety and effectiveness of nanoparticle-based therapies, preclinical assessment remains a fundamental step. In vitro, studies evaluate nanoparticle-cell interactions, cytotoxicity, and cellular uptake using cell cultures. In vivo, studies in animal models provide valuable information on nanoparticle biodistribution, clearance, and potential adverse effects. Long-term safety assessments are critical to understanding the impact of chronic exposure to nanoparticles.

2.2 Impact of particle size on biocompatibility

Particle size plays a crucial role in determining the biocompatibility of nanoparticles and their interactions with biological systems. Smaller nanoparticles tend to have longer circulation times in the bloodstream, thanks to their "stealth" behavior, which allows them to evade the immune system and avoid rapid clearance by the liver and spleen. This extended circulation time is advantageous for drug delivery applications, as it enhances the nanoparticles' ability to reach the target site and improve therapeutic efficacy [42–44]. Moreover, nanoparticle size influences their cellular uptake, with smaller nanoparticles being more readily taken up by cells through various endocytic pathways. Their small size allows for efficient interactions with cell membranes and internalization by cells. Conversely, larger nanoparticles may face challenges in cellular uptake due to size restrictions and potential steric hindrance. Efficient cellular uptake is critical for effective drug delivery, as it determines the intracellular delivery of therapeutic agents to target sites. Nanoparticle size also impacts their biodistribution within the body. Smaller nanoparticles tend to accumulate preferentially in tissues with leaky vasculature, such as tumor tissues, through the enhanced permeability and retention (EPR) effect. This phenomenon is beneficial for targeted drug delivery to tumors, as it allows nanoparticles to accumulate in the tumor microenvironment. However, size-dependent biodistribution can also lead to unintended off-target effects and non-specific accumulation in healthy tissues. Thus, careful consideration of nanoparticle size is necessary to achieve selective and effective targeting while minimizing unintended accumulation in non-target organs.

Furthermore, nanoparticle size influences their clearance from the body. Smaller nanoparticles can be cleared through the kidneys, leading to renal excretion. On the other hand, larger nanoparticles are typically cleared by the reticuloendothelial system (RES), primarily the liver and spleen. The clearance route can significantly impact the nanoparticles' behavior, biodistribution, and potential toxicity. For instance, nanoparticles cleared through the kidneys may undergo faster elimination from the body, reducing their accumulation in tissues and potential toxicity. In contrast, nanoparticles cleared through the RES may accumulate in these organs, raising concerns about long-term toxicity. To optimize nanoparticle biocompatibility and performance, researchers and scientists need to carefully consider and control particle size [45–49]. Smaller nanoparticles with extended circulation times are advantageous for drug delivery applications, while efficient cellular uptake is essential for effective intracellular drug delivery. Achieving selective targeting with size-dependent biodistribution requires a thoughtful design to maximize therapeutic efficacy and minimize off-target effects. Additionally, understanding the clearance route of nanoparticles helps in tailoring their behavior and potential toxicity. By optimizing particle size, scientists can enhance the safety and effectiveness of nanoparticle-based therapies, bringing us closer to the realization of precision medicine and personalized treatments.

2.3 Role of surface charge and functionalization in biocompatibility

The role of surface charge and functionalization in nanoparticle biocompatibility is crucial, as these factors significantly influence their interactions with biological systems. One key aspect is cellular uptake and internalization, where the surface charge of nanoparticles determines their efficiency in entering cells. Positively charged nanoparticles are more readily internalized due to electrostatic interactions with negatively charged cell membranes, while negatively charged nanoparticles may face reduced cellular uptake due to repulsive forces. Neutral or slightly negatively charged nanoparticles are often preferred to improve biocompatibility and minimize non-specific uptake. Another critical aspect is protein corona formation, which occurs when nanoparticles interact with proteins in biological fluids. The surface charge of nanoparticles influences the composition and properties of this protein corona. Positively charged nanoparticles can attract more opsonins, leading to enhanced phagocytosis and clearance by macrophages. To mitigate this issue, surface modification techniques, such as PEGylation, can be employed to introduce a hydrophilic and neutral charge, reducing protein adsorption and opsonization. This enhances nanoparticle circulation and biocompatibility by minimizing interactions with blood components.

Surface functionalization also plays a vital role in nanoparticle biocompatibility. Functionalization enables the attachment of targeting ligands to nanoparticles, improving their specificity and reducing off-target effects. Targeting ligands, such as antibodies or peptides, facilitate specific interactions with receptors on target cells or tissues, allowing for site-specific drug delivery and improved therapeutic outcomes. Functionalization offers a promising approach for personalized and targeted therapies, revolutionizing disease treatments. Furthermore, surface functionalization can impart stealth properties to nanoparticles. For instance, PEGylation involves attaching polyethylene glycol (PEG) chains to the nanoparticle surface, creating a hydrophilic and stealth-like surface. This modification reduces interactions with blood components and minimizes recognition by the immune system [50–55]. Stealth-functionalized nanoparticles are less likely to be recognized and cleared by the immune system, leading to prolonged circulation times. Enhanced circulation is particularly critical for nanoparticle-based drug delivery and imaging applications, as it allows nanoparticles to accumulate at the target site and improve therapeutic efficacy. In conclusion, the surface charge and functionalization of nanoparticles play fundamental roles in their biocompatibility and effectiveness in biomedical applications. Surface charge influences cellular uptake, protein corona formation, and interactions with biological systems. Careful selection and engineering of surface charge can lead to improved cellular interactions and reduced opsonization, enhancing nanoparticle circulation and biocompatibility. Moreover, surface functionalization enables targeted drug delivery, enhancing specificity and reducing off-target effects. Additionally, functionalization can impart stealth properties, prolonging circulation times and improving nanoparticle behavior in biological environments. By optimizing the surface charge and functionalization of nanoparticles, researchers can design safer and more effective nanoparticle-based therapies, contributing to the advancement of precision medicine and personalized treatments [56–58].

2.4 Immune responses to nanoparticles

Nanoparticles can elicit complex immune responses upon exposure to biological systems, and understanding these immune interactions is crucial for assessing their biocompatibility and potential therapeutic applications. The immune system plays a central role in recognizing and responding to foreign entities, including nanoparticles. Immune responses to nanoparticles involve various components of the immune system, including the innate and adaptive arms, each contributing to the nanoparticle's fate and behavior in the biological environment.

2.4.1 Innate immune responses

Upon encountering nanoparticles, immune cells of the innate immune system, such as macrophages and dendritic cells, can recognize them as foreign or non-self entities. This recognition triggers an immediate immune response, leading to the production of pro-inflammatory cytokines, such as interleukin-1β (IL-1β) and tumor necrosis factor-alpha (TNF-α). These cytokines are essential for initiating and amplifying the immune response and play significant roles in coordinating the recruitment and activation of other immune cells to the site of nanoparticle exposure. The activation of innate immune responses can result in local inflammation, which may influence the nanoparticles' behavior and interactions with cells and tissues.

2.4.2 Complement activation

The complement system is a critical component of the innate immune response and plays a significant role in the recognition and elimination of foreign entities, including nanoparticles. When nanoparticles interact with the complement system, they can trigger complement activation through various pathways. Complement activation can lead to opsonization, a process where the nanoparticles are coated with complement proteins and antibodies, marking them for recognition and clearance by phagocytic cells, such as macrophages. Additionally, complement activation can initiate an inflammatory response, further influencing the nanoparticles' fate in the body and their potential toxicity. Modulating complement activation is an important consideration in nanoparticle design to improve their biocompatibility and reduce undesired immune responses.

2.4.3 Adaptive immune responses

Prolonged or repeated exposure to nanoparticles can also lead to adaptive immune responses. The adaptive immune system recognizes specific antigens and generates tailored immune responses to combat them. Nanoparticles can act as antigens, stimulating the production of specific antibodies by B cells. The formation of nanoparticles-antibody complexes can lead to immune recognition and clearance, limiting their therapeutic efficacy upon repeated administration. Additionally, nanoparticle-specific T cells may be activated in certain instances, further contributing to adaptive immune responses. These adaptive immune reactions can influence the nanoparticles' pharmacokinetics, biodistribution, and overall biocompatibility. Understanding and controlling adaptive immune responses are essential for designing nanoparticle-based therapies with enhanced safety and efficacy.

2.4.4 Immunogenicity

The immunogenicity of nanoparticles refers to their ability to induce an immune response. Nanoparticles with specific physicochemical properties, such as size, charge, and surface functionalization, may be more immunogenic than others [59–64]. The immune system can recognize nanoparticles as foreign invaders and mount an immune response against them, potentially leading to adverse effects and reduced therapeutic efficacy. Reducing nanoparticle immunogenicity is a critical consideration in nanoparticle design to improve their biocompatibility and increase their chances of successful clinical translation. Strategies to minimize immunogenicity include employing biocompatible surface modifications and choosing materials that are less likely to trigger immune responses.

2.4.5 Regulatory implications

Understanding the immune responses to nanoparticles has significant regulatory implications for their translation from the laboratory to clinical applications. Regulatory agencies require a thorough assessment of the potential immunotoxicity of nanoparticles before they are approved for clinical use. This involves evaluating the nanoparticles' interactions with immune cells, cytokine production, complement activation, and potential for inducing adaptive immune responses. Preclinical studies are essential for identifying any potential immunotoxicity and guiding further optimization of nanoparticle design to ensure their safety and effectiveness in human applications. Comprehending the immune responses to nanoparticles is vital for designing safe and effective nanoparticle-based therapies [65,66]. Modulating nanoparticle interactions with the immune system can optimize their behavior, improve biocompatibility, and enhance their potential for therapeutic applications. In-depth studies on the immunological aspects of nanoparticle interactions are vital for advancing nanoparticle research and harnessing their full potential in medicine. By carefully considering and manipulating immune responses, researchers can develop nanoparticles that hold great promise for precision medicine, targeted drug delivery, and personalized therapeutic approaches. Moreover, regulatory agencies must remain vigilant in assessing the immunotoxicity of nanoparticles to ensure their safe clinical translation and contribute to the development of transformative nanoparticle-based therapies.

3.   Cardiovascular toxicity of nanoparticles

3.1   Effects of nanoparticles on heart function

The heart, as a crucial organ, plays a pivotal role in maintaining overall body homeostasis by pumping blood and ensuring oxygen and nutrient delivery to various tissues. However, nanoparticles can pose significant challenges to heart function, leading to cardiovascular toxicity. The effects of nanoparticles on heart function are multifaceted and can manifest in several ways. One of the primary concerns is myocardial toxicity, where nanoparticles accumulate in the heart tissue, particularly in cardiomyocytes, the cells responsible for heart contractions. This accumulation can disrupt cellular function, causing cellular damage or even cell death. Such adverse effects can lead to myocardial dysfunction, altering cardiac contractility, impairing electrical conduction, or inducing structural abnormalities [67]. Consequently, nanoparticles can contribute to the development of various cardiac diseases. Another critical aspect is the impact of nanoparticles on cardiac electrophysiology. Nanoparticles may interact with ion channels and ion transporters in cardiac cells, leading to changes in the electrical activity of the heart. Alterations in ion channel activity can result in irregular heartbeats, arrhythmias, and prolonged QT intervals. These abnormalities can pose a significant risk for life-threatening ventricular arrhythmias and may ultimately impair cardiac function, leading to an increased risk of sudden cardiac death. Furthermore, nanoparticles can influence cardiovascular hemodynamics, which refers to the study of blood flow dynamics and vascular resistance. Nanoparticles that accumulate in blood vessels may cause vascular constriction, affecting blood flow dynamics and increasing cardiac workload. Altered hemodynamics can result in hypertension and cardiovascular strain, which may exacerbate existing heart conditions, such as hypertension or heart failure.

Chronic exposure to nanoparticles can also induce cardiac remodeling, a process involving changes in the heart's structure and function in response to stress or injury. This can manifest as ventricular hypertrophy, fibrosis, and alterations in chamber dimensions. Cardiac remodeling negatively impacts heart function, potentially leading to heart failure or impairing the heart's ability to meet the body's demands. The effects of nanoparticles on heart function are complex and can be influenced by various factors, including the nanoparticle's physicochemical properties, route of exposure, and duration of exposure. Additionally, individual variability and pre-existing cardiovascular conditions can modulate the extent of nanoparticle-induced cardiac toxicity. To mitigate the potential adverse effects of nanoparticles on heart function, it is essential to gain a comprehensive understanding of their toxicological profile and establish safe exposure limits. Preclinical studies play a crucial role in elucidating the mechanisms underlying nanoparticle-induced cardiovascular toxicity and assessing potential risks. Additionally, appropriate dosing and careful consideration of nanoparticle properties can help minimize adverse effects while maximizing the benefits of nanoparticle-based therapies [68–70]. Moreover, researchers and regulatory bodies must consider the unique challenges posed by nanoparticles in heart-specific applications, such as targeted drug delivery to treat cardiovascular diseases. Ensuring the safety and efficacy of nanoparticle-based treatments requires a robust understanding of their interactions with cardiac tissues and the cardiovascular system as a whole. The effects of nanoparticles on heart function are a complex area of research with important implications for human health. As nanoparticles continue to find applications in various biomedical fields, including drug delivery and imaging, further investigation into their cardiovascular toxicity and potential strategies to enhance their safety profile will be critical to harness their full potential while safeguarding patient health. By advancing our knowledge of nanoparticle-cardiovascular interactions, we can pave the way for safer and more effective nanoparticle-based therapies that hold great promise in transforming healthcare.

3.2 Disruption of vascular integrity

The endothelium, a monolayer of cells lining the blood vessels, is a critical component of the cardiovascular system, responsible for maintaining vascular integrity and function. Nanoparticles, due to their unique properties and interactions with biological systems, have the potential to disrupt vascular integrity through various mechanisms. Endothelial dysfunction is one of the key consequences of nanoparticle exposure. Nanoparticles can directly affect endothelial cells, leading to the impairment of endothelial function. One significant aspect of endothelial dysfunction is the reduced production of nitric oxide (NO), a crucial signaling molecule that regulates vascular tone and blood flow. Diminished NO production can result in vascular constriction and increased vascular resistance, leading to hypertension and compromising blood flow to various organs.

Moreover, nanoparticles can stimulate the production of endothelin-1, a vasoconstrictor that further exacerbates endothelial dysfunction and contributes to vascular constriction. Nanoparticles can also disrupt vascular integrity by increasing vascular permeability. When nanoparticles interact with the endothelium, they can induce changes in cell signaling pathways, leading to alterations in the endothelial barrier function. As a result, vascular permeability increases, allowing plasma proteins and fluids to leak into the surrounding tissues. This increased vascular permeability can lead to tissue edema, impair blood flow, and compromise tissue function. The disruption of vascular integrity by nanoparticles can be particularly concerning in the context of diseases with altered endothelial barrier function, such as acute lung injury or sepsis.

Chronic exposure to nanoparticles has been associated with the promotion of atherosclerosis, a condition characterized by the buildup of lipid plaques in arterial walls. Nanoparticles can induce oxidative stress and inflammation in endothelial cells, initiating a cascade of events that promote the recruitment of immune cells to the arterial wall. The accumulation of immune cells and lipids leads to the formation of atherosclerotic plaques, which narrow and stiffen the arteries. The presence of atherosclerotic plaques can impede blood flow, increase the risk of thrombosis, and ultimately result in cardiovascular events, such as heart attacks and strokes. In addition to direct effects on the endothelium, nanoparticles can also impact vascular integrity indirectly through interactions with circulating blood components. For instance, nanoparticles can interact with platelets, leading to platelet activation and aggregation. Platelet activation can contribute to the formation of blood clots, which can further compromise blood flow and contribute to thrombotic events. Moreover, nanoparticles can modulate the behavior of immune cells and inflammatory mediators, influencing vascular inflammation and further disrupting vascular integrity. The effects of nanoparticles on vascular integrity can vary depending on factors such as nanoparticle size, surface charge, surface functionalization, and route of exposure. Additionally, individual susceptibility and pre-existing cardiovascular conditions can modulate the extent of vascular disruption caused by nanoparticles. Understanding the mechanisms by which nanoparticles disrupt vascular integrity is essential for the safe development and application of nanoparticle-based therapies. Researchers and regulatory bodies must consider the potential risks associated with nanoparticle exposure, especially in vulnerable populations, such as individuals with pre-existing cardiovascular diseases or other vascular disorders. Moreover, efforts should be made to optimize nanoparticle design and surface modification strategies to minimize their adverse effects on vascular integrity while maximizing their therapeutic benefits.

3.3 Nanoparticle-induced oxidative stress

Nanoparticle-induced oxidative stress is a critical mechanism that contributes to cardiovascular toxicity. Oxidative stress occurs when there is an imbalance between the production of reactive oxygen species (ROS) and the body's antioxidant defense system. Nanoparticles can trigger oxidative stress through various mechanisms, leading to cellular damage and potential adverse effects on the cardiovascular system. One of the primary ways nanoparticles induce oxidative stress is through electron transfer and Fenton reactions. Certain nanoparticles, particularly those containing transition metals like iron or copper, can undergo electron transfer reactions with oxygen, resulting in the production of superoxide radicals ($O_2^{\bullet-}$). Moreover, nanoparticles can catalyze Fenton reactions, where hydrogen peroxide (H_2O_2) reacts with transition metals to produce highly reactive hydroxyl radicals ($\bullet OH$). These ROS can cause cellular damage by oxidizing lipids, proteins, and DNA, leading to cellular dysfunction and potential long-term effects on cardiovascular health. Mitochondrial dysfunction is another mechanism by which nanoparticles induce oxidative stress. Mitochondria, the cellular organelles responsible for energy production, can accumulate nanoparticles upon exposure. The presence of nanoparticles in mitochondria can disrupt mitochondrial function, leading to impaired ATP production and increased ROS generation. Mitochondrial dysfunction can further contribute to oxidative stress, as ROS production occurs within mitochondria, creating a vicious cycle of cellular damage and dysfunction. In addition to directly generating ROS, some nanoparticles can also inhibit the activity of antioxidant enzymes, such as superoxide dismutase (SOD) and catalase. These enzymes play a crucial role in neutralizing ROS and maintaining redox balance within cells. When nanoparticles interfere with the activity of antioxidant enzymes, the body's ability to combat ROS is compromised, leading to heightened oxidative stress. Moreover, nanoparticle-induced oxidative stress can deplete cellular antioxidants, further exacerbating oxidative damage and increasing the susceptibility of cells to oxidative stress-induced injury.

The surface properties of nanoparticles, including their coatings and functionalization, can also influence the extent of oxidative stress they induce. Certain surface coatings, such as polyethylene glycol (PEG), have been shown to reduce ROS generation and mitigate oxidative stress. PEG coatings create a protective layer on the nanoparticle surface, shielding them from interactions with cellular components that could trigger oxidative stress. However, not all surface coatings have such protective effects. Some surface coatings may enhance ROS generation and exacerbate oxidative stress, particularly if they contain materials that can undergo redox reactions or catalyze ROS production. The impact of nanoparticle-induced oxidative stress on cardiovascular health is a complex and multifaceted area of research. Chronic exposure to nanoparticles with significant ROS-generating potential may lead to cumulative oxidative damage, increasing the risk of cardiovascular diseases such as atherosclerosis, hypertension, and heart failure. Additionally, the cardiovascular system's continuous exposure to ROS can lead to chronic inflammation, which is a key driver of cardiovascular diseases. Mitigating nanoparticle-induced oxidative stress is crucial for the safe and effective use of nanoparticles in various biomedical applications. Researchers must carefully consider nanoparticle design and surface modifications to minimize ROS generation and enhance biocompatibility. The development of nanoparticles with reduced ROS-generating potential and improved antioxidant properties can lead to safer and more effective nanoparticle-based therapies. Moreover, understanding the factors that influence nanoparticle-induced oxidative stress is essential for tailoring nanoparticles for specific applications. The choice of materials, surface coatings, and functionalization strategies should take into account the potential for oxidative stress induction and its implications for cardiovascular health.

3.4 Inflammation and endothelial dysfunction

Inflammation and endothelial dysfunction are interrelated processes that play a pivotal role in cardiovascular toxicity induced by nanoparticles. These processes can contribute to the development and progression of cardiovascular diseases, such as atherosclerosis, hypertension, and heart failure, which are associated with significant morbidity and mortality. Nanoparticles can trigger the release of pro-inflammatory cytokines, such as interleukin-6 (IL-6) and tumor necrosis factor-alpha (TNF-α), by immune cells and endothelial cells. Inflammatory cytokines are key mediators of the immune response, and their release in response to nanoparticle exposure can lead to vascular inflammation. Inflammation promotes the recruitment of immune cells to the site of nanoparticle exposure, amplifying the immune response and contributing to endothelial dysfunction. Additionally, inflammatory cytokines can activate endothelial cells, leading to increased expression of adhesion molecules and other inflammatory mediators. Nanoparticles can activate immune cells, such as macrophages and neutrophils, which are key players in the immune response. Activated immune cells can produce reactive oxygen species (ROS) and inflammatory cytokines, perpetuating the inflammatory response. ROS are highly reactive molecules that can cause cellular damage by oxidizing lipids, proteins, and DNA. ROS production by activated immune cells can contribute to endothelial dysfunction and impair vascular function.

Endothelial dysfunction induced by nanoparticles can lead to the upregulation of adhesion molecules on the surface of endothelial cells. Adhesion molecules, such as vascular cell adhesion molecule-1 (VCAM-1) and intercellular adhesion molecule-1 (ICAM-1), facilitate the adhesion and transmigration of immune cells into the vessel wall. Immune cell infiltration into the vessel wall promotes inflammation and the formation of atherosclerotic plaques. These plaques, composed of lipids, immune cells, and vascular smooth muscle cells, can narrow and stiffen arteries, compromising blood flow and increasing the risk of cardiovascular events. In addition to the generation of reactive oxygen species (ROS), nanoparticles can also induce the production of reactive nitrogen species (RNS), such as nitric oxide (NO) and peroxynitrite (ONOO$^-$). RNS can react with ROS to form highly reactive species, amplifying oxidative and nitrosative stress. Nitrosative stress can lead to nitration of proteins and modification of cellular signaling pathways, further exacerbating endothelial dysfunction and inflammation. The generation of RNS by nanoparticles contributes to the overall oxidative and nitrosative stress burden, which can have profound implications for cardiovascular health. The interplay between inflammation and endothelial dysfunction creates a vicious cycle that perpetuates cardiovascular toxicity induced by nanoparticles. Endothelial dysfunction leads to the upregulation of adhesion molecules, facilitating the infiltration of immune cells into the vessel wall. Immune cells, in turn, produce ROS, inflammatory cytokines, and RNS, leading to further endothelial dysfunction and inflammation. This cycle of inflammation and endothelial dysfunction can contribute to the development of atherosclerosis and other cardiovascular diseases.

The impact of inflammation and endothelial dysfunction on nanoparticle-induced cardiovascular toxicity is complex and influenced by various factors, such as nanoparticle properties, dose, and exposure duration. Chronic exposure to nanoparticles with pro-inflammatory properties or high oxidative potential can lead to persistent inflammation and endothelial dysfunction, increasing the risk of cardiovascular diseases. Additionally, individual susceptibility and pre-existing cardiovascular conditions can modulate the extent of inflammation and endothelial dysfunction induced by nanoparticles. Understanding the mechanisms underlying inflammation and endothelial dysfunction induced by nanoparticles is crucial for developing strategies to mitigate cardiovascular toxicity and enhance the safety of nanoparticle-based therapies. Researchers must consider nanoparticle design and surface modifications to minimize inflammation and endothelial dysfunction. The development of nanoparticles with reduced pro-inflammatory properties and enhanced biocompatibility can lead to safer and more effective nanoparticle-based therapies. Moreover, the crosstalk between inflammation and endothelial dysfunction highlights the importance of comprehensive evaluations of nanoparticle-induced cardiovascular toxicity. Researchers must consider multiple endpoints, including markers of inflammation, endothelial function, and vascular health, to fully understand the potential risks associated with nanoparticle exposure. By advancing our understanding of the interplay between inflammation and endothelial dysfunction, researchers can optimize the design and application of nanoparticle-based therapies, harnessing their potential for transformative advancements in cardiovascular medicine while safeguarding patient health.

4.   Strategies for enhancing nanoparticle biocompatibility

The success of nanoparticle-based applications in medicine and other fields heavily depends on their biocompatibility, which refers to their ability to interact safely with living organisms without inducing adverse effects. Nanoparticles have unique physicochemical properties that make them attractive for various applications, including drug delivery, imaging, and theranostics. However, their interactions with biological systems can elicit immune responses, cause toxicity, and limit their efficacy. To address these challenges, researchers have developed a range of strategies to enhance nanoparticle biocompatibility. This comprehensive discussion explores various approaches for improving nanoparticle biocompatibility, including surface modification techniques for improved circulation time, polyethylene glycol (PEG) functionalization, stealth nanoparticles, and biomimetic nanoparticles that mimic nature's biocompatibility.

4.1   Surface modification techniques for improved circulation time

Surface modification techniques play a crucial role in enhancing the circulation time and biocompatibility of nanoparticles, addressing one of the major challenges in nanoparticle-based therapeutics – their rapid clearance from the bloodstream by the mononuclear phagocyte system (MPS). The MPS, primarily composed of macrophages in the liver and spleen, recognizes and eliminates foreign particles, including nanoparticles, leading to their quick removal from circulation. This rapid clearance significantly limits the efficacy and therapeutic potential of nanoparticle-based drugs. To overcome this limitation, researchers employ various surface modification techniques to prolong circulation time and improve the biocompatibility of nanoparticles. One effective surface modification approach is ligand shielding. Ligand shielding involves the introduction of hydrophilic polymers or molecules on the nanoparticle surface, creating a protective layer that reduces interactions between nanoparticles and opsonins. Opsonins are proteins that mark foreign particles for phagocytosis by macrophages. By shielding the nanoparticles with hydrophilic polymers, such as polyethylene glycol (PEG), the ligand shielding technique helps "cloak" the nanoparticles, making them less recognizable to the MPS. As a result, the nanoparticles can evade recognition and phagocytosis by macrophages, leading to extended circulation time in the bloodstream. The ligand shielding approach improves nanoparticle biocompatibility by reducing their rapid clearance, allowing them to reach the target site and exert their therapeutic effect more effectively. Another surface modification technique is the use of zwitterionic coatings. Zwitterionic coatings contain both positive and negative charges on the nanoparticle surface, mimicking the charge distribution found in biological membranes. These coatings effectively reduce protein adsorption and opsonization, thus promoting nanoparticle stealth and biocompatibility. The zwitterionic coatings create a repulsive force that prevents the adhesion of opsonins to the nanoparticle surface, reducing macrophage recognition and clearance. The repulsive interactions between the zwitterionic-coated nanoparticles and blood components enable longer circulation times and improved biocompatibility. Zwitterionic coatings offer promising prospects for the development of long-circulating nanoparticles with reduced immunogenicity, as they provide a stealthy and biocompatible surface that minimizes interactions with the immune system.

Biomolecule-conjugated nanoparticles represent another surface modification strategy for enhancing nanoparticle biocompatibility. By attaching biomolecules, such as proteins, peptides, or antibodies, to the nanoparticle surface, researchers can create targeted nanoparticles that interact specifically with certain cells or tissues. The biomolecules serve as ligands that recognize receptors present on the surface of target cells. This targeted approach allows for specific interactions between the nanoparticles and the desired cells, reducing off-target effects and improving biocompatibility. Biomolecule-conjugated nanoparticles can be engineered to target specific tissues or disease sites, enabling precise drug delivery and enhancing therapeutic efficacy. Additionally, targeted interactions with specific cells or tissues can minimize nonspecific uptake and clearance, prolonging nanoparticle circulation time in the bloodstream and increasing their chances of reaching the target site. Surface modification techniques for improved circulation time and biocompatibility offer significant advantages for the development of nanoparticle-based therapeutics. By evading rapid clearance by the MPS, these modified nanoparticles can remain in circulation for extended periods, allowing more time for targeted drug delivery and enhancing treatment effectiveness. Furthermore, improved biocompatibility reduces the risk of adverse reactions and toxicity, making nanoparticle-based therapies safer for patients. Surface modification approaches, such as ligand shielding, zwitterionic coatings, and biomolecule conjugation, can be tailored to suit the specific requirements of different therapeutic applications, enabling the design of nanoparticles with optimized pharmacokinetics and enhanced therapeutic outcomes.

Despite their advantages, surface modification techniques also present challenges and considerations that must be carefully addressed. For example, the choice of surface coating and modification method can influence the overall stability and behavior of nanoparticles. Additionally, concerns related to potential immunogenicity, long-term safety, and the potential for immune responses to surface modifications must be thoroughly investigated. Researchers must strike a balance between optimizing nanoparticle circulation time and minimizing adverse effects associated with surface modifications.

4.2   Polyethylene glycol (PEG) functionalization

Polyethylene glycol (PEG) functionalization is a highly effective and widely used strategy for enhancing the biocompatibility of nanoparticles. PEG is a hydrophilic and biocompatible polymer that can be covalently attached to the nanoparticle surface, creating a steric barrier that reduces interactions with opsonins and immune cells. This surface modification technique, known as PEGylation, imparts "stealth" properties to the nanoparticles, enabling them to evade recognition by the immune system and prolong their circulation time in the bloodstream. The key to PEGylation's success lies in its ability to create a hydrophilic layer on the nanoparticle surface. The hydrophilic nature of PEG repels plasma proteins and immune cells, reducing the likelihood of opsonization, a process in which opsonins mark the nanoparticles for phagocytosis by macrophages [71,72]. By reducing opsonization and phagocytosis, PEGylation effectively enhances the nanoparticles' stealth properties. As a result, the PEGylated nanoparticles can circulate in the bloodstream for extended periods, increasing their chances of reaching the target site and exerting their therapeutic effect more effectively. The density and length of PEG chains can be tailored to fine-tune the biocompatibility and pharmacokinetics of the PEGylated nanoparticles. Higher PEG density and longer chains generally lead to increased stealth properties and longer circulation times. However, it is essential to strike a balance, as excessive PEGylation may hinder cellular uptake or interactions with target cells. The design of PEGylation must consider the specific application and the desired pharmacokinetic profile of the nanoparticles.

While PEGylation offers significant advantages for enhancing nanoparticle biocompatibility, it also has its challenges. One concern is that the presence of PEG on the nanoparticle surface may reduce nanoparticle-cell interactions. Cellular uptake of nanoparticles is often mediated by specific interactions between the nanoparticle surface and cell receptors. The presence of PEG may interfere with these interactions, potentially affecting the uptake of PEGylated nanoparticles by target cells. This limitation underscores the need for careful optimization of PEGylation to ensure that the nanoparticles retain their ability to interact with and enter target cells. Another challenge associated with PEGylation is the potential for anti-PEG antibody formation. Repeated administration of PEGylated nanoparticles can trigger an immune response, leading to the production of antibodies against PEG. These anti-PEG antibodies may recognize and clear PEGylated nanoparticles more rapidly upon subsequent administrations, reducing their circulation time and therapeutic efficacy [73,74]. To address this issue, researchers are exploring different strategies, such as using alternative polymers or designing nanoparticles with reduced immunogenicity, to improve the long-term stability and performance of PEGylated nanoparticles. Despite these challenges, PEGylation remains a valuable and versatile surface modification technique for enhancing nanoparticle biocompatibility. Its ability to create a protective steric barrier and confer stealth properties to nanoparticles makes PEGylation an attractive choice for various biomedical applications. PEGylated nanoparticles have been successfully employed in drug delivery systems, imaging agents, and therapeutic interventions, showing great promise in improving treatment outcomes and reducing side effects. As research in the field of nanoparticle-based therapies continues to advance, further optimization and refinement of PEGylation and other surface modification techniques will be essential. Understanding the intricate interactions between nanoparticles and biological systems will guide the development of nanoparticles with optimized pharmacokinetics, improved biocompatibility, and enhanced therapeutic efficacy. With continued progress, nanoparticle-based therapeutics hold the potential to revolutionize medicine, offering targeted and personalized treatments for a wide range of diseases and medical conditions.

4.3　Stealth nanoparticles and immune evasion

Stealth nanoparticles, a crucial advancement in nanomedicine, are designed to evade the immune system, allowing for prolonged circulation times and reduced non-specific interactions with immune cells [75–77]. By evading immune recognition, these nanoparticles can better reach their intended target sites and exert their therapeutic effects more effectively. One approach to achieving stealth properties involves the use of surface zwitterionic coatings. Zwitterionic coatings mimic the charge distribution found in biological membranes, creating a hydrophilic and neutral surface on the nanoparticles. This hydrophilic layer reduces the adsorption of plasma proteins, such as opsonins, which mark foreign particles for phagocytosis by macrophages. As a result, zwitterionic coatings help shield the nanoparticles from immune recognition, promoting their stealth behavior and improving biocompatibility. Another innovative strategy to promote immune evasion is through surface "self" signaling. Nanoparticles can be engineered to present signaling molecules on their surface that mimic the body's natural cellular signals. Immune cells recognize these "self" signals and interpret the nanoparticles as part of the body's own cells, rather than foreign invaders. By presenting such signals, nanoparticles can effectively evade immune recognition and reduce immune responses, leading to improved biocompatibility.

Furthermore, researchers are exploring the use of extracellular vesicles, such as exosomes, as a means of camouflaging nanoparticles. Extracellular vesicles are natural carriers in the body that facilitate intercellular communication. By coating nanoparticles with cell-derived extracellular vesicles, scientists can create "camouflaged" nanoparticles that closely resemble natural cell membranes. These camouflaged nanoparticles are less likely to be recognized and attacked by the immune system, contributing to enhanced stealth behavior and improved biocompatibility. Stealth nanoparticles offer great potential for revolutionizing various biomedical applications. In drug delivery systems, for example, stealth properties enable nanoparticles to avoid rapid clearance by the mononuclear phagocyte system (MPS), leading to prolonged circulation times and enhanced drug delivery to target tissues. By escaping immune recognition, these stealth nanoparticles can better reach their intended sites of action, ensuring more effective and targeted drug delivery. Moreover, the improved biocompatibility of stealth nanoparticles reduces the risk of immune reactions and adverse effects. This is particularly important for repeated administrations, as it reduces the likelihood of the immune system mounting a response against the nanoparticles, ensuring their long-term efficacy.

While stealth nanoparticles hold significant promise, challenges remain in their development and implementation. Achieving optimal stealth properties without compromising other crucial characteristics of nanoparticles, such as cellular uptake and targeting ability, requires careful design and optimization. Additionally, the long-term safety and potential immunogenicity of stealth nanoparticles need to be thoroughly assessed to ensure their clinical translation.

4.4 Biomimetic nanoparticles: Mimicking nature's biocompatibility

Biomimetic nanoparticles represent a promising approach in nanomedicine, as they seek to mimic natural structures and functions found in the body to enhance biocompatibility through natural recognition and interactions. Cell membrane-coated nanoparticles are a prominent example of biomimetic nanoparticles. These nanoparticles are created by fusing the surfaces of nanoparticles with cell membranes derived from specific cell types. The resulting cell membrane-coated nanoparticles retain the surface proteins and markers of the original cells, effectively acquiring "self" characteristics. This biomimicry enables the nanoparticles to evade immune recognition and reduce immunogenicity. Moreover, cell membrane-coated nanoparticles can interact with biological systems more effectively, leading to enhanced biocompatibility. These nanoparticles show great promise for applications in drug delivery, where their ability to mimic natural cells allows for targeted interactions with specific tissues and organs, leading to improved therapeutic outcomes.

Virus-like particles (VLPs) are another class of biomimetic nanoparticles. VLPs are non-infectious nanoparticles that resemble viruses in structure but lack viral genetic material. VLPs can be engineered to display specific surface proteins found on viruses, enabling them to interact with specific target cells [78–80]. This characteristic makes VLPs attractive for targeted drug delivery and vaccine development. By leveraging the natural ability of viruses to interact with host cells, VLPs can efficiently deliver therapeutic agents to specific cells or tissues, improving drug delivery efficacy while minimizing off-target effects. Biomimetic nanomaterials, such as proteins or nucleic acids, are designed to replicate the structure and properties of natural biomolecules. These nanomaterials exhibit enhanced biocompatibility as they closely resemble components already present in the body. By mimicking natural structures, biomimetic nanomaterials can interact with biological systems more efficiently, promoting compatibility and reducing immune responses.

This characteristic makes biomimetic nanomaterials valuable in various applications, including tissue engineering, where they can facilitate cellular adhesion and promote tissue regeneration. Additionally, biomimetic nanomaterials can be employed in targeted drug delivery systems, where they can be designed to specifically interact with certain receptors or cellular pathways, improving drug delivery precision. The concept of biomimetic nanoparticles has garnered significant interest in recent years due to its potential to revolutionize nanomedicine. By harnessing nature's biocompatibility and cellular recognition mechanisms, biomimetic nanoparticles offer a unique advantage over traditional nanoparticles. Their ability to interact with biological systems in a more natural and harmonious way reduces the risk of adverse reactions and immunogenicity, making them safer and more effective for various medical applications. However, challenges remain in the development and optimization of biomimetic nanoparticles. One major challenge is the scalability and reproducibility of cell membrane-coated nanoparticles and VLPs. These nanoparticles require careful preparation and purification processes to ensure their stability and consistency, which can be complex and time-consuming. Additionally, while biomimetic nanoparticles hold great potential, their long-term safety and potential immunogenicity need to be thoroughly investigated to ensure their clinical translation.

5. Targeted nanoparticles for specificity

Targeted nanoparticles have revolutionized the field of nanomedicine by enabling site-specific delivery of therapeutics and imaging agents. These nanoparticles are functionalized with ligands or molecules that recognize specific receptors or biomarkers on target cells or tissues, allowing for precise and selective interactions. Targeted nanoparticle design involves understanding the biology of the target site, selecting appropriate ligands, and optimizing the nanoparticle's physicochemical properties for effective delivery [81–84]. This comprehensive discussion explores the principles and mechanisms of targeted nanoparticles, including ligand-conjugated nanoparticles, active and passive targeting mechanisms, and the challenges and opportunities in their design.

5.1 Ligand-conjugated nanoparticles for site-specific delivery

Ligand-conjugated nanoparticles represent a powerful strategy for achieving site-specific delivery of therapeutic agents. By functionalizing nanoparticles with ligands that recognize specific receptors or biomarkers overexpressed on target cells or tissues, ligand-conjugated nanoparticles can improve the precision and efficiency of drug delivery. Several key aspects are crucial to the success of ligand-conjugated nanoparticles for site-specific delivery. The selection of ligands is a critical step in achieving specificity in targeting. Ligands can be antibodies, peptides, aptamers, or small molecules that selectively bind to receptors or biomarkers expressed on the target cells. The choice of ligands is determined by their affinity and specificity for the target, as well as their stability and biocompatibility. By selecting ligands that have a high affinity for the target, researchers can ensure that the nanoparticles will bind specifically to the desired cells or tissues, reducing off-target effects and enhancing therapeutic efficacy. Once the ligands are selected, they are conjugated to the surface of nanoparticles through various strategies. The conjugation chemistry must ensure the stability and activity of the ligands while preserving the integrity and biocompatibility of the nanoparticles. Covalent binding and non-covalent interactions are commonly employed methods for ligand conjugation. Additionally, some ligands can be encapsulated within the nanoparticle matrix [85]. The choice of conjugation strategy depends on the nature of the ligand and the nanoparticle, and it is essential to optimize the conjugation process to ensure the ligands retain their activity and targeting capabilities. One of the primary advantages of ligand-conjugated nanoparticles is their ability to facilitate enhanced cellular uptake through receptor-mediated endocytosis. The ligands recognize specific receptors on the surface of target cells, leading to specific binding and internalization of the nanoparticles. This targeted internalization improves drug delivery to the desired cells, increasing the concentration of the therapeutic agent at the target site and reducing exposure to healthy tissues. By bypassing non-targeted cells, ligand-conjugated nanoparticles minimize potential side effects and toxicity associated with conventional treatments.

Moreover, ligand-conjugated nanoparticles can overcome biological barriers that may impede drug delivery to the target site. For instance, the blood-brain barrier (BBB) is a highly selective barrier that prevents many drugs from entering the brain. By utilizing ligands that can interact with receptors expressed on the endothelial cells of the BBB, researchers can facilitate the transport of nanoparticles across this barrier and improve drug delivery to the brain. Similarly, in the context of cancer therapy, ligand-conjugated nanoparticles can exploit the unique features of the tumor microenvironment to target cancer cells specifically. Tumors often express specific receptors or markers that are absent or less abundant in healthy tissues, enabling the selective delivery of therapeutics to the tumor site. While ligand-conjugated nanoparticles offer great promise for site-specific delivery, challenges remain in their design and implementation. The complexity of ligand selection and conjugation requires careful consideration of factors such as ligand stability, binding kinetics, and potential immunogenicity. Moreover, achieving optimal targeting efficiency and minimizing off-target effects demand a deep understanding of the target's expression patterns and the biology of the delivery site. Additionally, the potential for ligand-mediated internalization raises questions about intracellular drug release and the potential for drug resistance. Researchers are actively exploring various strategies to address these challenges and improve the design and performance of ligand-conjugated nanoparticles.

5.2 Active and passive targeting mechanisms

Targeted nanoparticles utilize a combination of active and passive targeting mechanisms to achieve site-specific delivery, improving the efficacy and selectivity of drug delivery and imaging applications. Active targeting involves the use of ligands or molecules that specifically recognize receptors or biomarkers expressed on the surface of the target cells. Ligand-conjugated nanoparticles are designed to actively bind to these receptors, leading to specific uptake and accumulation at the target site [86,87]. This active targeting approach is advantageous because it allows for high specificity and selectivity in delivering therapeutic agents or imaging agents to the desired cells or tissues. In contrast, passive targeting relies on the enhanced permeability and retention (EPR) effect, which is a unique phenomenon observed in tumors and inflamed tissues. In regions with abnormal blood vessel structures, such as solid tumors, blood vessels are leaky, allowing nanoparticles to extravasate from the bloodstream into the tumor tissue. Furthermore, compromised lymphatic drainage in these regions leads to the retention of nanoparticles within the tumor microenvironment. As a result, nanoparticles can accumulate passively in the tumor tissue due to the EPR effect. Passive targeting is particularly useful for solid tumors, where nanoparticles can selectively accumulate in the tumor microenvironment, enhancing drug delivery and reducing systemic toxicity. By combining active and passive targeting strategies, researchers can achieve synergistic targeting, enhancing the specificity and effectiveness of nanoparticle delivery. In synergistic targeting, nanoparticles utilize the EPR effect for passive accumulation in the target tissue while also incorporating ligand-conjugated nanoparticles for additional active targeting. This combined approach aims to maximize nanoparticle accumulation and cellular uptake in the target tissue, further improving therapeutic outcomes. The EPR effect provides a broad distribution of nanoparticles in the tumor, while active targeting ensures precise targeting of specific cells within the tumor microenvironment. Synergistic targeting is particularly valuable for complex diseases like cancer, where heterogeneous cell populations and varying microenvironments necessitate multiple targeting approaches [88,89].

Various ligands can be used for active targeting, including antibodies, peptides, aptamers, and small molecules. The choice of ligands depends on the target cells or tissues and their expression patterns. Researchers must carefully select ligands with high affinity and specificity to ensure effective binding to the target receptors. Ligand selection is a crucial step in achieving successful active targeting and requires a deep understanding of the disease biology and the target's expression profile. The conjugation of ligands to the nanoparticle surface must be carefully optimized to preserve their activity while maintaining nanoparticle stability and biocompatibility. Conjugation chemistry can involve covalent binding, non-covalent interactions, or encapsulation within the nanoparticle matrix. The success of active targeting depends on the stability and integrity of the ligand-conjugated nanoparticles, as well as their ability to retain their targeting capabilities throughout the delivery process. Passive targeting through the EPR effect has been widely exploited for nanoparticle-based therapies. However, the EPR effect is not uniform across all tumors or inflamed tissues, and its magnitude can vary depending on factors such as tumor size, type, and location. Additionally, the EPR effect may be limited in some cases due to factors like increased interstitial pressure in tumors. To enhance the efficacy of passive targeting, researchers are investigating approaches to improve nanoparticle extravasation and retention within the target tissue. Strategies such as modifying the nanoparticle size, shape, or surface properties can potentially enhance passive targeting and improve nanoparticle accumulation in the desired tissue.

The combination of active and passive targeting mechanisms in synergistic targeting offers numerous advantages for nanoparticle-based therapies. By capitalizing on the benefits of both strategies, researchers can achieve enhanced drug delivery and imaging, leading to improved therapeutic outcomes. Synergistic targeting allows for precise delivery to target cells while also exploiting the EPR effect to passively accumulate in the tumor microenvironment. This approach can increase the therapeutic index of drugs by reducing systemic toxicity and minimizing off-target effects. Despite the potential benefits, challenges remain in the design and implementation of targeted nanoparticles [90,91]. The heterogeneity of diseases, such as cancer, poses a significant obstacle to achieving complete and uniform targeting. Additionally, nanoparticle stability, biocompatibility, and scalability are critical factors that must be addressed to ensure the successful translation of targeted nanoparticles from the laboratory to clinical applications. As researchers continue to explore innovative ligands, nanoparticle formulations, and targeting strategies, the field of targeted nanoparticles holds immense promise for revolutionizing precision medicine and personalized therapeutics. Through the continued advancement of active and passive targeting mechanisms, targeted nanoparticles will continue to play a pivotal role in enhancing drug delivery efficiency and improving patient outcomes in various disease settings.

Materials Research Forum LLC
https://doi.org/10.21741/9781644902998-7

5.3 Challenges and opportunities in targeted nanoparticle design

Targeted nanoparticle design presents both challenges and opportunities in the field of nanomedicine. One of the primary challenges is ligand selection and optimization for achieving specific and effective targeting. The choice of ligands must consider various factors, including receptor expression levels on target cells, ligand affinity, and ligand stability. Ligand-conjugated nanoparticles must be carefully optimized to ensure efficient targeting while minimizing off-target interactions. Another significant challenge is the potential immune response and immunogenicity associated with targeted nanoparticles. The introduction of foreign ligands on the nanoparticle surface may trigger immune recognition and lead to reduced efficacy upon repeated administration. Researchers must develop strategies to minimize immune responses and immunogenicity, allowing for long-term and repeated applications of targeted nanoparticles. Multifunctional nanoparticles that combine various functionalities, such as drug delivery, imaging, and therapeutic response monitoring, offer promising opportunities for personalized medicine and precision therapeutics. These multifunctional nanoparticles can be tailored to address individual patient needs based on their unique biomarkers and disease profiles. The ability to deliver drugs, monitor treatment response, and visualize disease sites in real time enhances the potential for improved therapeutic outcomes and reduced side effects.

The concept of precision medicine, enabled by targeted nanoparticles, holds immense potential for revolutionizing healthcare. Precision medicine aims to tailor treatments to individual patients, taking into account their genetic makeup, lifestyle, and environment. Targeted nanoparticles allow for specific and personalized drug delivery, optimizing treatment efficacy while minimizing adverse effects. By utilizing targeted nanoparticles, physicians can move towards more precise and individualized therapies, leading to better patient outcomes. However, the translation of targeted nanoparticles from the laboratory to clinical practice presents significant challenges. Regulatory hurdles, manufacturing processes, scalability, and cost-effectiveness are critical considerations for clinical applications. Ensuring the reproducibility and consistency of targeted nanoparticles is essential for obtaining regulatory approval and for widespread adoption in clinical settings. Additionally, scaling up the production of targeted nanoparticles to meet the demand for clinical trials and patient treatments requires robust and efficient manufacturing processes. Moreover, the safety and biocompatibility of targeted nanoparticles must be rigorously evaluated to ensure patient well-being. Preclinical studies are necessary to assess nanoparticle toxicity, biodistribution, and long-term effects. These studies provide valuable insights into potential risks and guide the optimization of nanoparticle formulations for clinical use.

In addition to challenges, there are several opportunities in targeted nanoparticle design. The development of novel ligands and targeting strategies allows for improved specificity and selectivity in delivering therapeutic agents to specific cells or tissues. The advancement of nanotechnology and nanomaterials has opened doors to innovative approaches for ligand-conjugated nanoparticles, enabling precise and efficient targeting. The integration of targeted nanoparticles with other therapeutic modalities, such as gene therapy and immunotherapy, offers exciting prospects for combination therapies [92–94]. Combining different treatment approaches can enhance treatment efficacy and overcome drug resistance, paving the way for more effective cancer therapies and treatments for other diseases. Furthermore, targeted nanoparticles provide opportunities for theranostics, where therapeutics and diagnostics are combined into a single platform. Theranostic nanoparticles enable simultaneous treatment and monitoring of disease

Nanoparticle Toxicity and Compatibility   Materials Research Forum LLC
Materials Research Foundations 161 (2024) 182-224   https://doi.org/10.21741/9781644902998-7

progression, offering real-time feedback on treatment response. Theranostic approaches enhance treatment efficiency, reduce treatment cycles, and promote patient well-being.

6. Stimuli-responsive nanoparticles

Stimuli-responsive nanoparticles are a cutting-edge class of nanomaterials that exhibit unique properties in response to specific environmental cues or triggers. These nanoparticles are designed to undergo controlled changes in their physicochemical properties, such as size, shape, surface charge, and drug release behavior, upon exposure to external stimuli. The ability of stimuli-responsive nanoparticles to intelligently respond to their surroundings has opened new avenues for precision medicine, targeted drug delivery, and disease-specific therapies. This comprehensive discussion explores the principles and applications of stimuli-responsive nanoparticles, focusing on their responsiveness to environmental cues, triggered drug release for precision medicine, and their potential to combat cardiovascular toxicity.

6.1 Smart nanoparticle systems: Responsive to environmental cues

Smart nanoparticle systems that respond to environmental cues offer numerous opportunities for precision medicine and targeted therapeutics. pH-responsive nanoparticles, for example, exploit the acidic tumor microenvironment to trigger drug release selectively. These nanoparticles remain stable at physiological pH but rapidly degrade in acidic environments, such as tumors or intracellular endosomes, leading to efficient drug release at the target site. By leveraging pH differences between healthy tissues and disease sites, pH-responsive nanoparticles enable targeted drug delivery and minimize off-target effects [95,96]. Temperature-responsive nanoparticles are another promising approach for controlled drug delivery. These nanoparticles change their size, shape, or drug-release behavior in response to specific temperatures. For instance, thermosensitive liposomes release their cargo upon exposure to mild hyperthermia, which can be achieved through external heating methods such as laser irradiation or magnetic fields. Temperature-responsive nanoparticles enable site-specific drug delivery to regions with elevated temperatures, such as tumors or inflamed tissues, while sparing healthy tissues. Enzyme-responsive nanoparticles exploit disease-associated enzymes for controlled drug release. These nanoparticles undergo structural changes or cargo release behavior upon exposure to specific enzymes present in pathological conditions. For example, matrix metalloproteinases (MMPs) are overexpressed in the tumor microenvironment and can be targeted to trigger drug release from enzyme-responsive nanoparticles. This targeted drug release strategy enhances therapeutic efficacy while reducing systemic toxicity.

Redox-responsive nanoparticles respond to changes in the cellular redox environment, which can be altered in various diseases, including cancer and inflammation. In the reducing intracellular environment, redox-responsive nanoparticles can undergo disulfide bond cleavage, leading to drug release. This mechanism allows for enhanced drug delivery to target cells with elevated levels of reducing agents, such as glutathione while maintaining stability in the extracellular environment. Light-responsive nanoparticles offer spatial and temporal control over drug release and other functionalities. These nanoparticles change their properties upon exposure to specific wavelengths of light. For instance, photodynamic therapy uses light-responsive nanoparticles that generate reactive oxygen species upon light activation, leading to localized cell death in tumors. Photoacoustic imaging utilizes light-responsive nanoparticles to convert absorbed light into acoustic signals, enabling non-invasive imaging of biological tissues. Light-responsive nanoparticles also find applications in optogenetics, where light is used to control cellular activity and signaling.

6.2 Triggered drug release for precision medicine

Stimuli-responsive nanoparticles represent a groundbreaking approach to precision medicine, as they offer precise control over drug release, allowing tailored treatments based on individual patient needs. These nanoparticles can be triggered to release drugs in response to specific internal or external cues, ensuring that therapeutic agents are delivered at the right time and in the right location. External triggering mechanisms involve the application of external stimuli, such as light, heat, ultrasound, or magnetic fields, to stimuli-responsive nanoparticles to induce drug release. For example, light-responsive nanoparticles can be activated by specific wavelengths of light to release drugs with high spatial and temporal precision [97,98]. Similarly, thermosensitive nanoparticles can release drugs upon exposure to mild hyperthermia, which can be achieved using external heating methods like laser irradiation or magnetic fields. External triggering allows researchers to precisely control drug release, enabling spatiotemporal drug delivery and reducing systemic side effects. Internal triggering mechanisms, on the other hand, utilize specific biological cues present at the target site to activate drug release. For instance, enzyme-responsive nanoparticles can be designed to release drugs in response to disease-associated enzymes, such as matrix metalloproteinases (MMPs) that are overexpressed in the tumor microenvironment. pH-responsive nanoparticles change their properties in response to variations in pH, allowing drug release in acidic environments, such as in tumors or intracellular endosomes. Redox-responsive nanoparticles, sensitive to changes in the cellular redox environment, can release drugs in the reducing intracellular environment, promoting enhanced drug delivery to target cells.

On-demand drug release is a significant advantage of stimuli-responsive nanoparticles. By engineering these nanoparticles to release drugs only in response to specific disease-related signals, therapeutic agents are released when and where they are needed most. This approach optimizes therapeutic efficacy, minimizing off-target effects and reducing the risk of adverse reactions. On-demand drug release also allows for personalized medicine, tailoring treatments to individual patient needs based on their disease status and response to therapy. Stimuli-responsive nanoparticles also enable combination therapies, where multiple drugs or therapeutic agents can be co-loaded and released in a sequential or synergistic manner. Combination therapies have demonstrated enhanced efficacy and can overcome drug resistance, particularly in diseases like cancer where tumor cells can develop resistance to single-agent treatments. By releasing different drugs in a controlled and coordinated manner, combination therapies can target multiple pathways involved in the disease process, increasing the likelihood of successful treatment outcomes.

6.3 Combating cardiovascular toxicity via stimuli-responsive strategies

Cardiovascular toxicity is a significant concern in the development of nanoparticle-based therapies, as nanoparticles can interact with the cardiovascular system, leading to adverse effects on heart function and vascular integrity. Stimuli-responsive strategies offer potential solutions to combat cardiovascular toxicity and improve the biocompatibility of nanoparticles in the cardiovascular system. Controlled drug release is a crucial aspect of combating cardiovascular toxicity using stimuli-responsive nanoparticles. By engineering nanoparticles to release drugs in a controlled manner, therapeutic agents can be delivered to the target site with precision, minimizing their systemic exposure and reducing the risk of cardiovascular side effects. Controlled drug release can also ensure that therapeutic concentrations are maintained at the target site for an extended period, improving treatment outcomes and reducing the need for frequent dosing. Biocompatible surface modifications are another essential strategy for reducing cardiovascular toxicity. Stimuli-responsive coatings, such as zwitterionic or PEGylated coatings, can be applied to the nanoparticle surface to improve biocompatibility. These coatings create a hydrophilic and stealth-like layer that reduces interactions with blood components and immune cells, extending nanoparticle circulation time and reducing the risk of cardiovascular toxicity.

Cardiac targeting is a promising approach to minimize cardiovascular toxicity and enhance therapeutic efficacy. By functionalizing stimuli-responsive nanoparticles with ligands that recognize specific receptors on cardiac cells or tissues, researchers can achieve site-specific drug delivery to the heart. This targeted approach ensures that therapeutic agents are delivered specifically to the heart, reducing the risk of cardiovascular toxicity in other organs. Optimizing nanoparticle clearance and biodistribution is critical for reducing cardiovascular toxicity. Stimuli-responsive nanoparticles can be engineered to optimize their clearance routes and biodistribution profiles. For example, adjusting nanoparticle size or surface charge can influence their clearance mechanism, allowing nanoparticles to be excreted through the kidneys instead of accumulating in the liver or spleen. Furthermore, ligand-mediated targeting can promote the preferential accumulation of nanoparticles in the target tissue, minimizing off-target effects in the cardiovascular system. Stimuli-responsive nanoparticles can also be combined with imaging agents to monitor their biodistribution and therapeutic efficacy in real time. Imaging and monitoring provide valuable insights into nanoparticle behavior in vivo, enabling researchers to evaluate the effectiveness of targeted drug delivery and make adjustments to optimize therapeutic outcomes and reduce cardiovascular toxicity [99–101].

7. Preclinical and clinical perspectives on biocompatibility and cardiovascular toxicity

The translation of nanoparticle-based therapies from preclinical studies to clinical trials requires a comprehensive understanding of their biocompatibility and potential cardiovascular toxicity. Preclinical assessment plays a crucial role in evaluating nanoparticle biocompatibility, identifying potential adverse effects, and optimizing therapeutic efficacy. However, despite rigorous preclinical testing, unexpected cardiovascular adverse events can still occur during clinical trials. This comprehensive discussion explores the preclinical and clinical perspectives on biocompatibility and cardiovascular toxicity of nanoparticles, including the assessment of biocompatibility in preclinical studies, the evaluation of cardiovascular safety in preclinical models, and the analysis of nanoparticle-related cardiovascular adverse events in clinical trials.

7.1 Preclinical assessment of nanoparticle biocompatibility

Preclinical assessment of nanoparticle biocompatibility is a critical and necessary step in the development of nanoparticle-based therapies. Before these novel therapies can be tested in humans, it is essential to understand how nanoparticles interact with biological systems and to evaluate their safety and efficacy. The preclinical assessment involves a comprehensive evaluation of various aspects of nanoparticle behavior in vitro and in vivo, providing valuable insights into their potential impact on human health. In vitro biocompatibility studies are typically the first step in preclinical assessment. These studies involve exposing nanoparticles to different cell types and tissues in laboratory settings. Researchers use techniques such as cell viability assays, flow cytometry, and immunofluorescence to assess the cytotoxicity of nanoparticles and their cellular uptake. These experiments help identify any potential toxic effects on cells and provide information on the nanoparticles' interactions with the cellular machinery. In vitro studies also evaluate the potential immunogenicity of nanoparticles, assessing whether they induce an immune response in cells. Following in vitro studies, in vivo experiments are conducted in animal models to evaluate the biodistribution and clearance of nanoparticles. These studies help understand how nanoparticles are distributed throughout the body, where they accumulate, and how they are eliminated from the system. Biodistribution studies are critical for identifying potential off-target effects, which can guide the selection of appropriate therapeutic doses. The clearance of nanoparticles from the body is an important consideration for their safety and efficacy, as prolonged retention may lead to toxic accumulation in organs.

Pharmacokinetic studies are an integral part of the preclinical assessment, providing insights into the systemic behavior of nanoparticles. These studies evaluate the absorption, distribution, metabolism, and excretion of nanoparticles in animals. Understanding the pharmacokinetics of nanoparticles is crucial for optimizing their dosing regimens and predicting how they will behave in humans. Additionally, pharmacodynamic studies are conducted to evaluate the therapeutic effects and potential adverse effects of nanoparticles in preclinical models. One critical aspect of nanoparticle biocompatibility is their immunological response. The immune system plays a significant role in recognizing and clearing foreign substances from the body, including nanoparticles[102,103]. Immunological studies assess the activation of immune cells and the release of cytokines or other inflammatory mediators in response to nanoparticle exposure. Understanding the immune response to nanoparticles is essential for predicting potential immunogenicity and inflammation in human subjects. Long-term safety assessments are crucial in preclinical studies to evaluate the impact of nanoparticles over extended periods. Chronic exposure studies help researchers understand the long-term effects of nanoparticles on cardiovascular health and other organs. These studies are particularly important for assessing the potential accumulation of nanoparticles in tissues and organs over time, as chronic exposure can lead to unforeseen adverse effects. Another critical consideration in preclinical assessment is the relevance of preclinical findings to human responses. Animal models used in preclinical studies may not fully represent human physiology, and species differences can impact the predictive validity of preclinical data. Researchers must carefully consider the similarities and differences between preclinical models and humans to ensure that the findings can be translated into clinical trials successfully.

In summary, preclinical assessment of nanoparticle biocompatibility is a multifaceted and comprehensive process that involves in vitro and in vivo studies. Evaluating the cytotoxicity, cellular uptake, immunogenicity, and pharmacokinetics of nanoparticles is essential for understanding their behavior in biological systems. Furthermore, long-term safety assessments and considerations of species differences contribute to the successful translation of nanoparticle-based therapies from preclinical studies to clinical trials. Through rigorous preclinical assessment, researchers can identify potential risks and optimize nanoparticle design for improved safety and efficacy in human subjects.

7.2   Cardiovascular safety evaluation in preclinical studies

Cardiovascular safety evaluation in preclinical studies is of utmost importance due to the sensitivity of the cardiovascular system to nanoparticle exposure. Rigorous assessment of cardiovascular safety provides valuable insights into the potential adverse effects of nanoparticles on heart function and vascular dynamics. Several key aspects are considered during cardiovascular safety evaluation in preclinical models to ensure the safety and efficacy of nanoparticle-based therapies. Hemodynamic assessment is a fundamental component of cardiovascular safety evaluation. Monitoring blood pressure and heart rate provides information on the cardiovascular effects of nanoparticles. Electrocardiograms (ECGs) are essential for evaluating the electrical activity of the heart and identifying potential arrhythmias or other cardiac abnormalities. Changes in hemodynamic parameters and ECG patterns can indicate cardiovascular toxicity induced by nanoparticle exposure [104,105]. Cardiac imaging techniques play a pivotal role in non-invasively assessing cardiac structure and function following nanoparticle exposure. Echocardiography, a widely used imaging modality, allows the visualization of the heart's anatomy and the assessment of its contractile function. Magnetic resonance imaging (MRI) provides detailed images of cardiac tissues and can detect subtle changes in cardiac structure and function. Cardiac imaging is instrumental in detecting any structural or functional alterations induced by nanoparticles. Biomarker analysis is a valuable tool for identifying potential cardiovascular injury or dysfunction caused by nanoparticle exposure. Biomarkers such as troponins and brain natriuretic peptide (BNP) are indicative of cardiac injury and can be measured in blood samples to assess nanoparticle-induced cardiovascular effects. Additionally, the analysis of inflammatory cytokines can provide insights into the inflammatory response triggered by nanoparticles in the cardiovascular system.

Histopathological examination of cardiac tissues is crucial for understanding the morphological changes induced by nanoparticle exposure. Microscopic evaluation of cardiac tissues allows the identification of tissue damage, inflammation, and structural alterations. The histopathological analysis complements other cardiovascular safety assessments and provides detailed information on the cellular and tissue-level effects of nanoparticles on the heart. Oxidative stress and inflammation are key mechanisms underlying cardiovascular toxicity induced by nanoparticles. Evaluating oxidative stress markers, such as reactive oxygen species (ROS) and antioxidant enzyme levels, can indicate the extent of oxidative damage caused by nanoparticles in the cardiovascular system. Furthermore, the measurement of inflammatory cytokines can provide insights into the inflammatory response elicited by nanoparticles, which can contribute to cardiovascular dysfunction. Reproductive and developmental toxicity evaluations are essential, especially if nanoparticle exposure occurs during early developmental stages. The cardiovascular system undergoes critical development during embryonic and fetal stages, and disruptions in this process can have long-term consequences on heart function. Assessing potential reproductive and developmental toxicity helps identify any cardiovascular impacts that may arise from nanoparticle exposure during crucial developmental periods.

7.3   Clinical trials and nanoparticle-related cardiovascular adverse events

Clinical trials play a pivotal role in evaluating the safety and efficacy of nanoparticle-based therapies in humans. Despite comprehensive preclinical assessments, unexpected cardiovascular adverse events can arise during clinical trials. Several key aspects are critical in understanding and managing nanoparticle-related cardiovascular adverse events in clinical settings. Safety monitoring is of paramount importance during clinical trials. Continuous surveillance and monitoring of patients allow for the timely detection and management of cardiovascular adverse events. Close monitoring enables researchers and healthcare providers to intervene promptly if any unexpected cardiovascular effects occur. Patient selection is a crucial factor in clinical trials, especially in the context of nanoparticle-based therapies targeting cardiovascular diseases. Considering individual cardiovascular risk factors and pre-existing conditions is essential when enrolling patients in clinical trials. Proper patient selection ensures that individuals with higher susceptibility to cardiovascular adverse events are appropriately assessed and accounted for during the trial [106].

Accurate and timely reporting of adverse events is essential in clinical trials. Adverse event reporting enables researchers to assess the cardiovascular safety profile of nanoparticle-based therapies rigorously. Transparent reporting helps inform regulators, healthcare providers, and patients about potential cardiovascular risks associated with the therapy. Understanding the underlying mechanisms of nanoparticle-related cardiovascular adverse events is essential for refining nanoparticle design and improving patient safety. Mechanistic insights enable researchers to identify the specific factors contributing to adverse events and develop strategies to mitigate them. A comprehensive risk-benefit analysis is necessary to make informed decisions during clinical trials. Evaluating the potential therapeutic benefits of nanoparticle-based therapies against their cardiovascular safety profile is essential. Researchers and regulators must weigh the risks and benefits of the therapy to ensure patient safety and therapeutic efficacy. Post-marketing surveillance is essential for monitoring nanoparticle-related cardiovascular adverse events after the therapy's approval and widespread use. Post-marketing surveillance provides real-world data on the therapy's safety and effectiveness in a larger patient population. Continuous monitoring ensures that any potential cardiovascular adverse events are detected promptly and managed appropriately.

8.   Regulatory aspects and future directions

As nanoparticle-based therapies continue to advance, regulatory considerations play a critical role in ensuring their safe and effective translation from research to clinical applications. Regulatory guidelines provide a framework for evaluating nanoparticle biocompatibility, safety, and efficacy, safeguarding patient welfare. Moreover, future directions in nanoparticle research focus on enhancing biocompatibility, refining therapeutic strategies, and optimizing clinical translation. This comprehensive discussion explores the current regulatory guidelines for nanoparticles in medicine, prospects for enhancing nanoparticle biocompatibility, and the challenges and opportunities in translating nanoparticle research into clinical applications.

8.1   Current regulatory guidelines for nanoparticles in medicine

Regulatory agencies worldwide have recognized the importance of comprehensive preclinical assessment for nanoparticle-based therapies in medicine. Preclinical evaluation includes in vitro studies, in vivo animal models, and detailed toxicity assessments to understand nanoparticle biocompatibility, safety, and potential adverse effects. These assessments provide valuable data on how nanoparticles interact with biological systems and help identify any potential risks associated with their use. Understanding nanoparticle biodistribution, accumulation, and clearance from the body is crucial for predicting their safety and optimizing therapeutic efficacy.

Biodistribution studies in relevant animal models provide valuable data on where nanoparticles are distributed in the body and how long they remain at specific sites. This information helps in determining potential off-target effects and guiding dosing strategies to maximize therapeutic benefits while minimizing adverse effects[107–109]. Assessing the immunological response to nanoparticles is another critical aspect of regulatory guidelines. Immunogenicity studies help determine the potential for immune reactions, inflammation, and hypersensitivity to nanoparticle-based therapies. Understanding the immunological response is essential for predicting potential adverse events and developing strategies to mitigate them.

Detailed toxicity and safety profiles are essential for evaluating potential risks associated with nanoparticle exposure. These profiles inform risk-benefit analyses and guide the selection of appropriate therapeutic doses. Regulatory agencies require thorough toxicological assessments to ensure that nanoparticles do not cause harm to patients and that their benefits outweigh the risks. Consistent manufacturing processes and quality control are vital for nanoparticle-based therapies. Ensuring the reproducibility and uniformity of nanoparticles is critical for clinical translation. Regulatory agencies emphasize the importance of adhering to good manufacturing practices (GMP) to maintain high-quality standards and ensure patient safety. Clinical trial design for nanoparticle-based therapies must adhere to ethical principles and comply with regulatory requirements. Rigorous study protocols, informed consent procedures, and patient safety monitoring are fundamental to ensure the validity and reliability of clinical trial results. Regulatory agencies review and approve clinical trial protocols to ensure that they meet the necessary scientific and ethical standards. Post-marketing surveillance involves continuous monitoring of nanoparticle-based therapies after approval to detect and manage potential adverse effects in real-world settings. This surveillance allows regulatory agencies to collect additional safety data and make informed decisions regarding the ongoing use of nanoparticle therapies. Harmonization of regulatory guidelines among different countries and regions is essential for facilitating global acceptance and collaboration in the development and evaluation of nanoparticle-based therapies. By harmonizing guidelines, regulatory agencies can ensure consistency in the evaluation and approval process, promoting the safe and efficient use of nanoparticle therapies worldwide.

8.2   Future prospects for enhancing nanoparticle biocompatibility

Future prospects for enhancing nanoparticle biocompatibility and improving nanoparticle-based therapies are promising and multifaceted. Targeted delivery and precision medicine are among the key advancements that offer significant potential in nanoparticle research. Targeted delivery systems allow for the specific release of therapeutic agents at the site of action, reducing off-target effects and enhancing therapeutic efficacy [110,111]. This approach holds particular importance in the treatment of diseases with complex pathologies, such as cancer, where precise drug delivery is crucial to avoid damage to healthy tissues. Personalized medicine is another exciting avenue in nanoparticle-based therapies. Personalized nanoparticle-based treatments involve tailoring therapies to individual patients based on their unique biomarkers, genetic makeup, and disease characteristics. By customizing treatment strategies, personalized medicine optimizes therapeutic outcomes and reduces the risk of adverse effects. Biomimetic nanoparticles are designed to replicate natural structures and functions found in the body. By mimicking the body's natural components, biomimetic nanoparticles offer improved biocompatibility and reduced immunogenicity. These nanoparticles are more likely to be recognized as "self" by the body, reducing immune responses and enhancing their compatibility with biological systems.

Smart nanoparticle systems are engineered to respond to specific environmental cues, enabling intelligent and controlled drug release. These nanoparticles can be designed to release therapeutic agents in response to changes in pH, temperature, or the presence of specific enzymes, providing precise and on-demand drug delivery. Smart nanoparticle systems offer the potential to enhance therapeutic precision and reduce potential adverse effects. Combination therapies, where multiple therapeutic agents are co-loaded within a single nanoparticle or administered sequentially, present exciting opportunities to enhance therapeutic efficacy. Combination therapies can target multiple pathways involved in disease progression, leading to synergistic effects and potentially overcoming drug resistance. Gene editing and therapeutics represent a groundbreaking area in nanoparticle research. Nanoparticles can serve as carriers for gene editing tools and therapeutic molecules, allowing for precise targeting of genetic mutations and the treatment of genetic diseases at the molecular level. This technology has the potential to revolutionize the treatment of genetic disorders, offering hope for patients with currently untreatable conditions.

Theranostic nanoparticles, which combine therapeutic and diagnostic functionalities, enable real-time monitoring of treatment responses and disease progression. These nanoparticles can simultaneously deliver therapeutic agents while providing diagnostic information, allowing clinicians to monitor treatment efficacy and make timely adjustments as needed. Non-invasive imaging and monitoring techniques play a crucial role in advancing nanoparticle-based therapies. These imaging methods allow researchers and clinicians to visualize nanoparticle biodistribution, drug release, and therapeutic responses without the need for invasive procedures. Real-time monitoring provides valuable data to optimize treatment strategies and ensure patient safety. While the future prospects for enhancing nanoparticle biocompatibility are promising, there are still challenges to address. As nanoparticle-based therapies advance toward clinical use, it is essential to ensure their safety and effectiveness in humans. Comprehensive preclinical assessment and adherence to regulatory guidelines are critical to identify potential risks and mitigate adverse effects.

One challenge is the potential for immune responses to nanoparticles, leading to inflammation and reduced therapeutic efficacy. Understanding the immunological response to nanoparticles is essential for predicting potential adverse events and developing strategies to minimize immune reactions. Long-term safety is another area of concern. While preclinical studies provide valuable data on short-term toxicity and biocompatibility, long-term safety data are necessary to ensure the safety of nanoparticle-based therapies over extended periods of use. Chronic exposure studies in preclinical models are vital for understanding the long-term impact of nanoparticles on human health [112,113]. Additionally, the potential for nanoparticle accumulation in the cardiovascular system remains a concern. As nanoparticles are designed to circulate in the bloodstream, their interaction with the cardiovascular system requires careful evaluation. Hemodynamic assessments, cardiac imaging, and biomarker analysis can provide valuable insights into the impact of nanoparticles on heart function and vascular dynamics. Moreover, translating nanoparticle-based therapies from bench to bedside involves addressing challenges in manufacturing scalability, quality control, and cost-effectiveness. Ensuring consistent manufacturing processes and high-quality standards is crucial for the clinical translation of nanoparticle therapies.

8.3   Translating nanoparticle research into clinical applications

Translating nanoparticle research into clinical applications is a complex and challenging process that requires careful consideration of regulatory, manufacturing, cost-effectiveness, safety, and clinical aspects. While nanoparticle-based therapies hold immense promise for revolutionizing medicine, several hurdles must be overcome to ensure successful translation into clinical practice. One of the primary challenges in translating nanoparticle research into clinical applications is obtaining regulatory approval and compliance. Regulatory agencies around the world, such as the Food and Drug Administration (FDA) in the United States and the European Medicines Agency (EMA) in Europe, have specific guidelines and requirements for evaluating nanoparticle-based therapies. Researchers and developers must navigate these regulatory pathways, submit comprehensive data on nanoparticle biocompatibility, safety, and efficacy, and demonstrate that the benefits of the therapies outweigh the potential risks. The regulatory approval process can be time-consuming and resource-intensive, but it is essential to ensure patient safety and the ethical use of nanoparticle-based therapies in clinical settings. The scalability of manufacturing is another crucial aspect to consider when translating nanoparticle research into clinical applications. Efficient and scalable manufacturing processes are necessary to produce nanoparticle-based therapies in large quantities for widespread use. Manufacturing techniques must ensure the reproducibility and consistency of nanoparticle products, adhering to high-quality standards. Cost-effective manufacturing processes are also vital to make nanoparticle-based therapies accessible and affordable to patients, healthcare systems, and insurance providers.

Ensuring the safety and efficacy of nanoparticle-based therapies in clinical settings is paramount. Preclinical studies provide valuable insights into nanoparticle biocompatibility and potential adverse effects. However, clinical trials in human subjects are necessary to evaluate the therapies' safety and effectiveness in real-world scenarios. Rigorous clinical trial design, including randomized controlled trials and appropriate endpoints, is essential to generate robust and reliable data on nanoparticle-based therapies' clinical outcomes. Patient safety monitoring throughout the clinical trial process is crucial to detect and manage any unexpected adverse events promptly. Patient recruitment and informed consent procedures are critical components of translating nanoparticle research into clinical applications. Identifying suitable patients for clinical trials, particularly for targeted therapies, is essential to ensure the therapies' efficacy and safety. Informed consent procedures must be transparent and thorough, providing patients with comprehensive information about the potential risks and benefits of participating in the trials. Market access and reimbursement are essential considerations for the successful commercialization and adoption of nanoparticle-based therapies. Securing market access involves navigating healthcare systems and obtaining approvals for the therapies to be covered by insurance providers or national healthcare programs. Reimbursement for nanoparticle-based therapies is critical to ensure their affordability to patients and healthcare systems [114–116]. Demonstrating the cost-effectiveness and clinical value of nanoparticle-based therapies is crucial for gaining market access and reimbursement approval. Collaboration and interdisciplinary research play a pivotal role in translating nanoparticle research into clinical applications. Nanoparticle research involves expertise from various fields, including chemistry, materials science, biology, medicine, and engineering. Collaboration among researchers, clinicians, industry stakeholders, and regulatory agencies fosters a synergistic approach and accelerates the translation of nanoparticle research into clinical applications. Interdisciplinary collaboration enables the integration of diverse perspectives and facilitates the development of innovative solutions to address the challenges in nanoparticle-based therapies.

Conclusion

Nanoparticle-based therapies offer promising opportunities in medicine, with applications ranging from targeted drug delivery to diagnostics. These nanoparticles possess unique properties that improve treatment outcomes and reduce side effects, particularly in cancer therapy, where they can selectively target tumors. However, challenges related to biocompatibility, toxicity, and immunogenicity must be addressed for successful clinical translation. Biocompatibility is crucial in nanoparticle applications, ensuring safe interactions with living organisms. Particle size, surface charge, and functionalization play key roles in determining biocompatibility. Poor biocompatibility can lead to adverse immune reactions and inflammation, necessitating safer nanoparticle designs. Cardiovascular toxicity is a significant concern in nanoparticle-based therapies, impacting heart function and vascular integrity. Thorough preclinical assessments are essential to identify potential adverse effects and optimize therapeutic strategies. Despite rigorous preclinical testing, unexpected cardiovascular adverse events can still occur during clinical trials, requiring continuous safety monitoring and risk-benefit analyses. Stimuli-responsive nanoparticles offer intelligent and tailored approaches for precision medicine. They can respond to specific environmental cues or triggers, enabling controlled drug release and minimizing off-target effects. Smart nanoparticle systems provide spatiotemporal control over drug release, improving therapeutic precision. The future prospects for stimuli-responsive nanoparticles involve targeted delivery, personalized therapies, and biomimetic nanoparticles.

Regulatory guidelines are crucial in guiding nanoparticle development and evaluation. Rigorous preclinical assessments and careful clinical trial design ensure patient welfare. Harmonizing regulatory guidelines globally facilitates collaboration in nanoparticle development. Future prospects focus on enhancing nanoparticle biocompatibility and refining therapeutic strategies, such as personalized and precision medicine. Collaboration among researchers, clinicians, industry stakeholders, and regulatory agencies is vital for successful clinical translation. Post-marketing surveillance continuously evaluates nanoparticle safety and efficacy in real-world settings. In conclusion, nanoparticles hold great promise for revolutionizing medical treatments. Addressing challenges, adhering to regulatory guidelines, and embracing future directions will pave the way for safer and more effective nanoparticle-based therapies, bringing us closer to the vision of precision and personalized healthcare.

References

[1]　Z. Malanchuk, V. Moshynskyi, Y. Malanchuk, V. Korniienko, M. Koziar, Results of research into the content of rare earth materials in man-made phosphogypsum deposits, Key Eng. Mater. 844 (2020) 77–87. https://doi.org/10.4028/www.scientific.net/KEM.844.77

[2]　R. Szőllősi, Á. Molnár, S. Kondak, Z. Kolbert, Dual effect of nanomaterials on germination and seedling growth: Stimulation vs. phytotoxicity, Plants. 9 (2020) 1–30. https://doi.org/10.3390/plants9121745

[3]　W. Yin, L. Zhou, Y. Ma, G. Tian, J. Zhao, L. Yan, X. Zheng, P. Zhang, J. Yu, Z. Gu, Y. Zhao, Phytotoxicity, Translocation, and Biotransformation of NaYF4 Upconversion Nanoparticles in a Soybean Plant, Small. 11 (2015) 4774–4784. https://doi.org/10.1002/smll.201500701

[4]   M. Alipanah, D.M. Park, A. Middleton, Z. Dong, H. Hsu-Kim, Y. Jiao, H. Jin, Techno-Economic and Life Cycle Assessments for Sustainable Rare Earth Recovery from Coal Byproducts using Biosorption, ACS Sustain. Chem. Eng. 8 (2020) 17914–17922. https://doi.org/10.1021/acssuschemeng.0c04415

[5]   W. Gwenzi, L. Mangori, C. Danha, N. Chaukura, N. Dunjana, E. Sanganyado, Sources, behaviour, and environmental and human health risks of high-technology rare earth elements as emerging contaminants, Sci. Total Environ. 636 (2018) 299–313. https://doi.org/10.1016/j.scitotenv.2018.04.235

[6]   V. Harish, D. Tewari, M. Gaur, A.B. Yadav, S. Swaroop, M. Bechelany, A. Barhoum, Review on Nanoparticles and Nanostructured Materials: Bioimaging, Biosensing, Drug Delivery, Tissue Engineering, Antimicrobial, and Agro-Food Applications, Nanomaterials. 12 (2022). https://doi.org/10.3390/nano12030457

[7]   T.N.V.K.V. Prasad, P. Sudhakar, Y. Sreenivasulu, P. Latha, V. Munaswamy, K. Raja Reddy, T.S. Sreeprasad, P.R. Sajanlal, T. Pradeep, Effect of nanoscale zinc oxide particles on the germination, growth and yield of peanut, J. Plant Nutr. 35 (2012) 905–927. https://doi.org/10.1080/01904167.2012.663443

[8]   C. Panneerselvam, K. Murugan, M. Roni, A.T. Aziz, U. Suresh, R. Rajaganesh, P. Madhiyazhagan, J. Subramaniam, D. Dinesh, M. Nicoletti, A. Higuchi, A.A. Alarfaj, M.A. Munusamy, S. Kumar, N. Desneux, G. Benelli, Fern-synthesized nanoparticles in the fight against malaria: LC/MS analysis of Pteridium aquilinum leaf extract and biosynthesis of silver nanoparticles with high mosquitocidal and antiplasmodial activity, Parasitol. Res. 115 (2016) 997–1013. https://doi.org/10.1007/s00436-015-4828-x

[9]   K. Shahane, M. Kshirsagar, S. Tambe, D. Jain, S. Rout, M.K.M. Ferreira, S. Mali, P. Amin, P.P. Srivastav, J. Cruz, R.R. Lima, An Updated Review on the Multifaceted Therapeutic Potential of Calendula officinalis L., Pharmaceuticals. 16 (2023) 611. https://doi.org/10.3390/ph16040611

[10]   S. Dev, A. Sachan, F. Dehghani, T. Ghosh, B.R. Briggs, S. Aggarwal, Mechanisms of biological recovery of rare-earth elements from industrial and electronic wastes: A review, Chem. Eng. J. 397 (2020) 124596. https://doi.org/10.1016/j.cej.2020.124596

[11]   C. Krishnaraj, E.G. Jagan, R. Ramachandran, S.M. Abirami, N. Mohan, P.T. Kalaichelvan, Effect of biologically synthesized silver nanoparticles on Bacopa monnieri (Linn.) Wettst. plant growth metabolism, Process Biochem. 47 (2012) 651–658. https://doi.org/10.1016/j.procbio.2012.01.006

[12]   S.T. Khan, S.F. Adil, M.R. Shaik, H.Z. Alkhathlan, M. Khan, M. Khan, Engineered nanomaterials in soil: Their impact on soil microbiome and plant health, Plants. 11 (2022) 1–25. https://doi.org/10.3390/plants11010109

[13]   C. Luo, Y. Deng, K. Inubushi, J. Liang, S. Zhu, Z. Wei, X. Guo, X. Luo, Sludge biochar amendment and alfalfa revegetation improve soil physicochemical properties and increase diversity of soil microbes in soils from a rare earth element mining wasteland, Int. J. Environ. Res. Public Health. 15 (2018). https://doi.org/10.3390/ijerph15050965

[14]   F. Nkinahamira, A. Alsbaiee, Q. Zeng, Y. Li, Y. Zhang, M. Feng, C.P. Yu, Q. Sun, Selective and fast recovery of rare earth elements from industrial wastewater by porous β-cyclodextrin and magnetic β-cyclodextrin polymers, Water Res. 181 (2020) 115857. https://doi.org/10.1016/j.watres.2020.115857

[15]   E.J. Lee, U. Song, M. Shin, G. Lee, J. Roh, Y. Kim, Functional analysis of TiO2 nanoparticle toxicity in three plant Species, Biol. Trace Elem. Res. 155 (2013) 93–103. https://doi.org/10.1007/s12011-013-9765-x

[16]   X.Y. He, C.L. Zheng, X. Sui, Q.G. Jing, X. Wu, J.Y. Wang, W.T. Si, X.F. Zhang, Biological damage to Sprague-Dawley rats by excessive anions contaminated groundwater from rare earth metals tailings pond seepage, J. Clean. Prod. 185 (2018) 523–532. https://doi.org/10.1016/j.jclepro.2018.03.074

[17]   J. Iqbal, B.A. Abbasi, T. Yaseen, S.A. Zahra, A. Shahbaz, S.A. Shah, S. Uddin, X. Ma, B. Raouf, S. Kanwal, W. Amin, T. Mahmood, H.A. El-Serehy, P. Ahmad, Green synthesis of zinc oxide nanoparticles using Elaeagnus angustifolia L. leaf extracts and their multiple in vitro biological applications, Sci. Rep. 11 (2021) 1–13. https://doi.org/10.1038/s41598-021-99839-z

[18]   T. Sabo-Attwood, J.M. Unrine, J.W. Stone, C.J. Murphy, S. Ghoshroy, D. Blom, P.M. Bertsch, L.A. Newman, Uptake, distribution and toxicity of gold nanoparticles in tobacco (Nicotiana xanthi) seedlings, Nanotoxicology. 6 (2012) 353–360. https://doi.org/10.3109/17435390.2011.579631

[19]   H. Fathollahzadeh, J.J. Eksteen, A.H. Kaksonen, E.L.J. Watkin, Role of microorganisms in bioleaching of rare earth elements from primary and secondary resources, Appl. Microbiol. Biotechnol. 103 (2019) 1043–1057. https://doi.org/10.1007/s00253-018-9526-z

[20]   W.A. Mohammad, S.M. Ali, N. Farhan, S.M. Said, The toxic effect of zinc oxide nanoparticles on the terrestrial slug Lehmannia nyctelia (Gastropoda-Limacidae), J. Basic Appl. Zool. 82 (2021) 1–9. https://doi.org/10.1186/s41936-021-00214-1

[21]   A. Hussain, M. Priyadarshi, S. Dubey, Experimental study on accumulation of heavy metals in vegetables irrigated with treated wastewater, Appl. Water Sci. 9 (2019) 1–11. https://doi.org/10.1007/s13201-019-0999-4

[22]   N. Saha, S. Dutta Gupta, Low-dose toxicity of biogenic silver nanoparticles fabricated by Swertia chirata on root tips and flower buds of Allium cepa, J. Hazard. Mater. 330 (2017) 18–28. https://doi.org/10.1016/j.jhazmat.2017.01.021

[23]   N. Musee, M. Thwala, N. Nota, The antibacterial effects of engineered nanomaterials: Implications for wastewater treatment plants, J. Environ. Monit. 13 (2011) 1164–1183. https://doi.org/10.1039/c1em10023h

[24]   J. Singh, V. Kumar, K.H. Kim, M. Rawat, Biogenic synthesis of copper oxide nanoparticles using plant extract and its prodigious potential for photocatalytic degradation of dyes, Environ. Res. 177 (2019) 108569. https://doi.org/10.1016/j.envres.2019.108569

[25]   A. Singh, N.B. Singh, I. Hussain, H. Singh, Effect of biologically synthesized copper oxide nanoparticles on metabolism and antioxidant activity to the crop plants Solanum

lycopersicum and Brassica oleracea var. botrytis, J. Biotechnol. 262 (2017) 11–27. https://doi.org/10.1016/j.jbiotec.2017.09.016

[26]   Y. Yang, L. Zhang, X. Huang, Y. Zhou, Q. Quan, Y. Li, X. Zhu, Response of photosynthesis to different concentrations of heavy metals in Davidia involucrata, PLoS One. 15 (2020) 1–16. https://doi.org/10.1371/journal.pone.0228563

[27]   K. Kornarzyński, A. Sujak, G. Czernel, D. Wiącek, Effect of Fe3O4 nanoparticles on germination of seeds and concentration of elements in Helianthus annuus L. under constant magnetic field, Sci. Rep. 10 (2020) 1–10. https://doi.org/10.1038/s41598-020-64849-w

[28]   R. Amooaghaie, M.R. Saeri, M. Azizi, Synthesis, characterization and biocompatibility of silver nanoparticles synthesized from Nigella sativa leaf extract in comparison with chemical silver nanoparticles, Ecotoxicol. Environ. Saf. 120 (2015) 400–408. https://doi.org/10.1016/j.ecoenv.2015.06.025

[29]   V. Shah, I. Belozerova, Influence of metal nanoparticles on the soil microbial community and germination of lettuce seeds, Water. Air. Soil Pollut. 197 (2009) 143–148. https://doi.org/10.1007/s11270-008-9797-6

[30]   M. Edahbi, B. Plante, M. Benzaazoua, Environmental challenges and identification of the knowledge gaps associated with REE mine wastes management, J. Clean. Prod. 212 (2019) 1232–1241. https://doi.org/10.1016/j.jclepro.2018.11.228

[31]   A.K. Sakr, M.M. Abdel Aal, K.A. Abd El-Rahem, E.M. Allam, S.M. Abdel Dayem, E.A. Elshehy, M.Y. Hanfi, M.S. Alqahtani, M.F. Cheira, Characteristic Aspects of Uranium(VI) Adsorption Utilizing Nano-Silica/Chitosan from Wastewater Solution, Nanomaterials. 12 (2022) 3866. https://doi.org/10.3390/nano12213866

[32]   W.S. Choi, H.J. Lee, Nanostructured Materials for Water Purification: Adsorption of Heavy Metal Ions and Organic Dyes, Polymers (Basel). 14 (2022) 1–26. https://doi.org/10.3390/polym14112183

[33]   A.M.E. Khalil, F.A. Memon, T.A. Tabish, B. Fenton, D. Salmon, S. Zhang, D. Butler, Performance evaluation of porous graphene as filter media for the removal of pharmaceutical/emerging contaminants from water and wastewater, Nanomaterials. 11 (2021) 1–24. https://doi.org/10.3390/nano11010079

[34]   M.Z.A. Zaimee, M.S. Sarjadi, M.L. Rahman, Heavy metals removal from water by efficient adsorbents, Water (Switzerland). 13 (2021). https://doi.org/10.3390/w13192659

[35]   A.T. Besha, Y. Liu, D.N. Bekele, Z. Dong, R. Naidu, G.N. Gebremariam, Sustainability and environmental ethics for the application of engineered nanoparticles, Environ. Sci. Policy. 103 (2020) 85–98. https://doi.org/10.1016/j.envsci.2019.10.013

[36]   R.D. Handy, N. Van Den Brink, M. Chappell, M. Mühling, R. Behra, M. Dušinská, P. Simpson, J. Ahtiainen, A.N. Jha, J. Seiter, A. Bednar, A. Kennedy, T.F. Fernandes, M. Riediker, Practical considerations for conducting ecotoxicity test methods with manufactured nanomaterials: What have we learnt so far?, Ecotoxicology, 21 (2012) 933–972. https://doi.org/10.1007/s10646-012-0862-y

[37]   N. Haque, A. Hughes, S. Lim, C. Vernon, Rare earth elements: Overview of mining, mineralogy, uses, sustainability and environmental impact, Resources. 3 (2014) 614–635. https://doi.org/10.3390/resources3040614

[38]   P. Pyšek, D.M. Richardson, Invasive species, environmental change and management, and health, Annu. Rev. Environ. Resour. 35 (2010) 25–55. https://doi.org/10.1146/annurev-environ-033009-095548

[39]   X. Yin, C. Martineau, I. Demers, N. Basiliko, N.J. Fenton, The potential environmental risks associated with the development of rare earth element production in Canada, Environ. Rev. 29 (2021) 354–377. https://doi.org/10.1139/er-2020-0115

[40]   G. V. Vimbela, S.M. Ngo, C. Fraze, L. Yang, D.A. Stout, Antibacterial properties and toxicity from metallic nanomaterials, Int. J. Nanomedicine. 12 (2017) 3941–3965. https://doi.org/10.2147/IJN.S134526

[41]   S. Muzammil, J. Neves Cruz, R. Mumtaz, I. Rasul, S. Hayat, M.A. Khan, A.M. Khan, M.U. Ijaz, R.R. Lima, M. Zubair, Effects of Drying Temperature and Solvents on In Vitro Diabetic Wound Healing Potential of Moringa oleifera Leaf Extracts, Molecules. 28 (2023) 710. https://doi.org/10.3390/molecules28020710

[42]   M.L. Carmo Bastos, J.V. Silva-Silva, J. Neves Cruz, A.R. Palheta da Silva, A.A. Bentaberry-Rosa, G. da Costa Ramos, J.E. de Sousa Siqueira, M.R. Coelho-Ferreira, S. Percário, P. Santana Barbosa Marinho, A.M. do R. Marinho, M. de Oliveira Bahia, M.F. Dolabela, Alkaloid from Geissospermum sericeum Benth. & Hook.f. ex Miers (Apocynaceae) Induce Apoptosis by Caspase Pathway in Human Gastric Cancer Cells, Pharmaceuticals. 16 (2023) 765. https://doi.org/10.3390/ph16050765

[43]   J. Wei, C. Wang, S. Yin, X. Pi, L. Jin, Z. Li, J. Liu, L. Wang, C. Yin, A. Ren, Concentrations of rare earth elements in maternal serum during pregnancy and risk for fetal neural tube defects, Environ. Int. 137 (2020) 105542. https://doi.org/10.1016/j.envint.2020.105542

[44]   M. Usman, M. Farooq, A. Wakeel, A. Nawaz, S.A. Cheema, H. ur Rehman, I. Ashraf, M. Sanaullah, Nanotechnology in agriculture: Current status, challenges and future opportunities, Sci. Total Environ. 721 (2020) 137778. https://doi.org/10.1016/j.scitotenv.2020.137778

[45]   A. Thakur, A. Kumar, S. Sharma, R. Ganjoo, H. Assad, Computational and experimental studies on the efficiency of Sonchus arvensis as green corrosion inhibitor for mild steel in 0.5 M HCl solution, Mater. Today Proc. 66 (2022) 609–621. https://doi.org/10.1016/j.matpr.2022.06.479

[46]   A. Thakur, A. Kumar, Recent trends in nanostructured carbon-based electrochemical sensors for the detection and remediation of persistent toxic substances in real-time analysis, Mater. Res. Express. 10 (2023) 034001. https://doi.org/10.1088/2053-1591/acbd1a

[47]   C. Dhonchak, N. Agnihotri, A. Kumar, A. Thakur, A. Kumar, Computational Insights in the Spectrophotometrically Analyzed Niobium (V)-3-Hydroxy-2-(4-methylphenyl)-4H-chromen-4-one Complex using DFT Method, Biointerface Res. Appl. Chem. 13 (2023) 357.

https://doi.org/10.33263/BRIAC134.357

[48]   A. Thakur, A. Kumar, R. Zhang, Alcoholic Beverage Purification Applications of Activated Carbon, in: C. Verma, M.A. Quraishi (Eds.), Act. Carbon, The Royal Society of Chemistry, 2023: pp. 152–178. https://doi.org/10.1039/bk9781839169861-00152

[49]   A. Thakur, S. Sharma, R. Ganjoo, H. Assad, A. Kumar, Anti-Corrosive Potential of the Sustainable Corrosion Inhibitors Based on Biomass Waste: A Review on Preceding and Perspective Research, J. Phys. Conf. Ser. 2267 (2022) 012079. https://doi.org/10.1088/1742-6596/2267/1/012079

[50]   A. Thakur, A. Kumar, Recent advances on rapid detection and remediation of environmental pollutants utilizing nanomaterials-based (bio)sensors, Sci. Total Environ. 834 (2022) 155219. https://doi.org/10.1016/j.scitotenv.2022.155219

[51]   G. Parveen, S. Bashir, A. Thakur, S.K. Saha, P. Banerjee, A. Kumar, Experimental and computational studies of imidazolium based ionic liquid 1-methyl- 3-propylimidazolium iodide on mild steel corrosion in acidic solution, Mater. Res. Express. 7 (2019) 016510. https://doi.org/10.1088/2053-1591/ab5c6a

[52]   A. Thakur, S. Kaya, A.S. Abousalem, S. Sharma, R. Ganjoo, H. Assad, A. Kumar, Computational and experimental studies on the corrosion inhibition performance of an aerial extract of Cnicus Benedictus weed on the acidic corrosion of mild steel, Process Saf. Environ. Prot. 161 (2022) 801–818. https://doi.org/10.1016/j.psep.2022.03.082

[53]   A. Thakur, S. Kaya, A.S. Abousalem, A. Kumar, Experimental, DFT and MC simulation analysis of Vicia Sativa weed aerial extract as sustainable and eco-benign corrosion inhibitor for mild steel in acidic environment, Sustain. Chem. Pharm. 29 (2022) 100785. https://doi.org/10.1016/j.scp.2022.100785

[54]   C. Dhonchak, N. Agnihotri, A. Kumar, R. Kamal, A. Thakur, A. Kumar, Spectrophotometric Investigation and Computational Studies of Zirconium(IV)-3-hydroxy-2-[1'-phenyl-3'-(p-methoxyphenyl)-4'-pyrazolyl]-4H-chromen-4-one Complex, J. Anal. Chem. 78 (2023) 856–865. https://doi.org/10.1134/S1061934823070055

[55]   D. Sharma, A. Thakur, M.K. Sharma, R. Sharma, S. Kumar, A. Sihmar, H. Dahiya, G. Jhaa, A. Kumar, A.K. Sharma, H. Om, Effective corrosion inhibition of mild steel using novel 1,3,4-oxadiazole-pyridine hybrids: Synthesis, electrochemical, morphological, and computational insights, Environ. Res. 234 (2023) 116555. https://doi.org/10.1016/j.envres.2023.116555

[56]   A. Thakur, S. Kaya, A. Kumar, Recent Innovations in Nano Container Based Self-Healing Coatings in the Construction Industry, Curr. Nanosci. 18 (2021) 203–216. https://doi.org/10.2174/1573413717666210216120741

[57]   S. Kaya, A. Thakur, A. Kumar, The role of in Silico/DFT investigations in analyzing dye molecules for enhanced solar cell efficiency and reduced toxicity, J. Mol. Graph. Model. 124 (2023) 108536. https://doi.org/10.1016/j.jmgm.2023.108536

[58]   M.H. Sarfraz, M. Zubair, B. Aslam, A. Ashraf, M.H. Siddique, S. Hayat, J.N. Cruz, S.

Muzammil, M. Khurshid, M.F. Sarfraz, A. Hashem, T.M. Dawoud, G.D. Avila-Quezada, E.F. Abd_Allah, Comparative analysis of phyto-fabricated chitosan, copper oxide, and chitosan-based CuO nanoparticles: antibacterial potential against Acinetobacter baumannii isolates and anticancer activity against HepG2 cell lines, Front. Microbiol. 14 (2023) 1188743. https://doi.org/10.3389/fmicb.2023.1188743

[59]   A. Thakur, A. Kumar, S. Kaya, D.V.N. Vo, A. Sharma, Suppressing inhibitory compounds by nanomaterials for highly efficient biofuel production: A review, Fuel. 312 (2022) 122934. https://doi.org/10.1016/j.fuel.2021.122934

[60]   C. Verma, A. Thakur, R. Ganjoo, S. Sharma, H. Assad, A. Kumar, M.A. Quraishi, A. Alfantazi, Coordination bonding and corrosion inhibition potential of nitrogen-rich heterocycles: Azoles and triazines as specific examples, Coord. Chem. Rev. 488 (2023) 215177. https://doi.org/10.1016/j.ccr.2023.215177

[61]   D. Sharma, A. Thakur, M.K. Sharma, K. Jakhar, A. Kumar, A.K. Sharma, O.M. Hari, Synthesis, Electrochemical, Morphological, Computational and Corrosion Inhibition Studies of 3-(5-Naphthalen-2-yl-[1,3,4]oxadiazol-2-yl)-pyridine against Mild Steel in 1 M HCl, Asian J. Chem. 35 (2023) 1079–1088. https://doi.org/10.14233/ajchem.2023.27711

[62]   A. Thakur, A. Kumar, S. Kaya, R. Marzouki, F. Zhang, L. Guo, Recent Advancements in Surface Modification, Characterization and Functionalization for Enhancing the Biocompatibility and Corrosion Resistance of Biomedical Implants, Coatings. 12 (2022) 1459. https://doi.org/10.3390/coatings12101459

[63]   A. Thakur, A. Kumar, Computational insights into the corrosion inhibition potential of some pyridine derivatives: A DFT approach, Eur. J. Chem. 14 (2023) 246–253. https://doi.org/10.5155/eurjchem.14.2.246-253.2408

[64]   A. Thakur, S. Kaya, A. Kumar, Recent Trends in the Characterization and Application Progress of Nano-Modified Coatings in Corrosion Mitigation of Metals and Alloys, Appl. Sci. 13 (2023) 730. https://doi.org/10.3390/app13020730

[65]   K. Maguire, G. Sheriff, Comparing distributions of environmental outcomes for regulatory environmental justice analysis, Int. J. Environ. Res. Public Health. 8 (2011) 1707–1726. https://doi.org/10.3390/ijerph8051707

[66]   T.E. Novotny, K. Lum, E. Smith, V. Wang, R. Barnes, Cigarettes butts and the case for an environmental policy on hazardous cigarette waste, Int. J. Environ. Res. Public Health. 6 (2009) 1691–1705. https://doi.org/10.3390/ijerph6051691

[67]   K. Sexton, S.H. Linder, The role of cumulative risk assessment in decisions about environmental justice, Int. J. Environ. Res. Public Health. 7 (2010) 4037–4049. https://doi.org/10.3390/ijerph7114037

[68]   J. Spengler, G. Adamkiewicz, Indoor air pollution: An old problem with new challenges, Int. J. Environ. Res. Public Health. 6 (2009) 2880–2882. https://doi.org/10.3390/ijerph6112880

[69]   J. Stenlid, J. Oliva, J.B. Boberg, A.J.M. Hopkins, Emerging diseases in european forest ecosystems and responses in society, Forests. 2 (2011) 486–504.

https://doi.org/10.3390/f2020486

[70]   M.S. Hossain, M.A. Hossain, M.A. Rahman, M.M. Islam, M.A. Rahman, T.M. Adyel, Health risk assessment of pesticide residues via dietary intake of market vegetables from Dhaka, Bangladesh, Foods. 2 (2013) 64–75. https://doi.org/10.3390/foods2010064

[71]   D. Siniscalco, A. Cirillo, J.J. Bradstreet, N. Antonucci, Epigenetic findings in autism: New perspectives for therapy, Int. J. Environ. Res. Public Health. 10 (2013) 4261–4273. https://doi.org/10.3390/ijerph10094261

[72]   R. Arora, Nanocomposite polyaniline for environmental and energy applications, Mater. Today Proc. 44 (2021) 633–636. https://doi.org/10.1016/j.matpr.2020.10.603

[73]   K. Wan, Y. Li, Y. Wang, G. Wei, Recent advance in the fabrication of 2d and 3d metal carbides-based nanomaterials for energy and environmental applications, Nanomaterials. 11 (2021) 1–34. https://doi.org/10.3390/nano11010246

[74]   P.M.J. Bos, S. Gottardo, J.J. Scott-Fordsm, M. van Tongeren, E. Semenzin, T.F. Fernandes, D. Hristozov, K. Hund-Rinke, N. Hunt, M.A. Irfan, R. Landsiedel, W.J.G.M. Peijnenburg, A.S. Jiménez, P.C.E. van Kesteren, A.G. Oomen, The MARINA risk assessment strategy: A flexible strategy for efficient information collection and risk assessment of nanomaterials, Int. J. Environ. Res. Public Health. 12 (2015) 15007–15021. https://doi.org/10.3390/ijerph121214961

[75]   M. Sarfraz, W. Qun, L. Hui, M.I. Abdullah, Environmental risk management strategies and the moderating role of corporate social responsibility in project financing decisions, Sustain. 10 (2018). https://doi.org/10.3390/su10082771

[76]   Z. Feng, W. Chen, Environmental regulation, green innovation, and industrial green development: An empirical analysis based on the spatial Durbin model, Sustain. 10 (2018). https://doi.org/10.3390/su10010223

[77]   J.Y. Heurtebise, Sustainability and ecological civilization in the age of Anthropocene: An epistemological analysis of the psychosocial and "culturalist" interpretations of global environmental risks, Sustain. 9 (2017). https://doi.org/10.3390/su9081331

[78]   R. Licker, B. Ekwurzel, S.C. Doney, S.R. Cooley, I.D. Lima, R. Heede, P.C. Frumhoff, Attributing ocean acidification to major carbon producers, Environ. Res. Lett. 14 (2019) 124060. https://doi.org/10.1088/1748-9326/ab5abc

[79]   A. Bhatnagar, F. Kaczala, W. Hogland, M. Marques, C.A. Paraskeva, V.G. Papadakis, M. Sillanpää, Valorization of solid waste products from olive oil industry as potential adsorbents for water pollution control-a review, 2014. https://doi.org/10.1007/s11356-013-2135-6

[80]   R.H. Bradley, R.F. Corwyn, H.P. McAdoo, C. García Coll, The home environments of children in the United States Part I: Variations by age, ethnicity, and poverty status, Child Dev. 72 (2001) 1844–1867. https://doi.org/10.1111/1467-8624.t01-1-00382

[81]   H. Peng, Y. Jia, C. Tague, P. Slaughter, An eco-hydrological model-based assessment of the impacts of soil and water conservation management in the Jinghe River Basin, China, Water

(Switzerland). 7 (2015) 6301–6320. https://doi.org/10.3390/w7116301

[82]   M. Cook, M. Webber, Food, fracking, and freshwater: The potential for markets and cross-sectoral investments to enable water conservation, Water (Switzerland). 8 (2016). https://doi.org/10.3390/w8020045

[83]   R. Kumar, A. Verma, A. Shome, R. Sinha, S. Sinha, P.K. Jha, R. Kumar, P. Kumar, Shubham, S. Das, P. Sharma, P.V.V. Prasad, Impacts of plastic pollution on ecosystem services, sustainable development goals, and need to focus on circular economy and policy interventions, Sustain. 13 (2021) 1–40. https://doi.org/10.3390/su13179963

[84]   J.P. Juanga-Labayen, I. V. Labayen, Q. Yuan, A Review on Textile Recycling Practices and Challenges, Textiles. 2 (2022) 174–188. https://doi.org/10.3390/textiles2010010

[85]   X. Zhang, F. Sun, H. Wang, Y. Qu, Green biased technical change in terms of industrial water resources in china's yangtze river economic belt, Int. J. Environ. Res. Public Health. 17 (2020). https://doi.org/10.3390/ijerph17082789

[86]   A.K. Chapagain, A.Y. Hoekstra, The water footprint of coffee and tea consumption in the Netherlands, Ecol. Econ. 64 (2007) 109–118. https://doi.org/10.1016/j.ecolecon.2007.02.022

[87]   A.K. Chapagain, S. Orr, An improved water footprint methodology linking global consumption to local water resources: A case of Spanish tomatoes, J. Environ. Manage. 90 (2009) 1219–1228. https://doi.org/10.1016/j.jenvman.2008.06.006

[88]   W. Gerbens-Leenes, A.Y. Hoekstra, The water footprint of sweeteners and bio-ethanol, Environ. Int. 40 (2012) 202–211. https://doi.org/10.1016/j.envint.2011.06.006

[89]   J. Huang, H.L. Zhang, W.J. Tong, F. Chen, The impact of local crops consumption on the water resources in Beijing, J. Clean. Prod. 21 (2012) 45–50. https://doi.org/10.1016/j.jclepro.2011.09.014

[90]   M.J. Blair, B. Gagnon, A. Klain, B. Kulišić, Contribution of biomass supply chains for bioenergy to sustainable development goals, Land. 10 (2021) 1–28. https://doi.org/10.3390/land10020181

[91]   B. Wu, P. Wang, S. Xiao, X. Yu, W. Shu, H. Zhang, M. Ding, Effects of Soil and Water Conservation Measures on Soil Bacterial Community Structure in Citrus Orchards, Res. Environ. Sci. 34 (2021) 419–430. https://doi.org/10.13198/j.issn.1001-6929.2021.01.02

[92]   A.Y. Hoekstra, A.K. Chapagain, P.R. van Oel, Advancing water footprint assessment research: Challenges in monitoring progress towards sustainable development goal 6, Water (Switzerland). 9 (2017). https://doi.org/10.3390/w9060438

[93]   C. Guo, J. Gao, B. Zhou, J. Yang, Factors of the ecosystem service value in water conservation areas considering the natural environment and human activities: A case study of Funiu mountain, China, Int. J. Environ. Res. Public Health. 18 (2021). https://doi.org/10.3390/ijerph182111074

[94]   L. Fang, F. Wu, Can water rights trading scheme promote regional water conservation in

china? Evidence from a time-varying DID analysis, Int. J. Environ. Res. Public Health. 17 (2020) 1–14. https://doi.org/10.3390/ijerph17186679

[95]　U. Shabbir, M. Rubab, A. Tyagi, D.H. Oh, Curcumin and its derivatives as theranostic agents in alzheimer's disease: The implication of nanotechnology, Int. J. Mol. Sci. 22 (2021) 1–23. https://doi.org/10.3390/ijms22010196

[96]　Q. Wang, Y. Zhu, B. Song, R. Fu, Y. Zhou, The In Vivo Toxicity Assessments of Water-Dispersed Fluorescent Silicon Nanoparticles in Caenorhabditis elegans, Int. J. Environ. Res. Public Health. 19 (2022) 4101. https://doi.org/10.3390/ijerph19074101

[97]　M.N. Hassan, K.V.R. Corresponding, S.O. Abah, E.I. Ohimain, B. Nagaraju, P. Gv, P. Ds, S. Mp, S. Sp, Treatment and Disposal Technologies for Medical Wastes in Developing Countries Where do We Start ?, Int. J. Med. Biomed. Res. 3 (2013) 82–88

[98]　V. Thakur, R. Anbanandam, Healthcare waste management: an interpretive structural modeling approach, Int. J. Health Care Qual. Assur. 29 (2016) 559–581. https://doi.org/10.1108/IJHCQA-02-2016-0010

[99]　P. Wang, M. Wang, F. Zhou, G. Yang, L. Qu, X. Miao, Development of a paper-based, inexpensive, and disposable electrochemical sensing platform for nitrite detection, Electrochem. Commun. 81 (2017) 74–78. https://doi.org/10.1016/j.elecom.2017.06.006

[100]　C. Zhu, A. Hu, J. Cui, K. Yang, X. Zhu, Y. Liu, G. Deng, L. Zhu, A lab-on-a-chip device integrated DNA extraction and solid phase PCR array for the genotyping of high-risk HPV in clinical samples, Micromachines. 10 (2019). https://doi.org/10.3390/mi10080537

[101]　J. Wu, M. Dong, S. Santos, C. Rigatto, Y. Liu, F. Lin, Lab-on-a-chip platforms for detection of cardiovascular disease and cancer biomarkers, Sensors (Switzerland). 17 (2017). https://doi.org/10.3390/s17122934

[102]　M.Z. Hua, S. Li, S. Wang, X. Lu, Detecting chemical hazards in foods using microfluidic paper-based analytical devices (μPADs): The real-world application, Micromachines. 9 (2018). https://doi.org/10.3390/mi9010032

[103]　R. Mazurczyk, J. Vieillard, A. Bouchard, B. Hannes, S. Krawczyk, A novel concept of the integrated fluorescence detection system and its application in a lab-on-a-chip microdevice, Sensors Actuators, B Chem. 118 (2006) 11–19. https://doi.org/10.1016/j.snb.2006.04.069

[104]　C.A. Başar, Applicability of the various adsorption models of three dyes adsorption onto activated carbon prepared waste apricot, J. Hazard. Mater. 135 (2006) 232–241. https://doi.org/10.1016/j.jhazmat.2005.11.055

[105]　M. Lewoyehu, Comprehensive review on synthesis and application of activated carbon from agricultural residues for the remediation of venomous pollutants in wastewater, J. Anal. Appl. Pyrolysis. 159 (2021) 105279. https://doi.org/10.1016/j.jaap.2021.105279

[106]　I. Safarik, K. Horska, K. Pospiskova, M. Safarikova, Special Section on Open Door Initiative (ODI)-2 nd edition Magnetically Responsive Activated Carbons for Bio-and Environmental Applications, Int. Rev. Chem. Eng. 4 (2012) 346–352

[107] K.Y. Foo, B.H. Hameed, The environmental applications of activated carbon/zeolite composite materials, Adv. Colloid Interface Sci. 162 (2011) 22–28. https://doi.org/10.1016/j.cis.2010.09.003

[108] K.P. Gopinath, D.V.N. Vo, D. Gnana Prakash, A. Adithya Joseph, S. Viswanathan, J. Arun, Environmental applications of carbon-based materials: a review, Environ. Chem. Lett. 19 (2021) 557–582. https://doi.org/10.1007/s10311-020-01084-9

[109] J. Bedia, M. Peñas-Garzón, A. Gómez-Avilés, J.J. Rodriguez, C. Belver, Review on Activated Carbons by Chemical Activation with FeCl3, C — J. Carbon Res. 6 (2020) 21. https://doi.org/10.3390/c6020021

[110] H.A.M. Bacelo, S.C.R. Santos, C.M.S. Botelho, Tannin-based biosorbents for environmental applications - A review, Chem. Eng. J. 303 (2016) 575–587. https://doi.org/10.1016/j.cej.2016.06.044

[111] M. Vohra, M. Al-Suwaiyan, M. Hussaini, Gas phase toluene adsorption using date palm-tree branches based activated carbon, Int. J. Environ. Res. Public Health. 17 (2020) 1–19. https://doi.org/10.3390/ijerph17249287

[112] A. Macías-García, J.P. Carrasco-Amador, V. Encinas-Sánchez, M.A. Díaz-Díez, D. Torrejón-Martín, Preparation of activated carbon from kenaf by activation with H3PO4. Kinetic study of the adsorption/electroadsorption using a system of supports designed in 3D, for environmental applications, J. Environ. Chem. Eng. 7 (2019) 103196. https://doi.org/10.1016/j.jece.2019.103196

[113] S. Banerjee, B. De, P. Sinha, J. Cherusseri, K.K. Kar, Applications of supercapacitors, 2020. https://doi.org/10.1007/978-3-030-43009-2_13

[114] R. Guillossou, J. Le Roux, R. Mailler, E. Vulliet, C. Morlay, F. Nauleau, J. Gasperi, V. Rocher, Organic micropollutants in a large wastewater treatment plant: What are the benefits of an advanced treatment by activated carbon adsorption in comparison to conventional treatment?, Chemosphere. 218 (2019) 1050–1060. https://doi.org/10.1016/j.chemosphere.2018.11.182

[115] A.J. Alkhatib, K. Al Zailaey, Medical and environmental applications of activated charcoal: review article, Eur. Sci. J. 11 (2015) 50–56

[116] X. Wang, N. Zhu, B. Yin, Preparation of sludge-based activated carbon and its application in dye wastewater treatment, J. Hazard. Mater. 153 (2008) 22–27. https://doi.org/10.1016/j.jhazmat.2007.08.011

Nanoparticle Toxicity and Compatibility   Materials Research Forum LLC
Materials Research Foundations 161 (2024) 225-244   https://doi.org/10.21741/9781644902998-8

Chapter 8

Ecotoxicology of Nanoparticles

Muhammad Zubair[1], Noor Fatima[1], Habibullah Nadeem[1], Sabir Hussain[2], Tanvir Shahzad[2], Muhammad Afzal[1], Sana Fatima[1], Muhammad Imran[3], Muhammad Hussnain Siddique[1]*

[1]Department of Bioinformatics and Biotechnology, Government College University Faisalabad, Pakistan

[2]Department of Environmental Sciences, Government College University Faisalabad, Pakistan

[3]Department of Environmental Sciences, COMSATS Institute of Information Technology, Vehari Campus, Pakistan

mhs1049@gmail.com

Abstract

The escalating concern within the field of ecotoxicology pertains to the burgeoning impacts of nanoparticles (NPs) on the environment. Owing to their distinctive physicochemical attributes, engineered nanomaterials have found pervasive use in many different fields, causing their dispersion into natural ecosystems. It summarizes nanoparticle ecotoxicology, concentrating on their environmental impacts. This provides NP bioavailability, toxicity processes, and exposure routes. It emphasizes risk assessment and sustainable nanomaterial innovation to reduce environmental damage. Addressing nanoparticle ecotoxicity is essential for nanotechnology's long-term sustainability and ecosystem protection. Furthermore, the chapter uncovering the toxicity mechanisms and implementing regulations to protect ecosystems from nanoparticle induced environmental hazards.

Keywords

Nanoparticles, Nanomaterials, Ecotoxicology, Reactive Oxygen Species, Bottom-Up, Top-Down

Contents

1.   Introduction

In Greek, 'nano' denotes 'dwarf' or 'very small' signifying a scale of one thousand millionth of a meter (10^{-9}m). Nanoparticles (NP) are encompassing materials of natural, incidental, or engineered origin, characterized by particles existing in a state of disassociation, as well as in aggregate and agglomerate configurations. Notably, in the number of size distribution, 50% or more of these particles processes external dimensions ranging from 1 nm to 100 nm [1, 2]. Despite bulk materials, nanomaterials (NMs) physical characteristics are determined by their shape and size [3]. The consequential impact of size on the physicochemical properties of materials including optical, electrical, mechanical and chemical attributes imparts significance to nanomaterials (NMs). Over the past two decades, customizing and engineering nanomaterials with unique physicochemical features has enabled many electronic, chemical, biological, and medicinal applications. Engineered nanomaterials (ENM) have different chemical properties from bulk materials because their small size causes them to have a high surface area. Many different types of consumer goods are starting to use ENM to improve their quality, including paintings, medications, and cosmetics. By 2015, the ENM-based product market is expected to be worth $100 billion annually worldwide [4].

Active use of nanoparticles poses environmental safety concerns. After 15 years of nanotechnology expansion, the first paper on engineered nanomaterials environmental implications occurred [5, 6]. Nanoecotoxicology is a novel field that focuses on how nanoscale materials affect the world. Nanomaterial behavior and fate in complicated environmental matrices is difficult to evaluate. The prospective liberation of nanoparticles within the ecological milieu should be evaluated throughout the products full life cycle including production, consumption and disposal. Nanoparticles may have not only favorable but also distinct harmful features from bulk compounds. Potential negative impacts should be considered from the cellular to the ecosystem level.

Three essential attributes of toxicity screening methodologies for nanoparticles have been identified by [7].

1. The scrutiny involves Physicochemical characterization encompassing dimensions, surface area, morphology, solubility, and aggregation, coupled with the explication of biological effects.

2. *in vitro* and

3. *in vivo* studies

These three fundamental components have been delineated concerning the potential human effects of nanoparticles (NPs). Ecosystem-wide issues are more complicated.

Objectives

- Analyze the toxicity of nanoparticles to various species in terrestrial and aquatic environments.

- Examine the dynamics and ultimate trajectory of nanoparticles within the context of natural ecosystem (such as soil, water, sediment).

- Evaluate the enduring repercussions of nanoparticle exposure on the structural and functional aspects of ecosystems.

- Establish risk assessment methods and recommendations for sustainable nanoparticle usage and management.

- Encourage cross-disciplinary collaborative research to tackle the complexities of nanoparticle ecotoxicity.

1.1 Nanomaterials

Nanomaterials (NMs) are define as materials wherein at least one dimension falls within the range of 1 to 100 nanometers [8].

1.2 Types of nanomaterials

Three distinctive categories of nanomaterials include discrete nanomaterials, nanoscale device materials and bulk nanomaterials [9].

1.2.1 Discrete nanomaterials

Discrete nanomaterials encompass substances exhibiting dimensions ranging from 1 to 10 nanometer along at least one axis, characterized by independence, and include nanoparticles and nanofibers such as carbon nanotubes). These discrete nanomaterials can be categorized as zero-dimensional (particles) or one-dimensional (fibers).

1.2.2 Nanoscale device materials

Material elements that are incorporated in devices at the nanoscale, known as nanoscale device materials, predominantly manifest as thin films. Broadly speaking, nanoscale device materials exhibit a two-dimensional morphology taking the form of thin sheets.

1.2.3 Bulk materials

Bulk materials refer to particles that possess diameters above 100 nm in all directions. Nanomaterials in discrete form or materials designed for nanoscale devices can be used to make bulk nanomaterials. bulk nanomaterials are three-dimensional.

2. Nanoparticles

Nanoparticles are entities characterized by dimensions ranging from 1 to 100 nanometers in size and may differ from bulk material [10]. Nanoparticles are drug delivery platforms that were employed in medical sciences at the start of 1990s [11]. Various classifications of nanoparticles exist, delineated by their structural attributes, dimensions, morphology, and functional characteristics such as metal nanoparticles, ceramic nanoparticles, Polymeric nanoparticles and fullerenes. Biogenic nanoparticles contribute significantly to diverse sectors, including industrial, medical, environmental applications and agricultural fields. Nanoparticles are used as a delivery method for controlling their surface properties and particle size. Metallic nanoparticles are extremely useful in a variety of industries including those that make optical devices, biological markers and medication delivery systems [12].

2.1 Synthesis methods

There are various methods employed for nanoparticles synthesis like "top-down" and "bottom-up" [13]. The top-down technique requires nanoparticles production via bulk and bottom-up method starts synthesis from atomic scale. The bottom-up process applies with chemical and plant-mediated synthesis but top-down strategy usually takes place through physical methods.

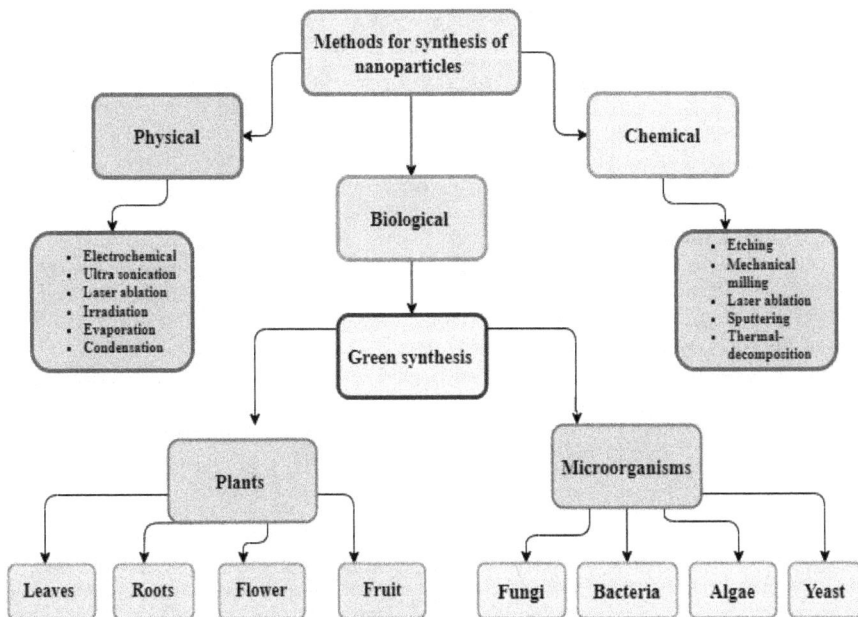

Figure 1. Green synthesis methods.

Table 1. Nanoparticles synthesis approaches

Bottom-up Approach	Bottom-up Approach	Top-down Approach
Green methodology	**Chemical methodology**	**Physical methodology**
This technique employs microorganisms, green plants and their extracted parts, yeast, organic molecules, and enzymes. All components in this process are non-hazardous and safe to use.	This technique employs a chemical reduction process. In addition, it involves light chemical, electrochemical, micro blend, pyrolysis, radio waves, salvo thermal and compound precipitate techniques.	This method involves dispersed pyrolysis, vaporized and gaseous states, lithography, arc extraction, and evaporation-condensation processes. The components of this technique are toxic.

3. Ecotoxicology

Ecotoxicology is a new field that studies biosphere contaminants and their impact on biome constituents particularly people [14]. In 1969, René Truhaut defined ecotoxicology as "the branch of toxicology that deals with the study of the harmful effects triggered by natural or synthetic substances, to the parts of ecosystems, animal (such as human), vegetable and microbial in a holistic context" [15]. The field of ecotoxicology primarily emerged as a means for assessing the toxicological impacts on aquatic environments. It is worth noting that the development of terrestrial ecotoxicological studies has been comparatively less extensive than that of aquatic studies. This discipline was established in 1989 by van Straalen and Denneman [16]. Nanoecotoxicology refers to the examination of the toxicological consequences that arising from the incorporation of nanoparticles within the ecological context, specifically in relation to their impact on humans, animals, plants, fungus, and other microbes. Both humans and animals may encounter with items containing nanoparticles through various means, such as contaminated water and air, or by consuming animals and plants that have been subjected to and amassed nanomaterials [17].

3.1 Ecotoxicity of nanoparticles

Nanoparticles ecotoxicology studies how nanoparticles, specially manufactured ones affect ecosystems and the environment. It emphasizes on understanding the potential adverse effects of nanoparticles on aquatic and terrestrial organisms as well as soil and water systems [18, 19]. Many nanoparticles have been released into the air and water due to nanotechnology. To safeguard human health and animals against the potential bad impacts of a wide spectrum of nanomaterials, more research are assessing the toxicity of industrial nanoparticles [20]. The initiation of oxidative stress stands out as a principal mechanism underlying the toxicity associated with nanoparticles. Nanoparticles can produce Reactive oxygen species (ROS) that may harm cells and then lipid peroxidation, oxidation of proteins and DNA strand breakage occur. Cancer, neurological and cardiovascular illnesses are linked to oxidative stress. These ailments can be caused by oxidative stress to nanoparticles, particularly those including iron or silver [21]. Many nanoparticle toxicity investigations have indicated ROS production by TiO_2 and fullerenes [22]. Nanoparticles cause inflammation in the body. Nanoparticles in tissues may trigger immune cells to secrete pro-inflammatory cytokines and chemokines. Chronic nanoparticle-induced inflammation can harm tissue and cause chronic illnesses. Nanoparticles also enhance inflammation. Nanoparticles affect cell membrane stability and function. This may spill the cell's contents and destroy neurons. Numerous factors influence the toxicity of nanomaterials.

- The impact of their dosage and exposure period
- The effect of the aggregation and concentration properties
- The effect of crystal size, shape, and structure
- The area influence and surface fictionalization [23].

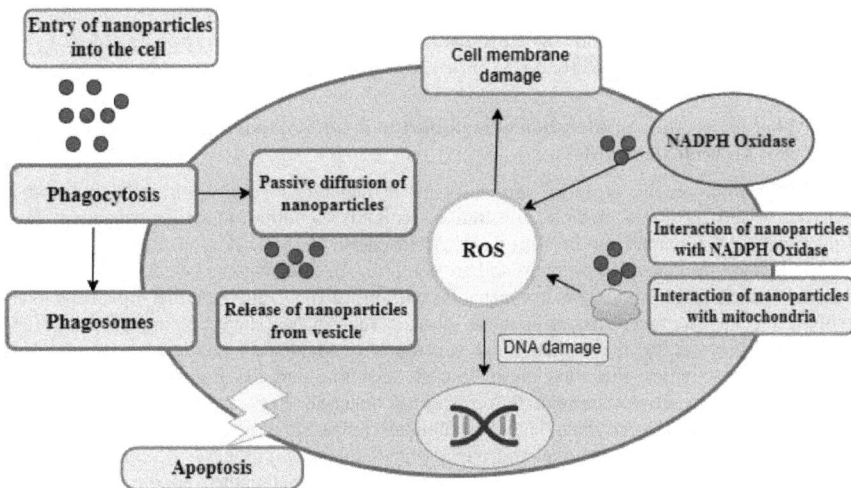

Figure 2. Mechanism of nanoparticle toxicity.

3.2   Ecotoxicity of metal nanoparticles

Solubility, binding selectivity to a biological location and other features all play a role in determining a metal's toxicity. Toxicity, specifically heavy metal exposure refers to any physiological or morphological modification in the body induced by chemicals and drugs that are ingested, injected, inhaled [24]. AgNPs are important in nanotechnology and nanomedicine. They have special physical, chemical and biological features that make them excellent and indispensable. Silver nanoparticles may kill numerous pathogens. In addition to manifesting antimicrobial activity, silver nanoparticles are demonstrating adverse and unacceptable effects on both human health and the environment [20].

3.3 Mechanisms of cytotoxicity induced by metal-based nanoparticles

Metal-based nanoparticles induce modifications in transmembrane and fundamental cellular functions, including differentiation, growth, apoptosis, and regulation of signal transduction pathways, having unique biological consequences [25]. However, no definitive results have been revealed about the cytotoxic mechanism generated by metal-based nanoparticles. Considerable focus has been directed towards oxidative stress and inflammation as primary contributors to cytotoxicity. Inflammation protects against infection, injury, and stress, but excessive inflammation can harm the body. Oxidative stress ensues from an overproduction of Reactive Oxygen Species (ROS) [26]. Lately, apoptosis and autophagy have been widely explored as metal-based NP cytotoxicity mechanisms. Different cytotoxic mechanisms are yet connected. For example, Inflammation and ROS production can be exacerbated by ROS activation and recruitment. Excess ROS may lead to mitochondrial malfunction, oxidative injury, apoptosis, and redox imbalance. Furthermore, cytotoxicity induced by metal-based NPs involves the interaction with numerous signaling networks. Furthermore, ZnO NPs can cause neurotoxicity by activating NF-κB, ERK, and p38 MAPK signaling pathways, while Cu NPs cause immunotoxicity by activating Nrf2, MAPK, and PI3K/AKT [27].

4. What makes the consideration of manufactured nanomaterials particularly significant for ecotoxicologists?

Within the environment, array of particles exist at nanometer scale, with colloids found in freshwater serving as a notable illustration (colloids denoting materials spanning from 1micrometer to 1 nanometer in size) [28]. Atmospheric volcanic dust particulates [29] and nanoscale particles generated throughs soil erosion [30]. An assertion can be advanced that these materials have existed in these environments for millions of years, implying that organisms likely evolved mechanism to coexist with these intrinsic substances. There are apprehensions and concern arise regarding the unintentional generation of nano-scale pollutants resulting from anthropogenic activities, including the release of air-borne particles from car exhausts or the formation of nanoparticles through the erosion of materials like car tyres persisting over a prolonged duration [31]. Attributable to their diminutive size and uniform composition, structure, or surface characteristics, manufactured nanoparticles (along with natural nanoparticles) often display distinctive physico-chemical properties and reactivities that set them apart from their large-scale counterparts. As an example, carbon fullerene nanoparticles (C60 particles),could manifest unique toxicity in contrast to fine graphite particles of 1 micrometer in size , even though both share a carbon [32].

5. The origin of nanoparticles in environmental context

5.1 Sources of nanoparticles derived from nature

In the natural environment, NPs emerge through diverse processes, encompassing photochemical reactions, volcanic eruptions, forest fires (potentially considered anthropogenic if intentionally started), and basic erosion. Natural events like dust storms, forest fires and volcanic eruptions have the capacity to generate significant amounts of nanoparticles, exerting a notable impact on the overall air quality worldwide. Approximately 10% of aerosols are attributed to human activities, with the remaining 90% originating naturally [33].

Nanoparticle Toxicity and Compatibility   Materials Research Forum LLC
Materials Research Foundations 161 (2024) 225-244   https://doi.org/10.21741/9781644902998-8

5.2   Human-generated (Anthropogenic) origins of nanoparticles

Anthropogenic refers to being human-made or stemming, whether by design inadvertently, from human activities, including the procedures like industrial production of nanoparticles, manufacturing, and the burning of fossil fuels. NPs release unintentionally from furnaces combusting hydrocarbon gases, even wax candles or wood [34, 35].

6.   The ecotoxicological impact of metal nps in soil

The presence of NPs in the soil may detrimentally affect the beneficial functions of indigenous microorganisms within the environment [36, 37]. NPs have the potential to discharged into the environment (aquatic, terrestrial, and atmospheric habitats) at various stages of their life cycle, generating serious concerns about potential ecological hazards Figure 3[38]. NPs can infiltrate soil via a variety of routes, including agricultural supplements of sewage sludge, air deposition, landfills or unintentional spills during industrial manufacturing [38]. Current models indicate that concentrations of TiO2 nanoparticles in soil treated with sewage sludge are projected to rise by 0.94 to 3.6 mg kg-1 annually. In contrast, the anticipated increase for AgNPs and fullerenes is more than 1000-fold lower. In this instance, concentration increases range from 0.09 to 0.65 g kg-1 and from 0.38 to 1.5 g kg-1 respectively [39].

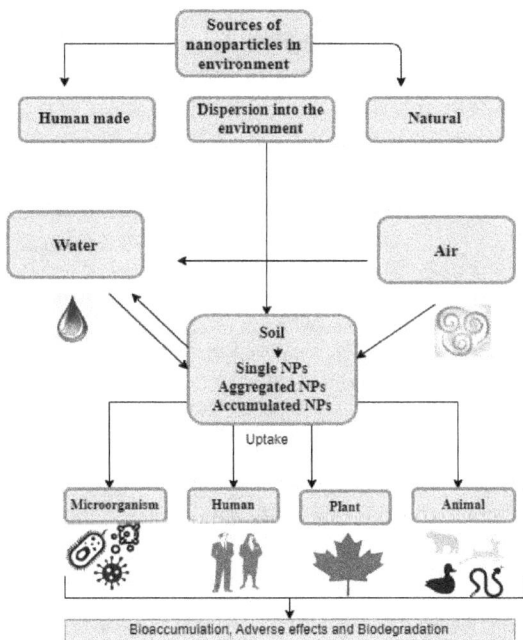

Figure 3. Diverse sources of NPs discharged into the environment and their ecological implications.

7. Impact of metal NPs on soil microorganism ecotoxicity

Soil microorganisms constitute integral elements of the soil ecosystem, exerting a pivotal role in the alteration and breakdown of diverse contaminants. This function contributes to the maintenance of soil productivity and ecological integrity [40]. They also possess the potential to influence both the developmental processes and the survival outcomes of plants. A wide number of microorganisms colonize terrestrial plant species in a variety of morphological locations, including internal tissues (endophytes), leaves (epiphytes), and the rhizosphere. Rhizosphere bacteria can boost nutrient intake, protect against infections, improve abiotic stress tolerance, and stimulate plant development [41]. However, some rhizosphere microbes can cause plant disease or impede plant growth. Hence, metal nanoparticles present in soil have the potential to adversely affect microbial pathogens while concurrently augmenting the beneficial activities of indigenous microorganisms within the environment, thereby implicating exposure to non-target species [36, 37]. The utilization of diverse metal nanoparticles, such as Ag NPs, ZnO, or TiO2, has been observed to influence vital plant symbionts, including arbuscular mycorrhizal fungi (AMF), which establish colonization in plant roots and contribute to heightened nutrient uptake. These alterations may have substantial and enduring implications for plant nutrient absorption and productivity within agro-ecosystems [42, 43]. Conversely, they have the capacity to mitigate metal toxicity in host plants through the regulation of interactions between metals and plant roots [44, 45].

8. Ecotoxicity of metal NPs in agricultural system (crops)

The toxicity of different nanoparticles on typical crop plants is reviewed. A variety of scientific evidences exist about the hazardous effects of NPs on plants belonging to plant families such as *Poaceae* (Gramineae), Leguminosae (*Fabaceae*), *Solanaceae*, *Amaryllidaceae*, *Cucurbitaceae*, and *Brassicaceae*.

8.1 Wheat (Triticum aestivum)

Wheat is one of the most investigated crops in terms of nanomaterial toxicity. The toxicity of Cu, CuO, Zn, Ag, TiO2, ZnO, Ag, and Fe_2O_3 NPs on wheat has received a lot of attention. CuO nanoparticles impede not only the growth of root and shoot, but they also have an effect on biomass. Furthermore, they have a negative impact on several processes of photosynthesis, such as lowering the chlorophyll content of tissues involved while raising the peroxidase and catalase activities in treated plants roots [46]. The hazardous potential of TiO2 has been demonstrated at the cell level by the production of micronuclei after exposure [47]. Fe2O3 has also been researched for its impact on seed germination, with larger concentrations reducing seed germination [48]. Silver nanoparticles have also been investigated for their impact on the growth characteristics of wheat. Silver nanoparticles have influenced the expression of numerous genes responsible for encoding proteins involved in primary metabolism and cellular defense at higher doses (10 mg/L) [49].

8.2 Rice (Oryza sativa)

When hydroponically grown rice is exposed to TiO_2 NPs, the amount of biomass decreases and the antioxidant defense system is harmed. It has also been shown that TiO_2 nanoparticles can change the levels of some metabolites inside cells, stop the process of making carbohydrates, and speed up the respiration routes [50]. When there are higher amounts of CeO_2 NPs, more H_2O_2 is made, which also causes more lipid peroxidation and electrolyte leaks in rice plants [51]. When seeds are handled with CuO NPs, they can slow down germination and growth and do a lot of damage to cells through oxidative stress [52]. It is possible for ZnO nanoparticles to damage the shoot and root growth of rice plants [53]. Ag NPs can cause changes in the shape of rice plant roots because they can damage the cell walls and vacuoles [54]. It has also been said that Ag NPs change the total amount of chlorophyll in plants and lower the number of carotenoids they have. It has also been seen that reactive oxygen species (ROS) production goes up in a way that depends on the concentration [55].

8.3   Maize (*Zea mays*)

It has been reported that TiO_2 nanoparticles accumulate in maize roots upon entry via the root pathway [56], which can elicit oxidative stress and inhibit root growth [57, 58]. The accumulation of Zn NPs within the roots was demonstrated by an analysis of Zn in the roots. It has also been revealed that root growth parameters, such as length, can be decreased [59]. According to [60], Seed germination in maize remains unaffected by CuO; however, the subsequent growth of seedlings is significantly impacted by the presence of CuO. Silver nanoparticles, as a distinct category of nanoparticles, impede the germination process of maize seedlings [61].

9.   Ecotoxicological effects on aquatic organisms

Aquatic creatures can absorb and accumulate nanoparticles primarily through direct passage (across the gills) into the body or by ingestion. However, they are also susceptible to endocytosis at the cellular level. Nanoparticles enter aquatic creatures via phagocytosis, endocytosis, and micropinocytosis. They may penetrate the cell membrane via transportation and absorption and are unbound particles in the cytoplasm. Studies reveal that nanoparticles easily impact fish gills. Fish consumption of fullerene can cause brain lipid peroxidation. Some C60 exposure causes mortality within 6-8 hours.

Microorganisms play a crucial role in aquatic ecosystems because they form the ecosystem's foundation and contribute to important processes including sustaining the nutritional cycle and decomposing waste. Silver and titanium dioxide nanoparticles have a negative impact on marine microorganisms. E. coli cell membranes communicate with silver nanoparticles to regulate transport and regulation. Titanium dioxide disinfects E. coli solely through photocatalytic activity and ROS. Since phytoplankton are the primary producers in the food chain, the growth and toxicity of nanoparticles on them can have a large impact on the entire marine ecosystem. Marine nanotoxicology studies indicate the effect of nanoparticles on bivalve mollusks immune function. Nanoparticles are harmful to aquatic ecosystems and have an effect on phagocytes, lysosomal activity, and enhanced apoptosis [62].

10.   Ecotoxicological effects on humans

Nanoparticles have been shown to enter the human body through the respiratory system, intestines or skin where they can be harmful to the brain, induce lung inflammation and cause heart problems. Because of their size and composition, several nanoparticles have been discovered to trigger irreversible cell death via organ damage and cellular inflammation. The toxicity of a NPs in an individual is determined by the individuals genetic make-up, that is dictated by the person's capacity to evolve and react to harmful compounds [23].

11. The ecotoxicological impact of alternative nanoparticles

Quantum dots (q-dots) constitute a diverse array of compounds characterized by distinct compositions, sizes, and surface coatings. While certain quantum dots exhibit pronounced cytotoxic effects, comprehensive ecotoxicity data is currently unavailable. The minimal amounts required for q-dot detection in medical imaging, together with rigorous regulations for medical waste recovery in hospitals, suggest that q-dot loss to the environment will be extremely small [63].

Dendrimers, organic molecules with many branches provide hollow cages for transporting drugs or therapeutic agents in nanomedicine. Studies indicate that cationic dendrimers demonstrate heightened cytotoxicity and hemolysis compared to their anionic or PEGylated counterparts. This is attributed to the ability of cationic molecules to destabilize cell membranes, leading to cell lysis. No ecotoxicity studies have been conducted on dendrimer NPs [64].

12. Challenges in nanoecotoxicology research

Behra and Krug (2008) published a paper in the "Nature Nanotechnology" journal, highlighting three significant challenges that require solutions in the forthcoming years [16].

- The selection of NPs for utilization in biological research and tests. In order to accurately identify nanoparticles during the course of tests, it is imperative to ascertain their physicochemical properties, as well as their potential for aggregation and sedimentation, among other pertinent characteristics. This comprehensive analysis is essential both prior to and subsequent to the experimental procedures.

- There is a requirement to investigate the mechanism by which organisms in various contexts capture manmade nanoparticles.

- The collection of organisms suitable for experimental purposes and the selection of measurement points available for utilization.

13. Assessment of environmental risks posed by nanoparticles

The risk assessment of NPs has commenced by first identifying the potential risks and exposure pathways for people inside a production environment. This particular setting is of utmost importance as it necessitates the resolution of safety concerns in order to facilitate progress in other domains of nanotechnology [65, 66]. Nanotechnology possesses the potential to yield numerous favorable outcomes in the realm of environmental sustainability, encompassing sustainable energy generation, remediation techniques, and enhanced product efficiency. Consequently, it is imperative to conduct a comprehensive risk assessment that carefully evaluates any potential adverse impacts and associated risks in light of these aforementioned benefits.

13.1 Origin of NPs in the environment

Natural and human-made (both deliberate and accidental) processes are both potential origins of NPs. Volcanic activity, forest fires, weathering, soil erosion by wind and water, creation from clay minerals, and dust storms in deserts are major natural processes that release nanoparticles into the atmosphere. Every year, atmospheric dust alone may contain several million tons of naturally occurring nanoparticles. Naturally occurring ambient nanoparticles exhibit significant heterogeneity in terms of size and possess the ability to undergo long-range transit spanning thousands of kilometers, while maintaining their suspended state in the atmosphere for extended periods of time. Unintentional discharges of nanoparticles into the surrounding ecosystem encompass several sources such as automobile emissions, fuel cell operations, and diverse industrial procedures [67]. Anthropogenic NPs are inadvertently or intentionally discharged into the environment through a range of industrial and mechanical procedures. Accurately estimating the annual release of designed NPs into the environment remains challenging due to the significant increase in manufacturing volumes. Due to the progress made in industrial processes and nanotechnologies, a significant quantity of engineered nanoparticles is currently being produced. Consequently, it is unavoidable that the release of nanoparticles into the air, water, and soil occurs, both purposefully and unintentionally, during the utilization of associated goods. Additionally, it encompasses methodologies such as drug administration, diagnostic procedures, biomedical imaging techniques, groundwater remediation, and other related applications [68].

13.2 Hazard assessment

During the hazard assessment, the evaluation of the nanoparticle's ability to induce harm is conducted. Various endpoints can be utilized to assess hazards, such as toxicity and ecotoxicity, encompassing physiological, genetic, or functional consequences, which might manifest as acute or chronic outcomes. However, it is important to acknowledge that in addition to toxicity, there may exist various other adverse environmental impacts that should be considered when conducting a risk assessment specific to nanoparticles. These impacts may include but are not limited to alterations in atmospheric and stratospheric processes, soil stability, and the availability of mineral nutrients [69].

13.3 Dose-Response assessment

Dose-response assessments are conducted subsequent to the hazard identification phase within the risk assessment process. The determination of dose-response relationships can be achieved by the execution of laboratory experiments or the utilization of mathematical models. However, the establishment of dose-response relationships for nanoparticles is not always straightforward. This is due to the fact that the dose, when determined by mass concentration, may be less pertinent compared to when it is determined by surface area. Additionally, variations in surface reactivity among various nanoparticle preparations can lead to disparities in toxicity [69].

13.4 Exposure assessment

The potential for encountering nanomaterials arises from their creation, similar to chemical compounds. Therefore, it is imperative to acquire a comprehensive understanding of the quantitative factors associated with distinct stages of production, encompassing purification, conditioning, functionalization, transport, and packaging. Equally significant is the knowledge of losses and waste streams related with each of these stages. In the context of environmental exposure, the acquisition of empirical data or the establishment of procedural frameworks is crucial for the accurate prediction of the persistence and mobility of substances in the air, soil, and water [69].

13.5 Risk characterization

Risk characterization represents the ultimate phase within the risk assessment process, whereby the amalgamation of data obtained from hazard identification, dosage response, and exposure stages is carefully examined to ascertain and effectively express the genuine probability of risk occurrence among populations who are exposed. The assessment of risk frequently involves the juxtaposition of the concentration of exposure with a presumed non-effectual exposure threshold [69].

14. Future research needs

Determining the environmental mobility and bioavailability of nanoparticles is the most important need for research in this field. These factors will determine whether or not nanoparticles can be absorbed by diverse creatures (including plants) and cause harm. Environmental degradation and public health impacts are possible due to contamination of drinking water and the food chain. Ongoing research on manufactured nanoparticles environmental impact faces challenges in detection, tracing and characterization as well as determining which to study. Nanoparticles are small and complex in their environments, such as water, sediments, soils and ecosystems. Additionally, there are many engineered nanoparticles and their derivatives like surface modifications of pristine materials [70].

Several factors should be considered seriously while studying the ecotoxicity of nanoparticles.

1. Choose suitable terrestrial or aquatic organisms and endpoints for nanoparticle ecotoxicity testing (or evaluate existing methodologies).

2. Creating realistic nanoparticle exposure protocols for ecotoxicity measurements.

3. Determining dose-response relationships, translocation, acute/chronic toxicity, toxicokinetic, excretion dynamics, and toxicity mechanisms.

To study nanoparticle environmental behavior, consider the following.

1. Nanoparticle mobility in soils, sediments and trash.

2. Adsorption/desorption behavior in respect to organic, biological, mineral, sediment and water components.

For nanoparticle metrology research, consider the following.

1. Developing nanoparticle detection and quantification methods for air, water, soils and trash.

2. Testing and modifying nanoparticle analysis procedures.

3. Standardized nanoparticle characterization requirements [71]

Conclusion

The study of nanoparticle ecotoxicology is an important field with broad consequences for the sustainability of the environment. As nanoparticles are integrated into more businesses and products, their ecological implications must be assessed. Exposure pathways, bioavailability and toxicity mechanisms have illuminated nanoparticle-environment interactions. This highlights the need for responsible nanomaterial development and risk assessment to reduce environmental harm. We can use nanotechnology while protecting our ecosystems by tackling nanoparticle ecotoxicity.

References

[1] E. Hood, Nanotechnology: Looking as we leap, Environ. Health Perspect. 112 (2004). https://doi.org/10.1289/ehp.112-a740.

[2] E.A.J. Bleeker, W.H. de Jong, R.E. Geertsma, M. Groenewold, E.H.W. Heugens, M. Koers-Jacquemijns, D. van de Meent, J.R. Popma, A.G. Rietveld, S.W.P. Wijnhoven, F.R. Cassee, A.G. Oomen, Considerations on the EU definition of a nanomaterial: Science to support policy making, Regul. Toxicol. Pharmacol. 65 (2013) 119–125. https://doi.org/10.1016/j.yrtph.2012.11.007.

[3] H. peng Feng, L. Tang, G. ming Zeng, Y. Zhou, Y. cheng Deng, X. Ren, B. Song, C. Liang, M. yun Wei, J. fang Yu, Core-shell nanomaterials: Applications in energy storage and conversion, Adv. Colloid Interface Sci. 267 (2019) 26–46. https://doi.org/10.1016/j.cis.2019.03.001.

[4] M.R. Wiesner, G. V. Lowry, P. Alvarez, D. Dionysiou, P. Biswas, Assessing the risks of manufactured nanomaterials, Environ. Sci. Technol. 40 (2006) 4336–4345. https://doi.org/10.1021/es062726m.

[5] V.L. Colvin, The potential environmental impact of engineered nanomaterials, Nat. Biotechnol. 21 (2003) 1166–1170. https://doi.org/10.1038/nbt875.

[6] M.H. Sarfraz, M. Zubair, B. Aslam, A. Ashraf, M.H. Siddique, S. Hayat, J.N. Cruz, S. Muzammil, M. Khurshid, M.F. Sarfraz, A. Hashem, T.M. Dawoud, G.D. Avila-Quezada, E.F. Abd_Allah, Comparative analysis of phyto-fabricated chitosan, copper oxide, and chitosan-based CuO nanoparticles: antibacterial potential against Acinetobacter baumannii isolates and anticancer activity against HepG2 cell lines, Front. Microbiol. 14 (2023) 1188743. https://doi.org/10.3389/fmicb.2023.1188743.

[7] G. Oberdörster, A. Maynard, K. Donaldson, V. Castranova, J. Fitzpatrick, K. Ausman, J. Carter, B. Karn, W. Kreyling, D. Lai, S. Olin, N. Monteiro-Riviere, D. Warheit, H. Yang, Principles for characterizing the potential human health effects from exposure to nanomaterials: Elements of a screening strategy, Part. Fibre Toxicol. 2 (2005). https://doi.org/10.1186/1743-8977-2-8.

[8] M.C. Roco, Nanotechnology: Convergence with modern biology and medicine, Curr. Opin. Biotechnol. 14 (2003) 337–346. https://doi.org/10.1016/S0958-1669(03)00068-5.

[9] S. Ammar, I.A.W. Ma, K. Ramesh, S. Ramesh, Polymers-based nanocomposite coatings, Nanomater. Coatings Fundam. Appl. (2019) 9–39. https://doi.org/10.1016/B978-0-12-815884-5.00002-8.

[10] D. Letchumanan, S.P.M. Sok, S. Ibrahim, N.H. Nagoor, N.M. Arshad, Plant-based biosynthesis of copper/copper oxide nanoparticles: An update on their applications in biomedicine, mechanisms, and toxicity, Biomolecules. 11 (2021) 564. https://doi.org/10.3390/biom11040564.

[11] A.C. Anselmo, S. Mitragotri, Nanoparticles in the clinic: An update, Bioeng. Transl. Med. 4 (2019). https://doi.org/10.1002/btm2.10143.

[12] H. Otsuka, Y. Nagasaki, K. Kataoka, PEGylated nanoparticles for biological and pharmaceutical applications, Adv. Drug Deliv. Rev. 55 (2003) 403–419. https://doi.org/10.1016/S0169-409X(02)00226-0.

[13] N.S. Alsaiari, F.M. Alzahrani, A. Amari, H. Osman, H.N. Harharah, N. Elboughdiri, M.A. Tahoon, Plant and Microbial Approaches as Green Methods for the Synthesis of Nanomaterials: Synthesis, Applications, and Future Perspectives, Molecules. 28 (2023) 463. https://doi.org/10.3390/molecules28010463.

[14] M.C. Newman, Y. Zhao, Ecotoxicology Nomenclature: LC, LD, LOC, LOEC, MAC, Encycl. Ecol. Five-Volume Set. (2008) 1187–1193. https://doi.org/10.1016/B978-008045405-4.00404-3.

[15] R. Truhaut, Ecotoxicology: Objectives, principles and perspectives, Ecotoxicol. Environ. Saf. 1 (1977) 151–173. https://doi.org/10.1016/0147-6513(77)90033-1.

[16] A. Kahru, H.C. Dubourguier, From ecotoxicology to nanoecotoxicology, Toxicology. 269 (2010) 105–119. https://doi.org/10.1016/j.tox.2009.08.016.

[17] A. Zielinska, F. Carreiró, A.M. Oliveira, A. Neves, B. Pires, D. Nagasamy Venkatesh, A. Durazzo, M. Lucarini, P. Eder, A.M. Silva, A. Santini, E.B. Souto, Polymeric Nanoparticles: Production, Characterization, Toxicology and Ecotoxicology, Molecules. 25 (2020) 3731. https://doi.org/10.3390/molecules25163731.

[18] F.S. Alves, J.N. Cruz, I.N. de Farias Ramos, D.L. do Nascimento Brandão, R.N. Queiroz, G.V. da Silva, G.V. da Silva, M.F. Dolabela, M.L. da Costa, A.S. Khayat, J. de Arimatéia Rodrigues do Rego, D. do Socorro Barros Brasil, Evaluation of Antimicrobial Activity and Cytotoxicity Effects of Extracts of Piper nigrum L. and Piperine, Separations. 10 (2023) 21. https://doi.org/10.3390/separations10010021.

[19] N.A. Monteiro-Riviere, R.J. Nemanich, A.O. Inman, Y.Y. Wang, J.E. Riviere, Multi-walled carbon nanotube interactions with human epidermal keratinocytes, Toxicol. Lett. 155 (2005) 377–384. https://doi.org/10.1016/j.toxlet.2004.11.004.

[20] L. Yang, D.J. Watts, Particle surface characteristics may play an important role in phytotoxicity of alumina nanoparticles, Toxicol. Lett. 158 (2005) 122–132. https://doi.org/10.1016/j.toxlet.2005.03.003.

[21] K. Fukushi, T. Sato, Using a surface complexation model to predict the nature and stability of nanoparticles, Environ. Sci. Technol. 39 (2005) 1250–1256. https://doi.org/10.1021/es0491984.

[22] C.M. Sayes, J.D. Fortner, W. Guo, D. Lyon, A.M. Boyd, K.D. Ausman, Y.J. Tao, B. Sitharaman, L.J. Wilson, J.B. Hughes, J.L. West, V.L. Colvin, The differential cytotoxicity of water-soluble fullerenes, Nano Lett. 4 (2004) 1881–1887. https://doi.org/10.1021/nl0489586.

[23] G. Pandey, P. Jain, Assessing the nanotechnology on the grounds of costs, benefits, and risks, Beni-Suef Univ. J. Basic Appl. Sci. 9 (2020). https://doi.org/10.1186/s43088-020-00085-5.

[24] N.R. Panyala, E.M. Peña-Méndez, J. Havel, Silver or silver nanoparticles: A hazardous threat to the environment and human health?, J. Appl. Biomed. 6 (2008) 117–129. https://doi.org/10.32725/jab.2008.015.

[25] P. Sharma, N.Y. Jang, J.W. Lee, B.C. Park, Y.K. Kim, N.H. Cho, Application of ZnO-based nanocomposites for vaccines and cancer immunotherapy, Pharmaceutics. 11 (2019) 493. https://doi.org/10.3390/pharmaceutics11100493.

[26] R.M. Touyz, F.J. Rios, R. Alves-Lopes, K.B. Neves, L.L. Camargo, A.C. Montezano, Oxidative Stress: A Unifying Paradigm in Hypertension, Can. J. Cardiol. 36 (2020) 659–670. https://doi.org/10.1016/j.cjca.2020.02.081.

[27] X. Zhou, L. Zhao, J. Luo, H. Tang, M. Xu, Y. Wang, X. Yang, H. Chen, Y. Li, G. Ye, F. Shi, C. Lv, B. Jing, The toxic effects and mechanisms of Nano-Cu on the spleen of rats, Int. J. Mol. Sci. 20 (2019) 1469. https://doi.org/10.3390/ijms20061469.

[28] J.R. Lead, K.J. Wilkinson, Aquatic colloids and nanoparticles: Current knowledge and future trends, Environ. Chem. 3 (2006) 159–171. https://doi.org/10.1071/EN06025.

[29] M. Ammann, H. Burtscher, H.C. Siegmann, Monitoring volcanic activity by characterization of ultrafine aerosol emissions, J. Aerosol Sci. 21 (1990) S275–S278. https://doi.org/10.1016/0021-8502(90)90237-R.

[30] S. Hasegawa, S. Wakamatsu, T. Ohara, Y. Itano, K. Saitoh, M. Hayasaki, S. Kobayashi, Vertical profiles of ultrafine to supermicron particles measured by aircraft over Osaka metropolitan area in Japan, Atmos. Environ. 41 (2007) 717–729. https://doi.org/10.1016/j.atmosenv.2006.09.031.

[31] R.D. Handy, B.J. Shaw, Toxic effects of nanoparticles and nanomaterials: Implications for public health, risk assessment and the public perception of nanotechnology, Heal. Risk Soc. 9 (2007) 125–144. https://doi.org/10.1080/13698570701306807.

[32] P.G. Barlow, K. Donaldson, J. MacCallum, A. Clouter, V. Stone, Serum exposed to nanoparticle carbon black displays increased potential to induce macrophage migration, Toxicol. Lett. 155 (2005) 397–401. https://doi.org/10.1016/j.toxlet.2004.11.006.

[33] D.A. Taylor, Dust in the wind, Environ. Health Perspect. 110 (2002). https://doi.org/10.1289/ehp.110-a80.

[34] P. Kumar, L. Pirjola, M. Ketzel, R.M. Harrison, Nanoparticle emissions from 11 non-vehicle exhaust sources - A review, Atmos. Environ. 67 (2013) 252–277. https://doi.org/10.1016/j.atmosenv.2012.11.011.

[35] R.B.M. de Almeida, D.B. Barbosa, M.R. do Bomfim, J.A.O. Amparo, B.S. Andrade, S.L. Costa, J.M. Campos, J.N. Cruz, C.B.R. Santos, F.H.A. Leite, M.B. Botura, Identification of a Novel Dual Inhibitor of Acetylcholinesterase and Butyrylcholinesterase: In Vitro and In Silico Studies, Pharmaceuticals. 16 (2023) 95. https://doi.org/10.3390/ph16010095.

[36] P. Gajjar, B. Pettee, D.W. Britt, W. Huang, W.P. Johnson, A.J. Anderson, Antimicrobial activities of commercial nanoparticles against an environmental soil microbe, Pseudomonas putida KT2440, J. Biol. Eng. 3 (2009). https://doi.org/10.1186/1754-1611-3-9.

Materials Research Forum LLC
https://doi.org/10.21741/9781644902998-8

[37] C.O. Dimkpa, D.E. Latta, J.E. McLean, D.W. Britt, M.I. Boyanov, A.J. Anderson, Fate of CuO and ZnO nano- and microparticles in the plant environment, Environ. Sci. Technol. 47 (2013) 4734–4742. https://doi.org/10.1021/es304736y.

[38] M. Simonin, A. Richaume, Impact of engineered nanoparticles on the activity, abundance, and diversity of soil microbial communities: a review, Environ. Sci. Pollut. Res. 22 (2015) 13710–13723. https://doi.org/10.1007/s11356-015-4171-x.

[39] T.Y. Sun, F. Gottschalk, K. Hungerbühler, B. Nowack, Comprehensive probabilistic modelling of environmental emissions of engineered nanomaterials, Environ. Pollut. 185 (2014) 69–76. https://doi.org/10.1016/j.envpol.2013.10.004.

[40] S.E. Smith, D.J. Read, Introduction, Mycorrhizal Symbiosis. (2002) 1–8. https://doi.org/10.1016/b978-012652840-4/50001-2.

[41] R.L. Berendsen, C.M.J. Pieterse, P.A.H.M. Bakker, The rhizosphere microbiome and plant health, Trends Plant Sci. 17 (2012) 478–486. https://doi.org/10.1016/j.tplants.2012.04.001.

[42] D.J. Burke, S. Zhu, M.P. Pablico-Lansigan, C.R. Hewins, A.C.S. Samia, Titanium oxide nanoparticle effects on composition of soil microbial communities and plant performance, Biol. Fertil. Soils. 50 (2014) 1169–1173. https://doi.org/10.1007/s00374-014-0938-3.

[43] Q. Wang, Z. Yang, Y. Yang, C. Long, H. Li, A bibliometric analysis of research on the risk of engineering nanomaterials during 1999-2012, Sci. Total Environ. 473–474 (2014) 483–489. https://doi.org/10.1016/j.scitotenv.2013.12.066.

[44] B.D. Chen, X.L. Li, H.Q. Tao, P. Christie, M.H. Wong, The role of arbuscular mycorrhiza in zinc uptake by red clover growing in a calcareous soil spiked with various quantities of zinc, Chemosphere. 50 (2003) 839–846. https://doi.org/10.1016/S0045-6535(02)00228-X.

[45] S. Muzammil, J. Neves Cruz, R. Mumtaz, I. Rasul, S. Hayat, M.A. Khan, A.M. Khan, M.U. Ijaz, R.R. Lima, M. Zubair, Effects of Drying Temperature and Solvents on In Vitro Diabetic Wound Healing Potential of Moringa oleifera Leaf Extracts, Molecules. 28 (2023) 710. https://doi.org/10.3390/molecules28020710.

[46] C.O. Dimkpa, J.E. McLean, D.E. Latta, E. Manangón, D.W. Britt, W.P. Johnson, M.I. Boyanov, A.J. Anderson, CuO and ZnO nanoparticles: Phytotoxicity, metal speciation, and induction of oxidative stress in sand-grown wheat, J. Nanoparticle Res. 14 (2012). https://doi.org/10.1007/s11051-012-1125-9.

[47] R. Rafique, Z. Zahra, N. Virk, M. Shahid, E. Pinelli, T.J. Park, J. Kallerhoff, M. Arshad, Dose-dependent physiological responses of Triticum aestivum L. to soil applied TiO2 nanoparticles: Alterations in chlorophyll content, H2O2 production, and genotoxicity, Agric. Ecosyst. Environ. 255 (2018) 95–101. https://doi.org/10.1016/j.agee.2017.12.010.

[48] H. Feizi, P. Rezvani Moghaddam, N. Shahtahmassebi, A. Fotovat, Impact of bulk and nanosized titanium dioxide (TiO2) on wheat seed germination and seedling growth, Biol. Trace Elem. Res. 146 (2012) 101–106. https://doi.org/10.1007/s12011-011-9222-7.

[49] C. Vannini, G. Domingo, E. Onelli, F. De Mattia, I. Bruni, M. Marsoni, M. Bracale, Phytotoxic and genotoxic effects of silver nanoparticles exposure on germinating wheat seedlings, J. Plant Physiol. 171 (2014) 1142–1148. https://doi.org/10.1016/j.jplph.2014.05.002.

[50] B. Wu, L. Zhu, X.C. Le, Metabolomics analysis of TiO2 nanoparticles induced toxicological effects on rice (Oryza sativa L.), Environ. Pollut. 230 (2017) 302–310. https://doi.org/10.1016/j.envpol.2017.06.062.

[51] C.M. Rico, M.I. Morales, R. McCreary, H. Castillo-Michel, A.C. Barrios, J. Hong, A. Tafoya, W.Y. Lee, A. Varela-Ramirez, J.R. Peralta-Videa, J.L. Gardea-Torresdey, Cerium oxide nanoparticles modify the antioxidative stress enzyme activities and macromolecule composition in rice seedlings, Environ. Sci. Technol. 47 (2013) 14110–14118. https://doi.org/10.1021/es4033887.

[52] A.K. Shaw, S. Ghosh, H.M. Kalaji, K. Bosa, M. Brestic, M. Zivcak, Z. Hossain, Nano-CuO stress induced modulation of antioxidative defense and photosynthetic performance of Syrian barley (Hordeum vulgare L.), Environ. Exp. Bot. 102 (2014) 37–47. https://doi.org/10.1016/j.envexpbot.2014.02.016.

[53] P. Boonyanitipong, B. Kositsup, P. Kumar, S. Baruah, J. Dutta, Toxicity of ZnO and TiO2 Nanoparticles on Germinating Rice Seed Oryza sativa L, Int. J. Biosci. Biochem. Bioinforma. (2011) 282–285. https://doi.org/10.7763/ijbbb.2011.v1.53.

[54] P. Zhang, Z. Guo, F.A. Monikh, I. Lynch, E. Valsami-Jones, Z. Zhang, Growing Rice (Oryza sativa) Aerobically Reduces Phytotoxicity, Uptake, and Transformation of CeO2Nanoparticles, Environ. Sci. Technol. 55 (2021) 8654–8664. https://doi.org/10.1021/acs.est.0c08813.

[55] P.M.G. Nair, I.M. Chung, Physiological and molecular level effects of silver nanoparticles exposure in rice (Oryza sativa L.) seedlings, Chemosphere. 112 (2014) 105–113. https://doi.org/10.1016/j.chemosphere.2014.03.056.

[56] C. Larue, J. Laurette, N. Herlin-Boime, H. Khodja, B. Fayard, A.M. Flank, F. Brisset, M. Carriere, Accumulation, translocation and impact of TiO2 nanoparticles in wheat (Triticum aestivum spp.): Influence of diameter and crystal phase, Sci. Total Environ. 431 (2012) 197–208. https://doi.org/10.1016/j.scitotenv.2012.04.073.

[57] M.R. Castiglione, L. Giorgetti, C. Geri, R. Cremonini, The effects of nano-TiO2 on seed germination, development and mitosis of root tip cells of Vicia narbonensis L. and Zea mays L, J. Nanoparticle Res. 13 (2011) 2443–2449. https://doi.org/10.1007/s11051-010-0135-8.

[58] Antimicrobial Activity of Titanium Dioxide Nanoparticles Biosynthesized by Enterococcus hirae, Curr. Sci. Int. (2022). https://doi.org/10.36632/csi/2022.11.4.30.

[59] R. Zhang, H. Zhang, C. Tu, X. Hu, L. Li, Y. Luo, P. Christie, Phytotoxicity of ZnO nanoparticles and the released Zn(II) ion to corn (Zea mays L.) and cucumber (Cucumis sativus L.) during germination, Environ. Sci. Pollut. Res. 22 (2015) 11109–11117. https://doi.org/10.1007/s11356-015-4325-x.

[60] Z. Wang, X. Xie, J. Zhao, X. Liu, W. Feng, J.C. White, B. Xing, Xylem- and phloem-based transport of CuO nanoparticles in maize (Zea mays L.), Environ. Sci. Technol. 46 (2012) 4434–4441. https://doi.org/10.1021/es204212z.

[61] L.R. Pokhrel, B. Dubey, Evaluation of developmental responses of two crop plants exposed to silver and zinc oxide nanoparticles, Sci. Total Environ. 452–453 (2013) 321–332. https://doi.org/10.1016/j.scitotenv.2013.02.059.

[62] E.Y. Krysanov, D.S. Pavlov, T.B. Demidova, Y.Y. Dgebuadze, Effect of nanoparticles on aquatic organisms, Biol. Bull. 37 (2010) 406–412. https://doi.org/10.1134/S1062359010040114.

[63] R. Hardman, A toxicologic review of quantum dots: Toxicity depends on physicochemical and environmental factors, Environ. Health Perspect. 114 (2006) 165–172. https://doi.org/10.1289/ehp.8284.

[64] R. Duncan, L. Izzo, Dendrimer biocompatibility and toxicity, Adv. Drug Deliv. Rev. 57 (2005) 2215–2237. https://doi.org/10.1016/j.addr.2005.09.019.

[65] I.N. Throbäck, M. Johansson, M. Rosenquist, M. Pell, M. Hansson, S. Hallin, Silver (Ag+) reduces denitrification and induces enrichment of novel nirK genotypes in soil, FEMS Microbiol. Lett. 270 (2007) 189–194. https://doi.org/10.1111/j.1574-6968.2007.00632.x.

[66] K. Shahane, M. Kshirsagar, S. Tambe, D. Jain, S. Rout, M.K.M. Ferreira, S. Mali, P. Amin, P.P. Srivastav, J. Cruz, R.R. Lima, An Updated Review on the Multifaceted Therapeutic Potential of Calendula officinalis L., Pharmaceuticals. 16 (2023) 611. https://doi.org/10.3390/ph16040611.

[67] S. Smita, S.K. Gupta, A. Bartonova, M. Dusinska, A.C. Gutleb, Q. Rahman, Nanoparticles in the environment: Assessment using the causal diagram approach, Environ. Heal. A Glob. Access Sci. Source. 11 (2012). https://doi.org/10.1186/1476-069X-11-S1-S13.

[68] S. Rana, P.T. Kalaichelvan, Ecotoxicity of Nanoparticles, ISRN Toxicol. 2013 (2013) 1–11. https://doi.org/10.1155/2013/574648.

[69] D.H. Oughton, T. Hertel-Aas, E. Pellicer, E. Mendoza, E.J. Joner, Neutron activation of engineered nanoparticles as a tool for tracing their environmental fate and uptake in organisms, Environ. Toxicol. Chem. 27 (2008) 1883–1887. https://doi.org/10.1897/07-578.1.

[70] C.A.M. van Gestel, Soil ecotoxicology: State of the art and future directions, Zookeys. 176 (2012) 275–296. https://doi.org/10.3897/zookeys.176.2275.

[71] R.D. Handy, R. Owen, E. Valsami-Jones, The ecotoxicology of nanoparticles and nanomaterials: Current status, knowledge gaps, challenges, and future needs, Ecotoxicology. 17 (2008) 315–325. https://doi.org/10.1007/s10646-008-0206-0.

Nanoparticle Toxicity and Compatibility
Materials Research Foundations 161 (2024) 245-266

Materials Research Forum LLC
https://doi.org/10.21741/9781644902998-9

Chapter 9

Toxicity of Nanoparticles in Biological Systems

Muhammad Umar Ijaz[1], Ali Akbar[1], Asma Ashraf[2], Muhammad Saqalein[3],
Hammad Ahmad Khan[1], Sumreen Hayat[3], Saima Muzammil[3]*

[1]Department of Zoology, Wildlife and Fisheries, University of Agriculture, Faisalabad, Pakistan

[2]Department of Zoology, Government college University Faisalabad, Pakistan

[3]Institute of Microbiology, Government college University Faisalabad, Pakistan

saimamuzammil83@gmail.com

Abstract

Nanoparticles (NPs) are minute particles with sizes ranging from 1 to 100 nanometers. They are characterized by their unique physiochemical properties due to their small size, which makes them valuable in several fields such as medicine, electronics, energy and material science. They are also used for drug delivery, imaging, energy applications and improving material properties. However, the potential toxicity of NPs to biological systems has raised concerns in recent years. The toxicity of NPs to biological systems arises from their physiochemical properties and interactions with biological components. These interactions can lead to the induction of various cellular pathways, including oxidative stress (OS), inflammation, genotoxicity and immune dysregulation among others, which can result in cellular damage and adverse effects on various organs and tissues. The toxic effects of NPs can vary with their size, shape, surface charge, concentration, mode of administration, as well as the type of biological system being exposed. Therefore, understanding NPs-instigated toxicity is critical for safe and effective use in various applications. This chapter presents an overview of the toxicological effects of NPs on biological systems with a particular focus on the liver, kidney, reproductive and cardiovascular systems along with mechanisms that trigger toxicity in these organs. Furthermore, the chapter discusses the various routes of exposure of NPs and their biodistribution within the body including the factors that influence their distribution and accumulation in different tissues.

Keywords

Nanoparticles, Toxicity, Nephrotoxicity, Hepatotoxicity, Reproductive Toxicity

Contents

1.   Introduction

The advent of nanotechnology has sparked a remarkable revolution in the industrial sector. Nanostructured materials, due to their unique physiochemical & electrical characteristics have garnered significant attention in the field of electronic, biotechnology & aerospace engineering. In the domain of medicine, NPs are considered as emerging drug delivery system for therapeutic agents, including proteins, DNA and monoclonal antibodies. This approach offers a novel means to mitigate the limitations of conventional drug delivery techniques i.e., poor bioavailability, rapid clearance and non-specific targeting, thereby enhancing the efficacy and safety of treatments [1-3]. The rapid advancement of nanotechnology in medical science has propelled the design and synthesis of nanomaterials for biomedical applications. NPs, with their unique physiochemical properties such as small size (1-100 nm in diameter), high surface area to volume ratio, distinct electronic, magnetic optical and mechanical properties as well as diverse shapes have garnered

Nanoparticle Toxicity and Compatibility　　　　　　　　Materials Research Forum LLC
Materials Research Foundations 161 (2024) 245-266　　　　https://doi.org/10.21741/9781644902998-9

immense interest in various fields. The properties of NPs can be fine-tuned by controlling their size, shape and surface chemistry, making them promising tools for targeted drug delivery, imaging and sensing applications. The unique features of NPs offer great potential for overcoming the limitations of conventional therapeutic & diagnostic approaches, leading to the development of more effective & personalized treatments [4-9].

NPs can arise from diverse natural & anthropogenic sources, including oceans, volcanic eruptions, glacial ice cores, surface water, physical and chemical weathering of rocks, ground water, diesel exhaust, treated drinking water, atmospheric water, electroplating and welding processes among others. These sources can generate a variety of NPs in the environment and their potential impacts on health and the ecosystem [10-13]. NPs can aggregate or persist in air, water, soil and biological system, affecting their exposure & potential impacts. [14]. Consumer products can also release NPs into the environment through continuous washing and use to their potential accumulation and persistence in different environment media [15].

Since geological times, the ecological system has been exposed to NPs through natural process such as dust storms [16], volcanic eruptions [17], erosions [18], forest fires [19] and sea sprays [20] among others leading to their potential accumulation and transformation in different environmental media. The growing use of nanotechnology and human activities has resulted in an increased environmental exposure to various shapes and sizes of NPs, posing potential risks to ecological safety & human health. The expanding use of NPs in consumer products, industrial processes and biomedical applications has raised concerns about their environmental impacts, particularly their persistence, transformation & toxicity in different environmental systems [20,15].

Reactive oxygen species (ROS) formation is considered as one of the underlying mechanisms of NPs toxicity, leading to OS, and inflammation. Therefore, evaluating toxicity of NPs in biological systems is very important for assessing their safety and efficacy in biomedical applications, including rug delivery, gene delivery & pharmacological applications. The understanding of the phytochemical properties of NPs and their interactions with biological systems can provide valuable insights into their potential risks, benefits & guide the development of safe and effective nanotherapeutics [21-24].

2.　　Routes of exposure

The current understanding of the intricate mechanisms underlying the exposure of the human body to NPs remains imprecise & it is the subject of intense exploration worldwide. Indeed, the routes by which NPs penetrate the human system, are taken up by cells and tissues, distributed & their potential health effects are still under investigation. There are three primary ways in which NPs can penetrate the human body: via skin, ingestion as well as inhalation. Once inside the body, NPs can travel through the bloodstream or respiratory system to reach organs [25-27] Fig.1.

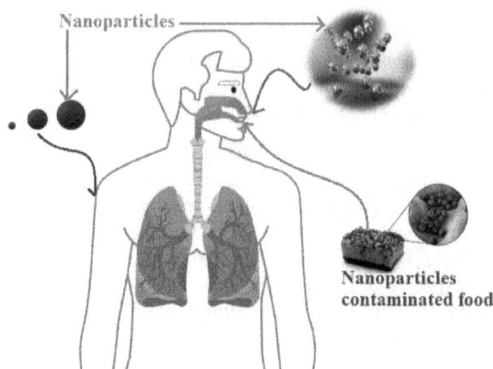

Fig.1 Primary routes of NPs exposure

3.   Nanoparticle biodistribution

NPs are able to enter the systemic circulation through various exposure routes and subsequently distribute to different body regions i.e., liver, kidney, lungs, testes, spleen, blood cells, thymus, heart, & brain [28-31]. To investigate the distribution & deposition of nanoceria particles (NCPs) in CD1 mice, the study utilized inductively coupled plasma mass spectrometry (ICP-MS). The rats were intraperitoneally exposed with NCPs and subsequently dosed weekly with 0.5 mgkg^{-1} NCPs for a duration of 2 or 5 weeks [32]. Results showed that NCPs were predominantly found in the spleen & liver with negligible concentrations detected in lungs, kidney & brain. The high concentration of NCPs in the spleen may be attributed to the greater number of macrophages that transported the NPs to the organ. On the other hand, the liver's high vascularization and detoxification role could also account for its NCPs deposition. Different distribution patterns were seen for the three administration routes. Oral administration had lower uptake due to the absence of gastrointestinal system involvement. In another study exploring gold-NPs (Au-NPs) with varying sizes (15, 50, 100 & 200 nm) in mice, it was observed that the smallest AuNPs (15 nm) exhibited rapid accumulation in vital body organs. Intriguingly, both 15 nm and 50 nm AuNPs were detected in the brain, suggesting their ability to cross the blood-brain barrier (BBB). Larger AuNPs (200 nm) expressed very minimal occurrence in different organs [33]. Xie et al. conducted a study on the nanotoxicity & biodistribution of 2 distinct sized silica NPs (SNPs) (20 nm and 80 nm) in mice for a period of 30 days [34]. He used 125I-SNPs to obtain numerical distribution in various organs, which revealed a similar distribution pattern to other studies. The high vascularity for the liver & spleen may have contributed to the observed deposition of SNPs in these organs. The transmission electron microscopic images of liver & spleen samples revealed that small NPs accumulated within the phagolysosomes of macrophages present in the tissues. Furthermore, the study observed that smaller NPs exhibited intense deposition in the spleen & liver compared to their larger counterparts. This observation could be attributed to the large surface potential of NPs, which increases protein binding efficiency & promotes recognition via surface receptors [35] Fig.2.

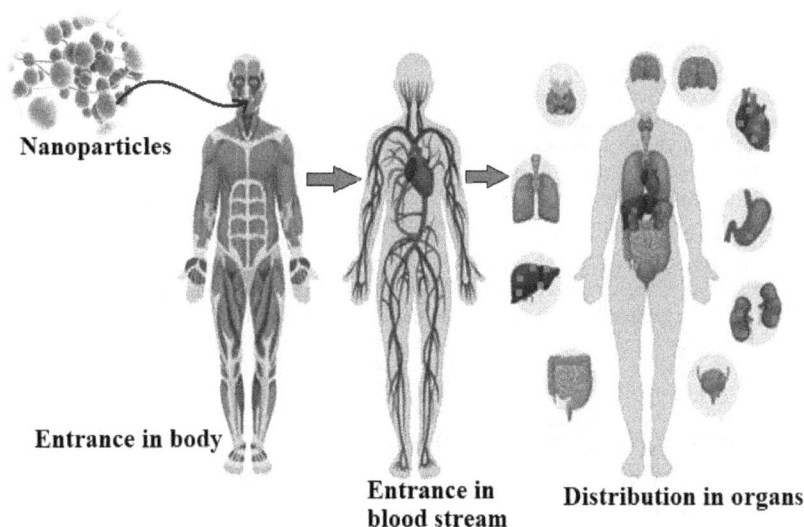

Fig. 2 Biodistribution of NPs

3.1 Correlation of NPs properties with biodistribution profile

Although the toxic characteristics of NPs are not yet fully understood, they are potentially associated with specific physicochemical properties. These attributes might encompass the surface catalytic activity associated with metallic or semiconducting properties, the uptake of NPs, or their dissolution. NPs can manifest diverse shapes & structures i.e., spheres, needles, tubes, plates, and various others. As particles are reduced to the nanoscale level, their surface area to volume ratio increases, resulting in a greater number of molecules being exposed on the NPs surface. This elevated surface area potentially augments the intrinsic toxicity of the NP [36] Table 1.

3.1.1 NPs size and structure configuration

Several studies have reported that NPs can cause more harm than their corresponding macro-sized particles. In order to test this, the cytotoxicity of Au particles Au NPs and Ag NPs with different sizes (2-4 nm, 5-7 nm and 20-40 nm) was evaluated in-*vitro* using J774 A1 macrophages. The results showed that the cytotoxicity increased as the size of the AuNPs decreased. Even though the small and medium-sized NPs only differ by about 3 nm in size, their surface characteristics could still be a factor in their potential toxicity. In the case of Ag NPs, the cytotoxicity was lower than that of AuNPs of different sizes. While the uptake of NPs by cells was evident through the presence of numerous vesicles in cells treated with both 10 ppm concentrations of NPs, it was notable that AuNPs had a more pronounced impact compared to Ag NPs [37]. Pan et al. reported that AuNPs with small sizes (1.2, 1.4 and 1.8 nm) could induce significantly different cell responses despite slight differences in their sizes [38]. The investigation by Pan et al. revealed that there were significant differences in the cellular response induced by small AuNPs (1.2, 1.4 and 1.8 nm) with a minute difference in their size. The 1.3 nm AuNPs were observed to be highly toxic while NPs larger than 15 nm were comparatively less toxic. Similarly, studies on TiO2 and carbon black NPs have shown that they induce greater inflammation and damage to epithelial cells in rat lungs when in the form of NPs, as compared to their larger counterparts. Furthermore, smaller Zinc oxide NPs (ZNPs) were found to be more toxic than their larger counterparts in human lung epithelial cell (A549) [39]. It could be inferred that cytotoxicity of NPs varies significantly with their size and the type of element they are composed of.

However, research has shown that NPs of the same size and surface area but with different shapes can have varying effects on cell viability. While there is not yet sufficient data on the toxicity of nanomaterials due to their various shapes it has been observed that rod-shape NPs can be more toxic than their corresponding spherical counterparts. For instance, in the case of ZNPs, at a fixed size and surface area, rod-shape ZNPs were found to be more cytotoxic than spherical ones in A549 cell. These findings suggest that both the size and shape of NPs can have considerable impact on their cytotoxicity [40].

3.1.2 Surface properties of NPs: composition and charges

Surface properties of NPs including their composition & charges have been found to play a remarkable role in their potential toxic effects [41]. Furthermore, intentional modification or functionalization of NPs surfaces can modulate their interactions with biological systems. Truel et al. conducted an extensive investigation into the intricate interactions among Ag & gold nanoparticles (Au NPs) of various functionalization, along with polystyrene NPs, in the presence of bovine serum albumin (BSA) [42]. It was revealed that the introduction of NPs into the protein solution led to the disruption of a portion of the protein's secondary structure, primarily caused by surface adsorption. The study indicated that citrate-coated Au & Ag NPs displayed considerably more strong interactions with BSA compared to polymeric or polymer-coated metallic NPs. This observation strongly suggested that the surface composition of NPs plays a dominant role in influencing their effect on BSA [42].

Chen et al. conducted a comprehensive assessment of how varying surface charges influence cell proliferation, viability & uptake behavior [43]. In investigating the interaction between hydroxyapatite NPs, which possess similar shape & size but different surface charges, it was observed that the uptake of positively charged NPs surpassed that of negatively charged particles. Intriguingly, neutral NPs were unable to traverse the cell membrane, likely attributed to their larger size [43]. In line with previous research, Goodman et al. conducted a study wherein they observed that positively charged Au NPs exhibited a moderate level of toxicity, while negatively charged Au NPs were deemed non-toxic. This investigation implies that the intracellular response and cytotoxicity of NPs are influenced by factors such as their composition and size, as well as the specific types of cells they interact with [44].

Nanoparticle Toxicity and Compatibility   Materials Research Forum LLC
Materials Research Foundations 161 (2024) 245-266   https://doi.org/10.21741/9781644902998-9

Table 1. Influence of nano-particles in correlation with their properties

Property	Influence	References
NPs size and structure configuration	Cytotoxicity of AuNPs and AgNPs increased as their size decreased.	[37]
	TiO2 and carbon black NPs induce more inflammation in lungs than larger particles.	[39]
	Rod-shaped NPs can be more toxic than spherical NPs of the same size.	[40]
Surface properties	Citrate-coated Au and Ag NPs have stronger interactions with BSA.	[42]
	Positively charged NPs are taken up more by cells than negatively charged ones.	[43]
	Positively charged Au NPs are moderately toxic; negatively charged ones are not.	[44]

Moreover, Gieser et al. reported the behavior of NPs that enter the small airways and alveoli. According to their findings, these NPs have the capability to immerse into the lining of alveoli, and it was observed that the reactivity of particle surfaces can lead to modifications in their biological effects [45]. For example, surface medications to SNPs can affect their fibrogenicity, inflammogenicity & cytotoxicity [46]. Liu et al. observed that NPs with a positive charge display a significant capacity to easily cross cell membranes. These positively charged NPs also have a strong affinity for negatively charged DNA molecules, which leads to detrimental outcomes, resulting in elevated toxicity when compared to their negatively charged and neutral counterparts. It was revealed that alterations in the surface chemistry of distinct NP types play a significant role in determining the extent of their toxic effects [47].

4.   Toxicity of silver nanoparticles (Ag NPs)

Several investigations have been conducted to improve the comprehension of the toxicity mechanism of Ag NPs, using both *in-vivo* models & animal cells [48, 9,50]. El Badawy et al. postulated that surface charge-related toxicity of *Bacillius sp.* can be observed with different Ag NPs exhibiting varying surface charges [51]. Wu et al. investigated the toxicity of three differently sized Ag NPs with different surface coatings (polythene glycol-coated Ag NPs [PEG-Ag NPs], silica-incorporated Ag NPs [Si-Ag NPs] and aminated-Si-AgNPs, it was observed that PEG-Ag NPs and Si-Ag NPs and Si-Ag NPs displayed greater toxicity as compared to Ami-Si-Ag NPs in *Chlamydomonas reinhardtii* and *E.colo* [52]. The properties of Ag NPs are influenced by factors i.e., shape, size, synthesis process & accumulation which affects their entrance in cell organelles, including mitochondria [53]. In certain instances, the coating of NPs with polysaccharides has been shown to exacerbate DNA damage when compared to the use of bare Ag NPs [54]. Several theories have been proposed to elaborate the mechanisms of Ag NP toxicity, including physical damage resulting from the association of Ag NPs with cell membrane [55] and dissolution Ag+ ions from Ag NPs resulting in ROS production & inhibition of enzyme. The communication between NPs & biological entities on the cell surface leads to the formation of a surface protein corona, which can be influenced by OS and surface receptors of cells, ultimately affecting the toxic effects of Ag NPs [56].

5. Toxicity of platinum nanoparticles (Pt NPs)

Pt NPs have gained attention as a promising therapeutic agent for cancer treatment, but they are still associated with some undesirable side effects [57]. Sørensen et al. evaluated the toxicity of pt NPs on green microalgae, *P. subcapitata* & found that the toxicity may be attributed to the favored surface binding of pt NPs to the polysaccharide-enriched cell wall of microalgae species [58]. In addition, it was reported that high dosages of pt NPs induced hepatocytes toxicity [59]. Furthermore, in a study by Konieczyn et al., it was observed that pt NPs elicited toxic effects on primary keratinocytes. These toxic effects led to a significant reduction in the cell metabolism process within the keratinocytes. [60].

6. Toxicity of gold nanoparticles (Au NPs)

Modifications of the surface charges of NPs may result in unfavorable interactions with biological systems, leading to altered molecular behavior & cellular toxicity [61]. Research suggests that cationic & anionic Au NPs can be toxic to cells by altering the mitochondrial membrane potential, resulting in OS, which is considered as the major mechanisms responsible for NPs toxicity [62]. Moreover, cationic and ionic surface charge modifications of Au NPs have been shown to stimulate the phagocytosis of lymphoid cell to a greater extent than neutral Au NPs [63]. Schaeublin et al. evaluated the cytotoxicity of 3 different Au NPs: positively charged, negatively charged and neutral. It was observed that all the NPs disrupted cell morphology & induced mitochondrial disturbance. However, DNA impairment was caused by all the NPs regardless of their surface charge. Notably, cationic Au NPs were found to be more cytotoxic compared to NPs *in-vitro* [63]. Furthermore, it is reported that all the Au NPs, whether carboxylated, PEGylated or uncoated induced genotoxicity in cells [64].

7. Toxicity of Copper nanoparticles (Cu NPs)

Cu NPs have been shown to induce various biochemical reactions upon exposure to cells. These reactions involve the oxidation of biomolecules i.e., lipids as well as proteins [63]. The conjugation between Cu NPs & proteins [64], dissolution of Cu NPs [65], photo-oxidation & the generation of ROS [66], as well as the oxidation-reduction cycling between $Cu2+$ & $Cu1+$ [67].

Many of the observed toxic effects and interactions between nanoparticles and biological systems involve the covalent modification of molecules following interaction with Cu NPs or CuO NPs, leading to OS in organisms. Furthermore, the liberation of copper ions from the particle's surface subsequent to their uptake by lysosomes can also trigger cellular stress [68]. Organisms possess intricate defense mechanisms to safeguard against Cu-induced toxicity, employing a range of regulated protein and enzyme-driven processes to maintain optimal homeostasis [69]. Cu is an indispensable metal in living organisms, with specific cell types, such as hepatocytes, adeptly sequestering Cu through binding with metallothionein. However, when facing Cu-induced toxicity, disruptions to homeostasis primarily arise from two well-established pathways: free radical-induced oxidative damage and exposure to excessive Cu ion concentrations. Lethal effects resulting from Cu-induced toxicity include hemolysis, GI tract inflammation, hepatocyte necrosis & altered lipid metabolism [69].

8. Toxicity of quantum dots (QDs)

QDs are semiconductor crystals renowned for their distinctive fluorescent properties, which stem from their exceptionally small particle size ranging between 1 and 10 nm [70]. Conventional Cd-based QDs, commonly referred as heavy metal QDs can release Cd ions into the environment through oxidation processes [71]. The accumulation of Cd in organisms can be important due to its long half-life, lasting approximately 25 to 30 years. Prolonged exposure to Cd can lead to serious health consequences, including an increased risk of cancer and damage to the reproductive system, especially affecting the testicles [72]. To mitigate the elevated toxicity associated with Cd-based QDs, researchers have successfully synthesized alternative QDs with reduced or non-toxic properties, suitable for both industrial and biological applications. Among the noteworthy alternatives are InP-QDs, Ag2S-QDs & carbon-CQDs [72-74]. The airborne concentrations of QDs (QDs) are of concern, particularly in occupational settings and environmental exposure. During printing processes that employ inks containing polyethylene glycol–cadmium telluride-QDs), there is a potential for the release and retention of these QDs in the air [75]. Inhalation represents a significant pathway for QD exposure. Upon inhalation, QD particles of various sizes are deposited at different locations within the respiratory system. NPs smaller than 100 nm tend to aggregate in the distal respiratory tract [76].

8.1 QDs instigated Pulmonary toxicity (PT)

Cd-based QDs have been found to induce toxic influence on the reduced cell viability, cell morphology, & disruption of integrity of cell membrane in pulmonary epithelial cells and alveolar macrophages. Cells serve as the fundamental structural and functional units of an organism. The observed morphological alterations in pulmonary cells provide evidence of the PT associated with QDs [77,78]. Cd-based QDs (QDs) can lead to granulomatous reactions & inflammation in the lungs [79]. The distribution of cadmium (Cd) in the pulmonary region is attributed to the accumulation of Cd-based QDs in the bronchioles [80].

Furthermore, Indium phosphide (InP)-QDs possess several properties i.e., large wide color tunability, absorption coefficients, high carrier mobility & low toxicity. These characteristics make InP-QDs suitable for various applications such as traffic lights, indoor & outdoor lighting as well as LED displays [81]. Pathologically, mice treated with InP-NH2 QDs exhibit notable congested alveolar septa. Extended congestion of these septa can result in the deterioration of elastic fibers, leading to reduced resilience, enlarged alveolar septa along with impaired pulmonary ventilation [82].

9.   Toxicity of zinc nano-particles (Zn NPs)

In an investigation carried out by De Souza et al. elucidated that a notable observation was made wherein mice were subjected to intraperitoneal treatment with Zn NPs (Zinc Nanoparticles) at two specific concentrations: an environmentally relevant dosage of 5.625×10^{-5} mg/kg, and a high concentration of 300 mg/kg. The outcome of this study revealed a significant augmentation in the accumulation of zinc (Zn) within the brain tissues of the treated mice [83]. ZnO NPs may be able to cross the BBB and translocate into the CNS through the circulatory system. Previous studies have also demonstrated the translocation of ZnO NPs into the CNS via the olfactory or taste nerve pathways. Rats and mice exposed to ZnO NPs through oral and intranasal instillation showed significant upregulation of Zn content in brain tissues [84,85]. The deposition of ZnO nanoparticles in the brain may lead to neurotoxicity through the stimulation of ROS production, triggering an inflammatory response & causing change in both ultrastructure and histopathology. Ultimately, these effects can result in neurodegeneration & functional impairments [86,87]. The CNS is considered more vulnerable to the adverse effects induced by ZnO NPs due to its high content of unsaturated lipids, low concentration of antioxidant enzymes along with limited cellular regeneration capacity in neural tissue [86,87,88]. Research has demonstrated that the oral administration of ZnO NPs in rats can elevate oxidative & nitrosative stresses, which are recognized to trigger for inflammation in the CNS, which activate glial cells, namely astrocytes and microglia, resulting in the release of pro-inflammatory mediators [89]. Increased levels of inflammatory cytokines have been reported in animals treated with ZnO NPs [90]. An upsurge in IL-1β levels in the brain has been associated with the emergence of acute neuronal lesions and neuronal cell death [91]. The activation and accumulation of astrocytes in the CNS, as evidenced by the upregulation of glial fibrillary acidic protein (GFAP), can impede axonal regeneration & growth, consequently contributing to neurodegeneration [92]. Moreover, it has been demonstrated that elevated concentrations of ZnO NPs can inhibit autophagy, leading to the formation of tau-protein aggregates in the brain. The accumulation of tau in the brain is believed to be indicative of axonal damage in brain tissue [93]. The neurotoxicity of ZnO NPs has also been evaluated by examining behavioral changes and cognitive function in different animal models. Mice exposed to ZnO NPs exhibited deficits in motor functions, as assessed by beam balance, pole, and footprint tests [94]. Similar effects on neuromuscular coordination and impaired crawling ability were observed in Drosophila melanogaster following oral administration of ZnO NPs [95]. Rats treated with ZnO NPs showed impaired learning and memory. The expression of memory & learning-related genes was found to be altered in these rats [95].

10.   Hepatotoxicity

The toxicity of NPs remains a major concern, especially in the hepatic system, which plays a critical function in the metabolism & elimination of xenobiotics from the body. The liver is also highly vascularized, rendering it susceptible to NPs accumulation due to its significant phagocytic activity [96]. Toxicity of NPs on liver cells is a multifactorial process, involving several mechanisms and several subcellular damages. Metallic NPs i.e., Ag, Au & Cu have been shown to induce subcellular damages in liver cells including mitochondrial dysfunctions, OS and inflammatory damages. These damages can ultimately result in cell death, fibrosis as well as hepatic cancer [97].

Mechanistically, the toxicity of metallic NPs on liver cells involves the generation of ROS, which can cause oxidative damages to cellular molecules, leading to DNA damage and cell death. Additionally, NPs can interfere with intracellular signaling pathways such as those involved in cell cycles regulation and apoptosis. Furthermore, NPs can activate pro-inflammatory cytokines pathways, leading to the initiation of an inflammatory cascade and subsequent damage [98].

NPs can also affect the activity of liver enzymes such as cytochrome P450 enzymes which are responsible for the metabolism and detoxification of foreign substances [99]. The toxicity of NPs on liver cells is further complicated by the heterogeneity of liver cell types i.e., Kupffer cells, hepatocytes & stellate cells. Different types of NPs can have distinct cellular uptake patterns and toxicological effects depending on the liver cell types [100]. For instance, Au NPs have been shown to induce mitochondrial damage & OS in hepatocytes, while Ag NPs can activate Kupffer cells and promote the release of pro-inflammatory cytokines [101] Fig.3.

11. Nephrotoxicity

Carbon-based NPs, such as carbon nanotubes and fullerenes, induce OS in renal cells by generating ROS and diminishing cellular antioxidants [102,103]. ROS are highly reactive molecules that can damage cellular components i.e., lipids, proteins & DNA [104]. In response to ROS, cell activate several signaling pathways i.e., NF-kB pathway, resulting in inflammatory injuries [105]. Metal-based NPs such as Ag, Au and titanium dioxide NPs induce OS, inflammation, genotoxicity and cytotoxicity in renal cells. The toxicity mechanisms of metal-based NPs involve the generation of ROS and induction of damage to the DNA leading to genotoxicity [106]. ROS generated by NPs can initiate a cascade of reactions that can result in oxidative damage in kidney involving DNA, lipids and protein dysfunction [104]. This damage can activate the NF-kB pathway and promote the expression of pro-inflammatory cytokines leading to inflammatory response in kidney [107]. Additionally, metal-based NPs can disrupt cellular membrane, leading to cytotoxicity [108].

Semiconductor-based NPs i.e., QDs, induce OS & genotoxicity in renal cells. Polymeric NPs and QDs could induce cytotoxicity in renal cells. The toxicity mechanism of polymeric NPs involves the disruption of cellular membrane and induction of apoptosis and inflammatory response via escalating the level of inflammatory and apoptotic proteins [109] Fig.3.

12. Reproductive toxicity

NPs are reported to cause reproductive toxicity [110]. Studies have demonstrated that NPs can adversely impact the reproductive system of mice *in-vivo* as well as *in-vitro*. On a cellular level, NPs can disrupt sperm activity, morphology, quality and quantity leading to infertility [110]. NPs have been observed to interfere with primary and secondary follicle development in female reproductive cells, leading to disruptions in cell arrangement and the formation of shapeless follicular antrum [111]. Furthermore, Ag NPs can affect Leydig cell proliferation, viability & gene expression [112] Table 2.

Table 2. Influence of nano-particles on reproductive system

Aspect	Effects	References
Sperm activity	Disruption of activity, morphology, and quality	[110]
Ovary & Follicle Development	Interference with primary and secondary follicle dev.	[111]
Leydig Cell	Impact on proliferation, viability, and gene expression	[112]

The reproductive toxicity of NPs is dependent on the dose, size and core structure as well as surface chemistry of the particles. NPs can accumulate in the ovary & testis, causing a decrease in testis and epididymis weight & changes in seminiferous tubule morphology [113,114]. Although changes in sexual behavior were not observed in male mice, a study reported a reduction in the number of regular estrous cycles in female mice when exposed to certain factors. However, the precise mechanisms responsible for these alterations in female reproductive cycles remain unclear [115].

OS-instigated apoptosis, caused by NP exposure is believed to be the major factor responsible for the reproductive & developmental toxicity of NPs. Studies have shown that NPs exposure increases the levels of glutathione, MDA & ROS indicating OS eventually triggering apoptotic and inflammatory response [115] Fig.3.

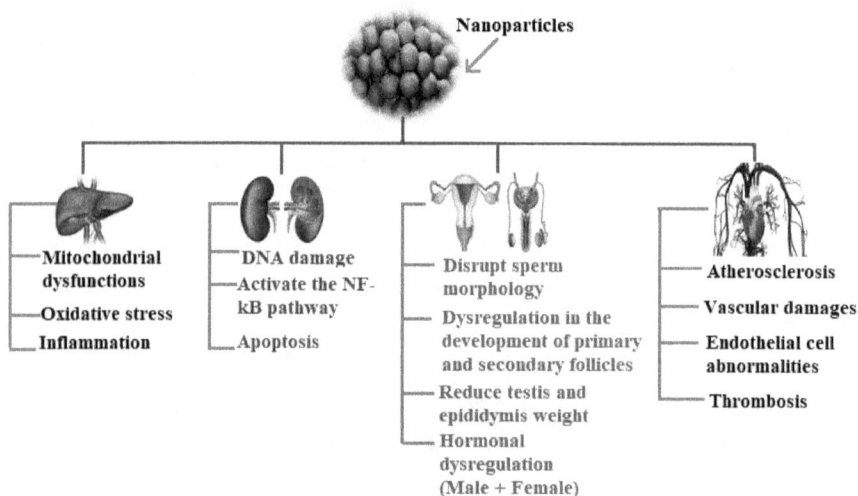

Fig. 3 Toxicological effects of NPs on liver, kidney, cardiovascular and reproductive system.

13. Cardiovascular toxicity

Exposure to a variety of engineered NPs has been demonstrated to instigate cardiovascular events as evidenced by *in-vitro* and *in-vivo* studies. These events include cardiac damage and dysfunction, abnormal angiogenesis, platelet activation, endothelial cell abnormalities, vascular damages, atherosclerosis & thrombosis [116-118]. These mechanism by which NP exposure induces cardiovascular toxicity are complex & multifunctional. The process of NP-triggered cardiovascular toxicity has been associated with OS, inflammatory responses, neural regulation & activation of ion channels. NPs hold potential for use in medical drug delivery systems, imaging & cell/organ-specific therapies within the cardiovascular system, current research on their cardiovascular toxicity is predominantly limited to a few experimental animal or cell culture model [119-121] Fig.3.

Conclusion

In conclusion, this chapter provides an overview of the toxicity of NPs on biological systems, particularly on liver, kidney reproductive as well as cardiovascular system. The unique physiochemical properties of NPs make them valuable in different fields, but their interactions with biological components can lead to adverse effects on organ, tissue as well as cellular level with major toxic effects such as inflammation, apoptosis as well as mechanism behind these toxicological damages including generation of ROS. Factors such as size, shape, surface charge, concentration, mode of administration and type of biological system being exposed influence NPs toxicity.

References

[1] Nowrouzi A, Meghrazi K, Golmohammadi T, Golestani A, Ahmadian S, Shafiezadeh M, Shajary Z, Khaghani S, Amiri AN. Cytotoxicity of subtoxic AgNP in human hepatoma cell line (HepG2) after long-term exposure. Iranian biomedical journal 2010; 14(1-2): 23-32.

[2] De Jong WH, Borm PJ. Drug delivery and nanoparticles: applications and hazards. International journal of nanomedicine 2008; 3(2): 133-149. https://doi.org/10.2147/IJN.S596

[3] Lewinski N, Colvin V, Drezek R. Cytotoxicity of nanoparticles. Small 2008; 4(1): 26-49. https://doi.org/10.1002/smll.200700595

[4] Farokhzad OC, Langer R. Nanomedicine: Developing smarter therapeutic and diagnostic modalities. Adv Drug Deliv Rev 2006;58(14):1456-9. https://doi.org/10.1016/j.addr.2006.09.011

[5] Roco MC. Nanotechnology: Convergence with modern biology and medicine. Curr Opin Biotechnol 2003;14(3):337-46. https://doi.org/10.1016/S0958-1669(03)00068-5

[6] Caruthers SD, Wickline SA, Lanza GM. Nanotechnological applications in medicine. Curr Opin Biotechnol 2007;18(1):26-30. doi: 10.1016/j.copbio.2007.01.006 https://doi.org/10.1016/j.copbio.2007.01.006

[7] Silva GA. Introduction to nanotechnology and its applications to medicine. Surg Neurol 2004;61(3):216-20. https://doi.org/10.1016/j.surneu.2003.09.036

[8] Singh M, Singh S, Prasad S, Gambhir IS. Nanotechnology in medicine and antibacterial effect of silver nanoparticles. Dig J Nanomater Bios 2008;3(3):115-22.

[9. Nie S, Xing Y, Kim GJ, Simons JW. Nanotechnology applications in cancer. Annu Rev Biomed Eng 2007;9:257-88. https://doi.org/10.1146/annurev.bioeng.9.060906.152025

[10] Rietmeijer F, Mackinnon I. Bismuth oxide nanoparticles in the stratosphere. J Geophys Res 1997;102:6621-7. https://doi.org/10.1029/96JE03989

[11] Cyrys J, Stölzel M, Heinrich J, Kreyling WG, Menzel N, Wittmaack K, Tuch T, Wichmann HE. Elemental composition and sources of fine and ultrafine ambient particles in Erfurt, Germany. Sci Total Environ 2003;305:143-56. https://doi.org/10.1016/S0048-9697(02)00494-1

[12] Hughes LS, Cass GR, Gone J, Ames M, Olmez I. Physical and chemical characterization of atmospheric ultrafine particles in the Los Angeles area. Environ Sci Technol 1998;32:1153-61. https://doi.org/10.1021/es970280r

[13] Siegmann P, Acevedo FJ, Siegmann K, Maldonado-Bascón S. A probabilistic source attribution model for nanoparticles in air suspension applied on the main roads of Madrid and Mexico City. Atmos Environ 2008;42:3937-48. https://doi.org/10.1016/j.atmosenv.2007.05.021

[14] Beduneau A, Ma Z, Grotepas CB, Kabanov A, Rabinow BE, Gong N, Mosley RL, Dou H, Boska MD, Gendelman HE. Facilitated monocyte-macrophage uptake and tissue distribution of superparmagnetic iron-oxide nanoparticles. PLoS One 2009;4:e4343. https://doi.org/10.1371/journal.pone.0004343

[15] Oberdörster G, Oberdörster E, Oberdörster J. Nanotoxicology: an emerging discipline evolving from studies of ultrafine particles. Environ Health Perspect 2005;113:823-39. https://doi.org/10.1289/ehp.7339

[16] Shi, Z., Shao, Longyi, 2005. Microscopy and mineralogy of airborne particles collected during severe dust storm episodes in Beijing, China. J. Geophys. Res. 110, D01303. https://doi.org/10.1029/2004JD005073

[17] Yano, E., Yokoyama, Y., Higashi, H., Nishii, S., Maeda, K., Koizumi, A., 1990. Health effects of volcanic ash: a repeat study. Arch. Environ. Health 45, 367-373. https://doi.org/10.1080/00039896.1990.10118757

[18] Taylor, D.A., 2002. Dust in the wind. Environ. Health Perspect. 110, A80-A87. https://doi.org/10.1289/ehp.110-a80

[19] Sapkota, A., Symons, J.M., Kleissl, J., Wang, L., Parlange, M.B., Ondov, J., Breysse, P.N., Diette, G.B., Eggleston, P.A., Buckley, T.J., 2005. Impact of the 2002 Canadian Forest fires on particulate matter air quality in Baltimore City. Environ. Sci. Technol. 39, 24-32. https://doi.org/10.1021/es035311z

[20] Buzea, C., Pacheco, I.I., Robbie, K., 2007. Nanomaterials and nanoparticles: sources and toxicity. Biointerphases 2. https://doi.org/10.1116/1.2815690

[21] Ahamed M, Karns M, Goodson M, Rowe J, Hussain SM, Schlager JJ, et al. DNA damage response to different surface chemistry of silver nanoparticles in mammalian cells. Toxicol Appl Pharmacol 2008;233(3):404-10. https://doi.org/10.1016/j.taap.2008.09.015

[22] Sharma V, Shukla RK, Saxena N, Parmar D, Das M, Dhawan A. DNA damaging potential of zinc oxide nanoparticles in human epidermal cells. Toxicol Lett 2009;185(3):211-8. https://doi.org/10.1016/j.toxlet.2009.01.008

[23] Wang F, Yu L, Monopoli MP, Sandin P, Mahon E, Salvati A, et al. The biomolecular corona is retained during nanoparticle uptake and protects the cells from the damage induced by cationic nanoparticles until degraded in the lysosomes. Nanomedicine 2013;9(8):1159-68. https://doi.org/10.1016/j.nano.2013.04.010

[24] Elsaesser A, Howard CV. Toxicology of nanoparticles. Adv Drug Deliv Rev 2012;64(2):129-37. https://doi.org/10.1016/j.addr.2011.09.001

[25] Slivka S, Landeen L, Zeigler F, Zimber M, Bartel R (1993) Characterization, barrier function, and drug metabolism of an in vitro skin model. J Invest Dermatol 100:40-46 https://doi.org/10.1111/1523-1747.ep12354098

[26] Brigger I, Dubernet C, Couvreur P (2002) Nanoparticles in cancer therapy and diagnosis. Adv Drug Deliv Rev 54:631-651 https://doi.org/10.1016/S0169-409X(02)00044-3

[27] Pattan G, Kaul G (2012) Health hazards associated with nanomaterials. Toxicol Ind Health 30:499-519. https://doi.org/10.1177/0748233712459900

[28] De Jong WH, Hagens WI, Krystek P, Burger MC, Sips AJ, Geertsma RE. Particle size-dependent organ distribution of gold nanoparticles after intravenous administration. Biomaterials 2008;29:1912-9. https://doi.org/10.1016/j.biomaterials.2007.12.037

[29] Akerman ME, Chan WC, Laakkonen P, Bhatia SN, Ruoslahti E. Nanocrystal argeting in vivo. PNAS 2002;99:12617-21 necessary. https://doi.org/10.1073/pnas.152463399

[30] Ballou B, Lagerholm BC, Ernst LA, Bruchez MP, Waggoner AS. Non-invasive imaging of quantum dots in mice. Bioconjugate Chem 2004;15:79-86. https://doi.org/10.1021/bc034153y

[31] Cagle DW, Kenmnel SJ, Mirzadeh S, Alford JM, Wilson LJ. In vivo studies of fullerene-based materials using endohedral metallofullerene radiotracers. PNAS 1999;96:5182-7. https://doi.org/10.1073/pnas.96.9.5182

[32] Hirst SM, Karakoti A, Singh S, Self W, Tyler R, Seal S, Reilly CM. Bio-distribution and in vivo antioxidant effects of cerium oxide nanoparticles in mice. Environ Toxicol 2013;28:107-18. https://doi.org/10.1002/tox.20704

[33] Sonavane G, Tomoda K, Makino K. Biodistribution of colloidal gold nanoparticles after intravenous administration: effect of particle size. Colloids Surf B 2008;66:274-80. https://doi.org/10.1016/j.colsurfb.2008.07.004

[34] Xie G, Jiao Sun, Gaoren Zhong, Liyi Shi, Dawei Zhang. Biodistribution and toxicity of intravenously administered silica nanoparticles in mice. Arch Toxicol 2010;84:183-90. https://doi.org/10.1007/s00204-009-0488-x

[35] Roy, R., Kumar, S., Tripathi, A., Das, M. and Dwivedi, P.D., 2014. Interactive threats of nanoparticles to the biological system. Immunology letters, 158(1-2), pp.79-87. https://doi.org/10.1016/j.imlet.2013.11.019

[36] Amato I. Making the right stuff. Sci News 1989;136:108-10. https://doi.org/10.2307/3973729

[37] Yen HJ, Hsu SH, Tsai CL. Cytotoxicity and immunological response of gold and silver nanoparticles of different sizes. Small 2009;5:1553-61. https://doi.org/10.1002/smll.200900126

[38] Pan Y, Neuss S, Leifert A, Fischler M, Wen F, Simon U, Schmid G, Brandau W, Jahnen-Dechent W. Size-dependent cytotoxicity of gold nanoparticles. Small 2007;3:1941-9. https://doi.org/10.1002/smll.200700378

[39] Hsiao IL, Huang YJ. Effects of various physicochemical characteristics on the toxicities of ZnO and TiO2 nanoparticles toward human lung epithelial cells. Sci Total Environ 2011;409:1219-28. https://doi.org/10.1016/j.scitotenv.2010.12.033

[40] Wilson MR, Lightbody JH, Donaldson K, Sales J, Stone V. Interactions Between ultrafine particles and transition metals in vivo and in vitro. Toxicol Appl Pharmacol 2002;184:172-279. https://doi.org/10.1006/taap.2002.9501

[41] Schellenberger EA, Reynolds F, Weissleder R, Josephson L. Surfacefunctionalized nanoparticle library yields probes for apoptotic cells. ChemBioChem 2004;5:275-9. https://doi.org/10.1002/cbic.200300713

[42] Treuel L, Malissek M, Gebauer JS, Zellner R. The influence of surface composition of nanoparticles on their interactions with serum albumin. ChemPhysChem 2010;11:3093-9. https://doi.org/10.1002/cphc.201000174

[43] Chen HW, Su SF, Chien CT, Lin WH, Yu SL, Chou CC, Chen JJ, Yang PC. Titanium dioxide nanoparticles induce emphysema-like lung injury in mice. FASEB J 2006;20:2393-5. https://doi.org/10.1096/fj.06-6485fje

[44] Goodman CM, McCusker CD, Yilmaz T, Rotello VM. Toxicity of gold nanoparticles functionalized with cationic and anionic side chains. Bioconjugate Chem 2004;15:897-900. https://doi.org/10.1021/bc049951i

[45] Geiser, M., Schurch, S., Gehr, P., 2003. Influence of surface chemistry and topography of particles on their immersion into the lung's surface-lining layer. J. Appl. Physiol. 94, 1793-1801. https://doi.org/10.1152/japplphysiol.00514.2002

[46] Schins, R.P., Duffin, R., Höhr, D., Knaapen, A.M., Shi, T., Weishaupt, C., Stone, V., Donaldson, K., Borm, P.J., 2002. Surface modification of quartz inhibits toxicity, particle uptake, and oxidative DNA damage in human lung epithelial cells. Chem. Res. Toxicol. 15, 1166-1173. https://doi.org/10.1021/tx025558u

[47] Liu, Y., Li, W., Lao, F., Liu, Y., Wang, L., Bai, R., Zhao, Y., Chen, C., 2011. Intracellular dynamics of cationic and anionic polystyrene nanoparticles without direct interaction with mitotic spindle and chromosomes. Biomaterials 32, 8291-8303. https://doi.org/10.1016/j.biomaterials.2011.07.037

[48] Hernández-Sierra, J.F., Ruiz, F., Pena, D.C., Martínez-Gutiérrez, F., Martínez, A.E., Guillén, A.D., Tapia-Pérez, H., Castañón, G.M., 2008. The antimicrobial sensitivity of Streptococcus mutans to nanoparticles of silver, zinc oxide, and gold. Nanomedicine 4, 237-240. https://doi.org/10.1016/j.nano.2008.04.005

[49] Ahamed, M., Karns, M., Goodson, M., Rowe, J., Hussain, S.M., Schlager, J.J., Hong, Y., 2008. DNA damage response to different surface chemistry of silver nanoparticles in mammalian cells. Toxicol. Appl. Pharmacol. 233, 404-410. https://doi.org/10.1016/j.taap.2008.09.015

[50] Kawata, K., Osawa, M., Okabe, S., 2009. In-vitro toxicity of silver nanoparticles at noncytotoxic doses to HepG2 human hepatoma cells. Environ. Sci. Technol. 43, 6046-6051. https://doi.org/10.1021/es900754q

[51] El Badawy, A.M., Silva, R.G., Morris, B., Scheckel, K.G., Suidan, M.T., Tolaymat, T.M., 2011. Surface charge-dependent toxicity of silver nanoparticles. Environ. Sci. Technol. 245, 283-287. https://doi.org/10.1021/es1034188

[52] Wu, F., Harper, B.J., Harper, S.L., 2017. Differential dissolution and toxicity of surface functionalized silver nanoparticles in small-scale microcosms: impacts of community complexity. Environ. Sci. Nano. 4, 359-372. https://doi.org/10.1039/C6EN00324A

[53] Aktera, M., Sikder, M.T., Rahman, M.M., Ullah, A.A., Hossain, K.F., Banik, S., Hosokawa, T., Saito, T., Kurasaki, M., 2018. A systematic review on silver nanoparticles-induced cytotoxicity: physicochemical properties and perspectives. J. Adv. Res. 9, 1-16. https://doi.org/10.1016/j.jare.2017.10.008

[54] Zhang, T., Wang, L., Chen, Q.C., 2014. Chen Cytotoxic potential of silver nanoparticles. Yonsei Med. J. 55, 283-291. https://doi.org/10.3349/ymj.2014.55.2.283

[55] Fabrega, J., Fawcett, S.R., Renshaw, J.C., Lead, J.R., 2009. Silver nanoparticle impact on bacterial growth: effect of pH, concentration, and organic matter. Environ. Sci. Technol. 43, 7285-7290. https://doi.org/10.1021/es803259g

[56] Navarro, E., Piccapietra, F., Wagner, B., Marconi, F., Kaegi, R., Odzak, N., Sigg, L., Behra, R., 2008. Toxicity of silver nanoparticles to Chlamydomonas reinhardtii. Environ. Sci. Technol. 42, 8959-8964. https://doi.org/10.1021/es801785m

[57] Azmi, M.A., Shad, K.F., 2017. Role of nanostructure molecules in enhancing the bioavailability of oral drugs. In: Nanostructures for Novel Therapy. pp. 375-407. https://doi.org/10.1016/B978-0-323-46142-9.00014-1

[58] Sørensen, S.N., Engelbrekt, C., Lützhøft, H.C.H., Jiménez-Lamana, J., Noori, J.S., Alatraktchi, F.A., Delgado, C.G., Slaveykova, V.I., Baun, A., 2016. A multimethod approach for investigating algal toxicity of platinum nanoparticles. Environ. Sci. Technol. 50, 10635-10643. https://doi.org/10.1021/acs.est.6b01072

[59] Labrador-Rached, C.J., Browning, R.T., Braydich-Stolle, L.K., Comfort, K.K., 2018. Toxicological implications of platinum nanoparticle exposure: stimulation of intracellular stress, inflammatory response, and akt signaling in-vitro. J. Toxicol. 1367801. https://doi.org/10.1155/2018/1367801

[60] Konieczny, P., Goralczyk, A.G., Szmyd, R., Skalniak, L., Koziel, J., Filon, F.L., Jura, J., 2013. Effects triggered by platinum nanoparticles on primary keratinocytes. Int. J. Nanomed. 8, 3963. https://doi.org/10.2147/IJN.S49612

[61] Malugin, A., Ghandehari, H., 2010. Cellular uptake and toxicity of gold nanoparticles in prostate cancer cells: a comparative study of rods and spheres. J. Appl. Toxicol. 30, 212-217. https://doi.org/10.1002/jat.1486

[62] De, M., Rotello, V.M., 2008. Synthetic "chaperones": nanoparticle-mediated refolding of thermally denatured proteins. Chem. Commun. (J. Chem. Soc. Sect. D) 30, 3504-3506. https://doi.org/10.1039/b805242e

[63] Sayes, C., A. Gobin, K. Ausman, J. Mendez, J. West and V. Colvin (2005). "Nano-C-60 cytotoxicity is due to lipid peroxidation." Biomaterials 26(36): 7587-7595. https://doi.org/10.1016/j.biomaterials.2005.05.027

[64] Deng, Z. J., G. Mortimer, T. Schiller, A. Musumeci, D. Martin and R. F. Minchin (2009). "Differential plasma protein binding to metal oxide nanoparticles." Nanotechnology 20(45): 455101. https://doi.org/10.1088/0957-4484/20/45/455101

[65] Auffan, M., J. Rose, M. R. Wiesner and J. Y. Bottero (2009). "Chemical stability of metallic nanoparticles: a parameter controlling their potential cellular toxicity in vitro." Environ Pollut 157(4): 1127-1133 https://doi.org/10.1016/j.envpol.2008.10.002

[66] Saison, C., F. Perreault, J. Daigle, C. Fortin, J. Claverie, M. Morin and R. Popovic (2010). "Effect of core-shell copper oxide nanoparticles on cell culture morphology and photosynthesis (photosystem II energy distribution) in the green alga, Chlamydomonas reinhardtii." Aquatic Toxicology 96(2): 109-114. https://doi.org/10.1016/j.aquatox.2009.10.002

[67] Hu, W., S. Culloty, G. Darmody, S. Lynch, J. Davenport, S. Ramirez-Garcia, K. Dawson, I. Lynch, J. Blasco and D. Sheehan (2014). "Toxicity of copper oxide nanoparticles in the blue mussel, Mytilus edulis: A redox proteomic investigation." Chemosphere 108: 289-299. https://doi.org/10.1016/j.chemosphere.2014.01.054

[68] Griffitt, R., R. Weil, K. Hyndman, N. Denslow, K. Powers, D. Taylor and D. Barber (2007). "Exposure to copper nanoparticles causes gill injury and acute lethality in zebrafish (Danio rerio)." Environmental Science & Technology 41(23): 8178-8186. https://doi.org/10.1021/es071235e

[69] Gaetke, L., H. Chow-Johnson and C. Chow (2014). "Copper: toxicological relevance and mechanisms." Archives of Toxicology 88(11): 1929-1938. https://doi.org/10.1007/s00204-014-1355-y

[70] C. Matea, T. Mocan, F. Tabaran, T. Pop, O. Mosteanu, C. Puia, C. Iancu, L. Mocan, Quantum dots in imaging, drug delivery and sensor applications, Int. J. Nanomed. 12 (2017) 5421-5431. https://doi.org/10.2147/IJN.S138624

[71] Marissa S. Giroux, Zahra Zahra, Omobayo A. Salawu, Robert M. Burgess, K.T. Ho, A.S. Adeleye, Assessing the environmental effects related to quantum dot structure, function, synthesis and exposure, ENVIRON SCI-NANO 9 (3) (2022) 867-910. https://doi.org/10.1039/D1EN00712B

[72] A. Goutam Mukherjee, U. Ramesh Wanjari, K. Renu, B. Vellingiri, A. Valsala Gopalakrishnan, Heavy metal and metalloid - induced reproductive toxicity, ENVIRON TOXICOL PHAR 92 (2022), 103859. https://doi.org/10.1016/j.etap.2022.103859

[73] G. Genchi, M.S. Sinicropi, G. Lauria, A. Carocci, A. Catalano, The effects of cadmium toxicity, Int. J. Environ. Res. Publ. Health 17 (11) (2020) 3782. https://doi.org/10.3390/ijerph17113782

[74] M.J. Molaei, Carbon quantum dots and their biomedical and therapeutic applications: a review, RSC Adv. 9 (12) (2019) 646-6481. https://doi.org/10.1039/C8RA08088G

[75] M. Bl'azquez S'anchez, I. Nelissen, V. Pomar-Portillo, A. Vílchez, J. Van Laer, A. Jacobs, E. Frijns, S. V'azquez-Campos, E. Fernandez-Rosas, Release and cytotoxicity screening of the

printer emissions of a CdTe quantum dots-based fluorescent ink, Toxicol. Lett. 347 (2021) 1-11. https://doi.org/10.1016/j.toxlet.2021.04.009

[76] K. B'eruB'e, D. Balharry, K. Sexton, L. Koshy, T. Jones, COMBUSTION-DERIVED nanoparticles: mechanisms of pulmonary toxicity, Clin. Exp. Pharmacol. P 34 (10) (2007) 1044-1050. https://doi.org/10.1111/j.1440-1681.2007.04733.x

[77] A.K. Jigyasu, S. Siddiqui, M. Lohani, I.A. Khan, M. Arshad, Chemically synthesized CdSe quantum dots inhibit growth of human lung carcinoma cells via ROS generation, EXCLI J 15 (2016) 54-63.

[78] K.G. Li, J.T. Chen, S.S. Bai, X. Wen, S.Y. Song, Q. Yu, J. Li, Y.Q. Wang, Intracellular oxidative stress and cadmium ions release induce cytotoxicity of unmodified cadmium sulfide quantum dots, Toxicol. Vitro 23 (6) (2009) 1007-1013. https://doi.org/10.1016/j.tiv.2009.06.020

[79] J.R. Roberts, J.M. Antonini, D.W. Porter, R.S. Chapman, J.F. Scabilloni, S. H. Young, D. Schwegler-Berry, V. Castranova, R.R. Mercer, Lung toxicity and biodistribution of Cd/Se-ZnS quantum dots with different surface functional groups after pulmonary exposure in rats, Part. Fibre Toxicol. 10 (2013) 5. https://doi.org/10.1186/1743-8977-10-5

[80] Y. Hsieh, H. Hsieh, H. Hsieh, T. Wang, C. Ho, P. Lin, C. Wang, Using laser ablation inductively coupled plasma mass spectrometry to characterize the biointeractions of inhaled CdSe quantum dots in the mouse lungs, J ANAL ATOM SPECTROM 28 (9) (2013) 1141-1396. https://doi.org/10.1039/c3ja50063b

[81] B. Chen, D. Li, F. Wang, InP quantum dots: synthesis and lighting applications, Small 16 (32) (2020), 2002454. https://doi.org/10.1002/smll.202002454

[82] G. Lin, T. Chen, Y. Pan, Z. Yang, L. Li, K. Yong, X. Wang, J. Wang, Y. Chen, W. Jiang, S. Weng, X. Huang, J. Kuang, G. Xu, Biodistribution and acute toxicity of cadmium-free quantum dots with different surface functional groups in mice following intratracheal inhalation, Nanotheranostics 4 (3) (2020) 173-183. https://doi.org/10.7150/ntno.42786

[83] de Souza, J. M., de Mendes, B. O., Guimarães, A. T. B., de Rodrigues, A. S. L., Chagas, T. Q., Rocha, T. L., & Malafaia, G. (2018). Zinc oxide nanoparticles in predicted environmentally relevant concentrations leading to behavioral impairments in male swiss mice. Science of the Total Environment, 613-614, 653-662. https://doi.org/10.1016/j.scitotenv.2017.09.051

[84] Attia, H., Nounou, H., & Shalaby, M. (2018). Zinc oxide nanoparticles induced oxidative DNA damage, inflammation and apoptosis in rat's brain after oral exposure. Toxics, 6(2). https://doi.org/10.3390/toxics6020029

[85] Aijie, C., Huimin, L., Jia, L., Lingling, O., Limin, W., Junrong, W., et al. (2017). Central neurotoxicity induced by the instillation of ZnO and TiO nanoparticles through the taste nerve pathway. Nanomedicine, 12(20), 2453-2470. https://doi.org/10.2217/nnm-2017-0171

[86] Salim, S. (2017). Oxidative stress and the central nervous system. Journal of Pharmacology and Experimental Therapeutics. https://doi.org/10.1124/jpet.116.237503

[87] Sayre, L. M., Perry, G., & Smith, M. A. (2008). Oxidative stress and neurotoxicity. Chemical Research in Toxicology. https://doi.org/10.1021/tx700210j

[88] Ortiz, G. G., González-Usigli, H., Pacheco-Moisés, F. P., Mireles-Ramírez, M. A., Sánchez-López, A. L., Torres-Sánchez, E. D., et al. (2017). Physiology and pathology of neuroimmunology: role of inflammation in Parkinson's disease. In Physiology and Pathology of Immunology. https://doi.org/10.5772/intechopen.70377

[89] Attia, H., Nounou, H., & Shalaby, M. (2018). Zinc oxide nanoparticles induced oxidative DNA damage, inflammation and apoptosis in rat's brain after oral exposure. Toxics, 6(2). https://doi.org/10.3390/toxics6020029

[90] Liu, H., Yang, H., Fang, Y., Li, K., Tian, L., Liu, X., et al. (2020). Neurotoxicity and biomarkers of zinc oxide nanoparticles in main functional brain regions and dopaminergic neurons. Science of the Total Environment, 705. https://doi.org/10.1016/j.scitotenv.2019.135809

[91] Fogal, B., & Hewett, S. J. (2008). Interleukin-1β: A bridge between inflammation and excitotoxicity? Journal of Neurochemistry. https://doi.org/10.1111/j.1471-4159.2008.05315.x

[92] Giovannoni, G. (2014). Cerebrospinal fluid analysis. In Handbook of Clinical Neurology. https://doi.org/10.1016/B978-0-444-52001-2.00029-7

[93] Llorens, F., Villar-Piqué, A., Candelise, N., Ferrer, I., & Zerr, I. (2019). Tau protein as a biological fluid biomarker in neurodegenerative dementias. In Cognitive Disorders. https://doi.org/10.5772/intechopen.73528

[94] Yaqub, A., Faheem, I., Anjum, K. M., Ditta, S. A., Yousaf, M. Z., Tanvir, F., & Raza, C. (2020). Neurotoxicity of ZnO nanoparticles and associated motor function deficits in mice. Applied Nanoscience (Switzerland), 10(1), 177-185. https://doi.org/10.1007/s13204-019-01093-3

[95] Sood, K., Kaur, J., & Khatri, M. (2017). comparative neurotoxicity evaluation of zinc oxide nanoparticles by crawling assay on Drosophila melanogaster. International Journal of Engineering Technology Science and Research, 4(4), 440-444.

[96] Schaeublin, N.M., Braydich-Stolle, L.K., Schrand, A.M., Miller, J.M., Hutchison, J., Schlager, J.J., Hussain, S.M., 2011. Surface charge of gold nanoparticles mediates mechanism of toxicity. Nanoscale 3, 410-420. https://doi.org/10.1039/c0nr00478b

[97] Vales, G., Suhonen, S., Siivola, K.M., Savolainen, K.M., Catalán, J., Norppa, H., 2020. Genotoxicity and cytotoxicity of gold nanoparticles in-vitro: role of surface functionalization and particle size. Nanomaterials 10, 271. https://doi.org/10.3390/nano10020271

[98. Patel, J. and Patel, A., 2015. Toxicity of Nanomaterials on the Liver, Kidney, and Spleen. Biointeractions of Nanomaterials, 1, pp.286-306.

[99] Boey, A. and Ho, H.K., 2020. All roads lead to the liver: metal nanoparticles and their implications for liver health. Small, 16(21), p.2000153. https://doi.org/10.1002/smll.202000153

[100] Yao, Y., Zang, Y., Qu, J., Tang, M. and Zhang, T., 2019. The toxicity of metallic nanoparticles on liver: the subcellular damages, mechanisms, and outcomes. International journal of nanomedicine, pp.8787-8804. https://doi.org/10.2147/IJN.S212907

[101] Pan, Y., Ong, C.E., Pung, Y.F. and Chieng, J.Y., 2019. The current understanding of the interactions between nanoparticles and cytochrome P450 enzymes-a literature-based review. Xenobiotica, 49(7), pp.863-876. https://doi.org/10.1080/00498254.2018.1503360

Materials Research Forum LLC
https://doi.org/10.21741/9781644902998-9

[102] Bartneck, M., Warzecha, K.T. and Tacke, F., 2014. Therapeutic targeting of liver inflammation and fibrosis by nanomedicine. Hepatobiliary surgery and nutrition, 3(6), p.364.

[103] Shvedova, A.A., Pietroiusti, A., Fadeel, B. and Kagan, V.E., 2012. Mechanisms of carbon nanotube-induced toxicity: focus on oxidative stress. Toxicology and applied pharmacology, 261(2), pp.121-133. https://doi.org/10.1016/j.taap.2012.03.023

[104] Oberdörster, E., 2004. Manufactured nanomaterials (fullerenes, C60) induce oxidative stress in the brain of juvenile largemouth bass. Environmental health perspectives, 112(10), pp.1058-1062. https://doi.org/10.1289/ehp.7021

[105] Juan, C.A., Pérez de la Lastra, J.M., Plou, F.J. and Pérez-Lebeña, E., 2021. The chemistry of reactive oxygen species (ROS) revisited: outlining their role in biological macromolecules (DNA, lipids and proteins) and induced pathologies. International Journal of Molecular Sciences, 22(9), p.4642. https://doi.org/10.3390/ijms22094642

[106] Ijaz, M.U., Mustafa, S., Batool, R., Naz, H., Ahmed, H. and Anwar, H., 2022. Ameliorative effect of herbacetin against cyclophosphamide-induced nephrotoxicity in rats via attenuation of oxidative stress, inflammation, apoptosis and mitochondrial dysfunction. Human & Experimental Toxicology, 41, p.09603271221132140. https://doi.org/10.1177/09603271221132140

[107] Sengul, A.B. and Asmatulu, E., 2020. Toxicity of metal and metal oxide nanoparticles: a review. Environmental Chemistry Letters, 18, pp.1659-1683. https://doi.org/10.1007/s10311-020-01033-6

[108] Ijaz, M.U., Jabeen, F., Ashraf, A., Imran, M., Ehsan, N., Samad, A., Saleemi, M.K. and Iqbal, J., 2022. Evaluation of possible protective role of Chrysin against arsenic-induced nephrotoxicity in rats. Toxin Reviews, 41(4), pp.1237-1245. https://doi.org/10.1080/15569543.2021.1993261

[109] Xiong, P., Huang, X., Ye, N., Lu, Q., Zhang, G., Peng, S., Wang, H. and Liu, Y., 2022. Cytotoxicity of Metal-Based Nanoparticles: From Mechanisms and Methods of Evaluation to Pathological Manifestations. Advanced Science, 9(16), p.2106049. https://doi.org/10.1002/advs.202106049

[110] Choi, A.O.; Brown, S.E.; Szyf, M.; Maysinger, D. Quantum dot-induced epigenetic and genotoxic changes in human breast cancer cells. J. Mol. Med. 2008, 86, 291-302. https://doi.org/10.1007/s00109-007-0274-2

[111] Singh, R.P. and Ramarao, P., 2013. Accumulated polymer degradation products as effector molecules in cytotoxicity of polymeric nanoparticles. toxicological sciences, 136(1), pp.131-143. https://doi.org/10.1093/toxsci/kft179

[112] Habas, K., Demir, E., Guo, C., Brinkworth, M.H. and Anderson, D., 2021. Toxicity mechanisms of nanoparticles in the male reproductive system. Drug Metabolism Reviews, 53(4), pp.604-617. https://doi.org/10.1080/03602532.2021.1917597

[113] Ajdary, M., Keyhanfar, F., Moosavi, M.A., Shabani, R., Mehdizadeh, M. and Varma, R.S., 2021. Potential toxicity of nanoparticles on the reproductive system animal models: A review. Journal of Reproductive Immunology, 148, p.103384. https://doi.org/10.1016/j.jri.2021.103384

[114] Garcia, T.X., Costa, G.M., França, L.R. and Hofmann, M.C., 2014. Sub-acute intravenous administration of silver nanoparticles in male mice alters Leydig cell function and testosterone levels. Reproductive Toxicology, 45, pp.59-70. https://doi.org/10.1016/j.reprotox.2014.01.006

[115] Schädlich A, Hoffmann S, Mueller T, et al. Accumulation of nano¬carriers in the ovary: a neglected toxicity risk? J Control Rel. 2012; 160(1):105-112. https://doi.org/10.1016/j.jconrel.2012.02.012

[116] Adebayo OA, Akinloye O, Adaramoye OA. Cerium oxide nanopar¬ticle elicits oxidative stress, endocrine imbalance and lowers sperm characteristics in testes of Balb/c mice. Andrologia. 2018;50(3):e12920. https://doi.org/10.1111/and.12920

[117] Wang, R., Song, B., Wu, J., Zhang, Y., Chen, A. and Shao, L., 2018. Potential adverse effects of nanoparticles on the reproductive system. International journal of nanomedicine, 13, p.8487. https://doi.org/10.2147/IJN.S170723

[118] Chen Z, Wang Y, Zhuo L, Chen S, Zhao L, Luan X, Wang H, Jia G. Effect of titanium dioxide nanoparticles on the cardiovascular system after oral administration. Toxicol Lett, 2015, 239(2):123-30. https://doi.org/10.1016/j.toxlet.2015.09.013

[119] Pryskoka AO. [Study of the effect of colloidal solution of silver nanoparticles on parameters of cardio- and hemo-dynamics in rabbits]. Lik Sprava, 2014, (5-6):146-50. https://doi.org/10.31640/LS-2014-(5-6)-21

[120] Chen Z, Meng H, Xing G, Yuan H, Zhao F, Liu R, Chang X, Gao X, Wang T, Jia G, Ye C, Chai Z, Zhao Y. Age-related differences in pulmonary and cardiovascular responses to SiO2 nanoparticle inhalation: nanotoxicity has susceptible population. Environ Sci Technol, 2008, 42(23):8985-92. https://doi.org/10.1021/es800975u

[121] Bourdon JA, Saber AT, Jacobsen NR, Williams A, Vogel U, Wallin H, Halappanavar S, Yauk CL. Carbon black nanoparticle intratracheal installation does not alter cardiac gene expression. Cardiovasc Toxicol, 2013, 13(4):406-12. https://doi.org/10.1007/s12012-013-9223-1

Keyword Index

b

www.ingramcontent.com/pod-product-compliance
Lightning Source LLC
Chambersburg PA
CBHW071337210326
41597CB00015B/1483